Gordon / Shaw

Computer Programs for Rheologists

SPE Books from Hanser Publishers

Bernhardt, Computer Aided Engineering for Injection Molding
Brostow/Corneliussen, Failure of Plastics
Chan, Polymer Surface Modification and Characterization
Charrier, Polymeric Materials and Processing – Plastics, Elastomers and Composites
Ehrig, Plastics Recycling
Gordon, Total Quality Process Control for Injection Molding
Gordon/Shaw, Computer Programs for Rheologists
Gruenwald, Plastics: How Structure Determines Properties
Macosko, Fundamentals of Reaction Injection Molding
Manzione, Applications of Computer Aided Engineering in Injection Molding
Matsuoka, Relaxation Phenomena in Polymers
Menges/Mohren, How to Make Injection Molds
Michaeli, Extrusion Dies for Plastics and Rubber
O'Brien, Applications of Computer Modeling for Extrusion and Other Continuous Polymer Processes
Progelhof/Throne, Polymer Engineering Principles
Rauwendaal, Polymer Extrusion
Saechtling, International Plastics Handbook for the Technologist, Engineer and User
Stoeckhert, Mold-Making Handbook for the Plastics Engineer
Throne, Thermoforming
Tucker, Fundamentals of Computer Modeling for Polymer Processing
Ulrich, Introduction to Industrial Polymers
Wright, Molded Thermosets: A Handbook for Plastics Engineers, Molders and Designers

Glenn V. Gordon / Montgomery T. Shaw

Computer Programs for Rheologists

Hanser Publishers, Munich Vienna New York

Hanser/Gardner Publications, Inc., Cincinnati

The Authors:
Dr. Glenn V. Gordon and Dr. Montgomery T. Shaw, Institute of Materials Science, University of Connecticut, 97 N. Eagleville Road, U-136, Storrs, CT 06269-3136, USA

Distributed in the USA and in Canada by
Hanser/Gardner Publications, Inc.
6600 Clough Pike, Cincinnati, Ohio 45244-4090, USA
Fax: +1 (513) 527-8950

Distributed in all other countries by
Carl Hanser Verlag
Postfach 86 04 20, 81631 München, Germany
Fax: +49 (89) 98 48 09

Library of Congress Cataloging-in-Publication Data
Gordon, Glenn V.
Computer programs for rheologists / Glenn V. Gordon, Montgomery T. Shaw.
p. cm.
Includes bibliographical references and index.
ISBN 1-56990-163-5
1. Polymers -- Rheology. I. Shaw, Montgomery T. II. Title.
QD381.9.R48G67 1994
620.1'9204292'02855362--dc20 93-48574

Die Deutsche Bibliothek – CIP-Einheitsaufnahme
Gordon, Glenn V.:
Computer programs for rheologists / Glenn V. Gordon; Montgomery T. Shaw. – Munich ; Vienna ; New York : Hanser ; Cincinnati : Hanser/Gardner Publ., 1994
ISBN 3-446-16290-9 (Hanser)
ISBN 1-56990-163-5 (Hanser/Gardner)
NE: Shaw, Montgomery T.:

Camera-ready copy prepared by the authors.
Printed and bound in Germany by Schoder Druck GmbH & Co. KG, Gersthofen

To Kara and Steven

Foreword

The Society of Plastics Engineers is pleased to sponsor and endorse this volume on "Computer Programs for Polymer Rheologists" by Glenn V. Gordon and Montgomery T. Shaw. The book, which brings together a wide range of analyses, is an ambitious presentation suitable for use as a university textbook as well as a professional resource for the plastics industry. The inclusion of the computer program on floppy disk with the printed volume, as recommended by SPE members, should be an invaluable aid to engineering users. By doing this, we take one small step into the electronic publishing age.

SPE, through its Technical Volumes Committee, has long sponsored books on various aspects of plastics. Its involvement has ranged from identification of needed volumes and recruitment of authors to peer review and approval and publication of new books.

Technical competence pervades all SPE activities, not only in the publication of books, but also in other areas such as sponsorship of technical conferences and educational programs. In addition, the Society publishes periodicals including – *Plastics Engineering, Polymer Engineering and Science, Polymer Processing and Rheology, Journal of Vinyl Technology* and *Polymer Composites* – as well as conference proceedings and other publications, all of which are subject to rigorous technical review procedures.

The resource of some 38,000 practicing plastics engineers, scientists, and technologists has made SPE the largest organization of its type worldwide. Further information is available from the Society at 14 Fairfield Drive, Brookfield, Connecticut 06804, U.S.A.

Eugene De Michele
Executive Director
Society of Plastics Engineers

Technical Volumes Committee
Claire Bluestein, Chairperson
Captan Associates, Inc.

Preface

All rheologists are faced with the task of getting as much information as possible from hard-won data. The standard texts carry descriptions of procedures that have been worked out over the years and are often applied simply because they "are there." While the theoretical assumptions behind these procedures are made quite clear in the more comprehensive textbooks, the numerical and statistical aspects are often neglected. As with many areas of science, rheology is still suffering from what we will call the "linearization syndrome" – that is, the propensity to manipulate the data until the plot is a straight line. The power-law viscosity function is a classic example. The reasons for this state of affairs is clear; in the precomputer days, any other analysis was prohibitively time consuming. However, many of these analyses do have limitations, including inadequate description of the results and poor statistical properties. But where does the rheologist turn? Faced with the task of developing custom software, most will choose to continue with the traditional and familiar methods.

The purpose of this volume is to provide the practicing rheologist with computer algorithms for performing these common computational tasks, but with due attention to the statistical aspects of the analysis. Emphasis is placed on determining the suitability of any empirical model and calculating the errors in the fitted parameters. Illustrative examples are provided to test the algorithms and serve as guides in the execution of the programs. Whenever possible, the programs were tested and compared with published data.

Assembled either for research work or instructional courses in polymer rheology at the University of Connecticut, the programs were written and modified as the need arose. The FORTRAN programs presented here conform to the American National Standards Institute (ANSI) standard X3.9-1978 and were compiled in Microsoft® FORTRAN (Version 5.1) in a PC-DOS environment with the floating-point option which uses a math coprocessor if one is installed in the computer. With slight modifications, the programs should be portable to any compiler implementing the full FORTRAN-77 standard. While the features of FORTRAN-90 would have been convenient, we appreciate that there are presently few compilers available for this new standard. All the algorithms have also been tested on an IBM ES/9000 mainframe with VS FORTRAN (Version 2 Release 5) in the Conversational Monitor System environment (CMS 5, VM/HPO Release 5 Level 50).

While it is probably safe to assume that similar rheological algorithms are in existence, the source codes are often not readily available. The advent of automated commercial rheometers has developed a market for software modules that have become prohibitively expensive or without the possibility for custom modification. In some instances, the use of "packaged" modeling functions (e.g., the polynomial function) in describing rheological data is often dangerous. On the assumption that other rheologists were in the same situation, we decided that it would be worthwhile to make these programs widely accessible. The scope of this volume does not cover the rheology of processing. With the possible exception of the single screw extruder, processing equipment is so individualized that off-the-shelf codes must either be extremely complex or require extensive customization.

The first half of the volume covers the analysis of mechanical and viscoelastic experiments on polymer solids and melts.

Chapter 1 deals with the mechanical tensile properties of plastics. Program `TENSIL` is provided to analyze data collected through any data acquisition system that outputs ASCII code. `TENSIL` can be used either when specimen extension is measured by the rate of grip separation or from an extension indicator. One feature of `TENSIL` is the capability to correct for the toe region at the start of the extension of the test specimen.

Chapter 2 introduces a method for calculating the equilibrium tensile properties for elastomers using the constant-force experiment. By modifying the phenomenological Mooney-Rivlin equation to introduce a third parameter, program `STRECH` circumvents the task of measuring the unstrained length of the elastomer; the result is a more precise determination of the Mooney-Rivlin constants. While the traditional Mooney plot analysis is included, we provide a much improved analysis which pays strict attention to the statistical aspects of the problem.

Chapter 3 provides two algorithms for describing the long-term physical relaxation process in elastomers. An important practical result of the proper treatment of relaxation data is the recovery of the equilibrium modulus along with its error. Program `CHATH3` is a direct application of the three-parameter Chasset-Thirion equation. We introduce a second method, `CHATH4`, that uses a fourth parameter to describe the response of very lightly cross-linked networks and directly calculates the entanglement network modulus.

Chapter 4 handles the conversion of viscosity data to the corresponding molecular weight distributions. This is an evolving area, with relatively little evidence concerning the universality of the chosen weighting functions, especially at high shear rates. Program `MWD` offers a new and very flexible treatment of the problem.

Chapter 5 addresses the time-temperature superposition principle. Data sets are examined as adjacent-temperature pairs, and the master curve is subsequently formed by the addition of the pairwise shifts. The two more prominent features of the program `SHIFTT` are its ability to handle widely spaced data and a provision to characterize the error involved in superposition using the Monte Carlo method. An auxiliary program, `CONVRT`, is provided to interpolate data obtained through a thermal scan of multiplexed time scales to fixed temperatures and create a file that is usable by `SHIFTT`.

The second half deals primarily with the flow of polymer melts.

Chapter 6 treats the problem of constant-radius or constant-volume squeeze-flow experiments run either at constant force or constant velocity. Program `SQUEEZ` provides a framework for handling data from any of these experiments, incorporating inertial and compliance corrections. From `SQUEEZ`, the user obtains estimates of the shear-rate-dependent viscosity and a characteristic time. `SQUEEZ` can also prepare the data for linear viscoelastic analysis using the program `FITSTR` available in Chapter 10.

Chapter 7 and program TWIST analyze both steady-state and start-up torsional flows in the parallel-plate geometry. The purpose of the start-up flow is to produce the transient function $\sigma^+(t, \dot{\gamma})$ which is important in interpreting the elastic properties of the fluid. With these analyses at hand the rheologist can freely use the parallel-plate geometry as an alternative to the traditional cone-and-plate without any sacrifice in the accuracy of the resulting material functions. Two additional programs are provided: STEADY calculates the torques for several viscosity models while TRANST does the same thing for transient shear flows, but for a nonlinear integral model.

Chapter 8 provides two algorithms for handling capillary rheometry data. Both methods address end, slip, and velocity-profile corrections in steady, isothermal flows in capillaries having circular cross sections. The traditional procedure, the sequential order of correcting capillary data, is accomplished in CAPSEQ. Data from at least two different sets of capillary sizes, each having at least two different aspect ratios, are required for CAPSEQ to execute properly. The alternative method, CAPSIM, simultaneously handles the three types of corrections by assuming functional forms for entrance pressure loss, slip velocity, and flow. CAPSIM is recommended if less than four, or an arbitrary collection, of capillary geometries are used, although the program can be applied to extensive data sets as well.

Chapter 9 deals with the modeling of nonlinear viscosity functions of melts with unimodal molecular weight distributions. While the most suitable function ultimately depends on the breadth of the molecular weight distribution, six viscosity models are provided to serve as guides to inserting any desired function. As written, VISFIT includes the two-parameter Bueche-Harding and Eyring models, the three-parameter Carreau, Cross, and Sutterby models, and the four-parameter generalized-rate model. The program structure makes it simple to add other models.

Chapter 10 uses nonlinear viscoelastic models for the prediction of the elements of the stress tensor in various deformations, including simple shear and extension, planar extension, and biaxial extension. Program STRESS includes six nonlinear models and is designed to predict stresses given the deformation history of the polymer. References are given that provide examples of how STRESS can be used to estimate the relative performance of polymers in selected fabrication processes. Included in Chapter 10 is the experimental program FITSTR designed to take stress-strain-time data and fit these data with a nonlinear viscoelastic model. The data need not be gathered using a steady deformation rate; they could, for example, be obtained from image analysis of a strip of polymer that is being stretched by a weight or pulled in a tensile testing machine. Once the model is established, the data can be transformed, using STRESS, to the closest constant-rate "experiment."

Every user will have an opinion about what a program should do and how it should operate. Some of the choices we have made have been influenced by the instruments we have in our laboratory, while other choices have been made somewhat arbitrarily. Features which the user may or may not find harmonious include the structure of the input data files, particularly their compatibility with the output formats defined by the software from one's favorite rheological instrument; the degree of terminal interaction to enhance control over the program's operation; and the type of information that appears in the output/results files, including a provision for graphical presentation. All 16 main programs in this volume produce ASCII files as their output. Often these files consist of functional, unadorned lists of data that require a minimum of editing prior to importing into a graphics program. The ASCII file is still the stan- dard information "interface" between sets of software and should provide the user maximum flexibility.

We have not ignored the question of program modification. For example, programs such as `CAPSIM` in Chapter 8 may require frequent alterations or additions as better empirical functions are introduced. The code is written to make these changes straightforward. Program `STRESS` in Chapter 10 is structured to encourage the user to add a favorite nonlinear constitutive model. Provided with the full listings, custom modification to these programs should be easy for FORTRAN programmers, even those with little experience. The modification of the fundamental algorithms is more difficult, but the present programs should provide a convenient standard against which the creators of alternatives can test their work.

For projects involving more extensive modification, many of the subprograms can be used intact. To aid the programmer, we have provided full listings of all but the published routines (*Numerical Recipes in* FORTRAN: *The Art of Scientific Computing* by Press et al., Copyright © Cambridge University Press 1986, 1992) and proprietary software (Copyright © 1992 Visual Numerics, Inc.) in Appendix B, while all the program dependencies are charted in Appendix A.

We hope the reader will find the programs in this volume useful. While most of the algorithms have been tested over the years with a variety of data, we occasionally encounter sets that will not run or execute properly. The reader can expect similar occurrences. While we will be pleased to make arrangements to help with modifications of the code to run such data, it should be understood that we are not obliged to do so. **The reader should also understand that the use of this, or any, software may result in subtle, hidden errors that are as difficult to anticipate as they are to detect. Thus, never use the results from these programs as the sole basis for any decisions of import.**

Storrs, Connecticut
July 1993

Glenn V. Gordon
Montgomery T. Shaw

Acknowledgments

During the preparation of the manuscript we were fortunate to have the comments and suggestions of several rheologists on various parts of the manuscript. We particularly wish to thank Professor Albert Co of the University of Maine, Dr. John Curro of Sandia National Laboratories, Professor John M. Dealy of McGill University, Professor James E. Mark of the University of Cincinnati, Dr. Shiro Matsuoka of AT&T Bell Laboratories, Professor Robert L. Powell of the University of California, Dr. William H. Tuminello of E. I. du Pont de Nemours, and Dr. Kurt F. Wissbrun (Hoechst Celanese, retired).

We are grateful to the publishers and authors for granting permission to publish several of the figures and data sets appearing in this volume.

We wish to acknowledge the authors of *Numerical Recipes in* FORTRAN: *The Art of Scientific Computing* (Cambridge University Press) and Visual Numerics, Inc. for permission to imbed in the executable codes routines from the Numerical Recipes Software (Copyright © 1986, 1992) and the IMSL Libraries (Copyright © 1992), respectively.

Table of Contents

Chapter 9 Modeling the Viscosity Function

Chapter 10 Predicting Stresses with Nonlinear Integral Models

Chapter 1

Tensile Stress-Strain Analysis

1.1 Introduction

The tensile stress-strain test is undoubtedly the most widely used and standardized viscoelastic characterization for solid polymers. A widely recognized standard is the ASTM D 638 test method for dumbbell-shaped specimens; the metric equivalent is the test method D 638M [1]. Incorporated in these test methods' documentation are references to other ASTM standards, including the test method D 882 for thin plastic sheets less than 1.0 mm thick [2]. Test method D 882 is particularly useful when testing polymeric materials of limited quantity. Shown in Figure 1.1 is a schematic representation of some typical responses for various classes of polymers ranging from brittle to rubbery behavior.

While simple in concept, the details of a test on a real polymeric material can be quite confusing. For example, tests on most polymers do not result in Hookean, elastic response at low strains. As illustrated in Figure 1.2, this situation often makes the initial modulus difficult to determine. Another area of confusion is the yield point, the determination of which can be somewhat arbitrary.

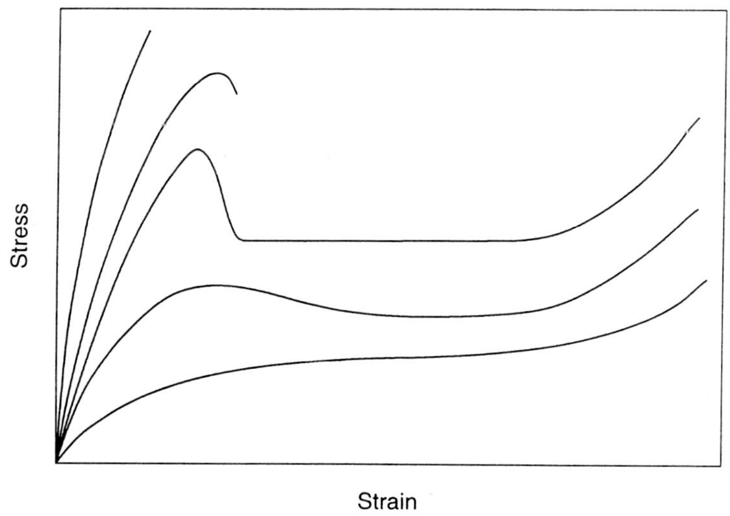

Figure 1.1 Schematic tensile stress-strain responses for various classes of polymers ranging from brittle to rubbery behavior.

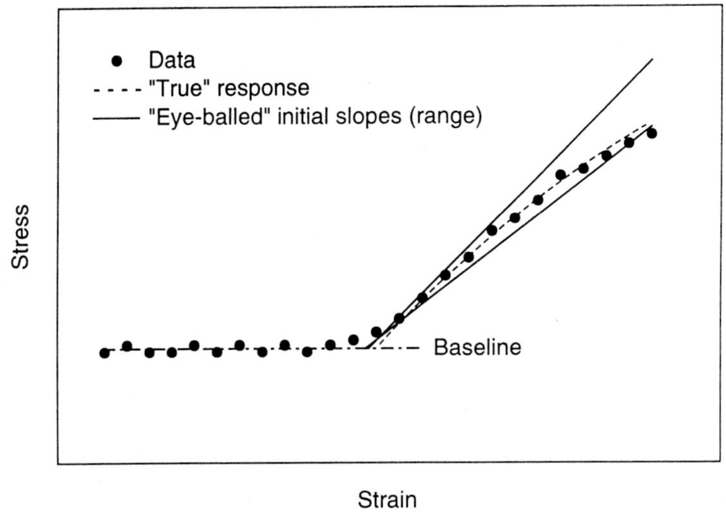

Figure 1.2 Stress-strain curve illustrating problems in determining the initial slope required for the modulus including a rounding of the initial portion of the stress-strain response (the region between the 11th and 14th point).

The program `TENSIL` is designed to reduce the difficulty of compensating for nonlinearity in the tensile stress-strain response; it does this by fitting an empirical, but realistic, model to the data. `TENSIL` also addresses problems associated with the takeup of slack and alignment of the test specimen at the start of a tensile experiment. The algorithm was originally conceived to analyze data from thin films of uniform original cross section where extension is based on the crosshead travel. Extensometer data can be handled as well. While `TENSIL` provides other pertinent mechanical information aside from the initial modulus, no direct determination is attempted for the yield point, the precise location of which is still subject to debate [3].

The modeling of the stress-strain response of a polymeric material should address the array of theoretical and practical aspects of the test. For one approach, the tensile stress-strain response at low strains for a viscoelastic material with a stress-relaxation modulus, $E(t)$, may be expressed using the Boltzmann equation

$$\sigma(t) \;\cong\; \frac{W(t)}{A_o} \;\cong\; \dot{\varepsilon}\int_0^t E(t-s)\,ds \tag{1.1}$$

where σ is the applied stress, W is the load, A_o is the initial cross-sectional area, $\dot{\varepsilon}$ is the initial strain rate, and t is time. When extensometer data is available, the initial strain rate is calculated from the initial slope of the strain-versus-time response [1]; otherwise, the strain rate is

$$\dot{\varepsilon} = \frac{V_{XH}}{L_o} \tag{1.2}$$

where V_{XH} is the constant crosshead speed of the testing machine and L_o is the initial gauge length.

A restriction to low strains (ε) permits the variable $\dot{\varepsilon}$ to be moved from under the integral sign in Equation 1.1. The tangent modulus, E_o, is defined as the slope of the stress-strain curve at zero strain:

$$E_o \equiv \lim_{\varepsilon \to 0} \frac{d\sigma}{d\varepsilon} = \frac{1}{\dot{\varepsilon}} \lim_{\varepsilon \to 0} \frac{d\sigma}{dt} \tag{1.3}$$

Applying Leibniz's rule to Equation 1.1 gives

$$\begin{aligned} E_o &= \lim_{t \to 0} \frac{d}{dt} \int_0^t E(t - s)\, ds \\ &= E(0) + \lim_{t \to 0} \int_0^t \frac{\partial}{\partial t} E(t - s)\, ds \end{aligned} \tag{1.4}$$

To eliminate the second term, the assumption has been made that the derivative of $E(t)$ is everywhere finite. Because of the nature of real materials, the remaining term, $E(0)$, is unfortunately still something of a concept rather than a measurable material property. In other words, the relaxation modulus at $t = 0$ that is derived from the initial slope depends on the particular derivative-finding numerical technique employed. Hence, there is the motivation to employ stress-strain data at higher strains for comparison with dynamic, creep, or stress relaxation tests [4-6].

Smith defined a modulus function, $F(t)$, for describing the linear viscoelastic response in a constant-strain-rate experiment (low strains) [4-6],

$$F(t) \equiv \frac{\sigma(t)}{\varepsilon(t)} = \frac{1}{t} \int_0^t E(t - s)\, ds \tag{1.5}$$

and has shown that

$$E(t) = F(t) \left[1 + \frac{d \ln F(t)}{d \ln t} \right] \tag{1.6}$$

Thus, values of the relaxation modulus, $E(t)$, can be calculated given stress-strain data gathered at a known, constant strain rate.

Based on a thermodynamic analysis of the glassy state, a particularly convenient empiricism for the stress-strain response of several polymers in tension has been used by Matsuoka [7]

$$\sigma(\varepsilon) = E_o \varepsilon \, e^{-\beta_3 \varepsilon} \tag{1.7}$$

where β_3 is a material constant dependent on strain rate. From Equation 1.7, the secant modulus, $E(\varepsilon)$, defined as the nominal stress-to-strain ratio at any specified point on the stress-strain response, can be determined. The Matsuoka model often provides good fits not only at low strains but through the yield point; however, necking must be absent [8]. The application of Equation 1.6 to Equation 1.7 gives an expression for the stress-relaxation modulus:

$$E(t) = E_o \, e^{-\beta_3 \dot{\varepsilon} t} (1 - \beta_3 \dot{\varepsilon} t) \tag{1.8}$$

The expression for the 1-s relaxation modulus, $E(1)$, is

$$E(1) = E_o \, e^{-\beta_3 \dot{\varepsilon}} (1 - \beta_3 \dot{\varepsilon}) \tag{1.9}$$

The practical aspects of analyzing stress-strain results provides additional motivation to automate the analysis. For example, a careful look at the stress-strain curve reveals a more or less pronounced rounding of the initial portion of the response referred to as the toe region [1-2]. This rounding, shown schematically in Figure 1.2 as the region between the 11th and 14th point, is usually an artifact due to the takeup of slack – a consequence of improper seating or alignment of the test specimen in the grips. Another factor contributing to the appearance of a toe region is the assumption of a step change in deformation rate from 0 to $\dot{\varepsilon}$ at the start of the run. This assumption is never exactly true.

To compensate for the toe region, the rheologist is forced to extrapolate the data, at higher strains, to zero strain. `TENSIL` handles this situation by assuming an appropriate stress-strain function. The function should be relatively insensitive to artifacts or faulty data at lower strains for accurate extrapolation back through the toe region.

From the preceding discussion, the Matsuoka model should be a reasonable choice for the extrapolation. Referring to Figure 1.2, Equation 1.7 may be modified to compensate for the toe region by introducing another parameter β_2:

$$\sigma(\varepsilon) = E_o(\varepsilon_a - \beta_2) \, e^{-\beta_3(\varepsilon_a - \beta_2)}, \qquad \varepsilon_a > \beta_2 \tag{1.10.1}$$

$$\sigma(\varepsilon) = 0, \qquad \varepsilon_a \leq \beta_2 \tag{1.10.2}$$

where the term $\varepsilon_a - \beta_2$ represents the actual strain as opposed to the apparent strain ε_a. Equation 1.10.1 is the basis for the objective function defined in the subroutine module FSS.

Another problem deals with the use of contact-type strain gauges. A review of the ASTM D 638 method by Staats-Westover and Matsuoka [9] points out that extensometer slippage and clamping force can be major concerns. A noncontact-type device would be preferable; one such device has been described by Staats-Westover and Matsuoka [9]. In elevated temperature environments, the use of any type of strain gauge may be impractical. Therefore, TENSIL will handle force data collected at constant time intervals under the assumption that the strain is spatially homogeneous; that is, the crosshead travel completely describes the strain in a test specimen of uniform original cross section. This assumption is particularly critical at low strains where the stress-strain function is to be fit.

As stated earlier, TENSIL does not attempt to look for a yield point. By definition, the yield point is the first point in the stress-strain response at which an increase in strain occurs without an increase in stress [1]. The definition implies that the condition $d\sigma/d\varepsilon = 0$ must be satisfied. However, the term "yield" has also been used to characterize the discontinuity from the Hookean region or the maximum point in the engineering stress-strain curve. Figure 1.3 illustrates the more common methods of obtaining the yield point. While curve *A* satisfies the definition of yield point, numerical data must be relatively precise before a derivative-finding routine such as SLOPE in Appendix B can be applied. More importantly, such a yield point is absent in many stress-strain curves. For the response represented by curve *B*, Birley et al. imply the use of "tangents" to extrapolate to the yield point [3]. A more common circumstance is illustrated by curve *C*, where the yield point is based on a specified offset strain [1]. Thus, the underlying message is that while graphical methods are available, it remains unclear how to approach a global definition for the yield point so as to code it in precise terms in the algorithm.

1.2 Description of the Algorithm TENSIL

Program TENSIL analyzes the tensile stress-strain response to find the tangent modulus at zero strain, the 1% secant modulus, the relaxation modulus at 1 s, the ultimate engineering stress, and the properties at break. The program is designed to handle force or force-extension data, in ASCII format, collected by any analog-to-digital (A/D) conversion system using up to two different acquisition rates. The use of multiple acquisition sections is appropriate when dealing with highly extensible materials ($\varepsilon > 1$). The program provides checks to aid the user in spotting data entry errors.

The algorithm will cope with noisy data up to a point; however, additional smoothing may become necessary. As written, SMOOTH calls the module MEDFIL

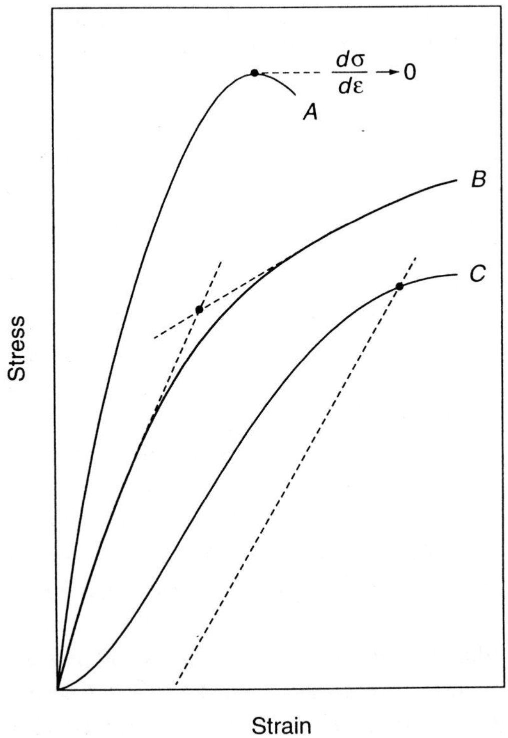

Figure 1.3 Schematic representations for some methods of obtaining the yield point (●). Refer to discussion in text for more information. Curve *B* adapted from Ref. 3. Copyright © 1992 Carl Hanser Verlag. Reprinted by permission.

which smooths an odd-number strip of points by replacing the midpoint of the strip with the median value of the points in the strip inclusive of the midpoint. `MEDFIL` will remove high-frequency noise. To handle coherent noise of lower frequencies, use low-pass filter algorithms such as `SMOOFT` available in *Numerical Recipes* [10]. Only a few simple changes are required to the subroutine `SMOOTH` imbedded in `TENSIL` to invoke one's favorite smoothing routine.

The program calls `CONVSI` to provide conversion factors to the SI system based on a selection of dimensional units available on most popular instruments (Table 1.1). The module `START` is called next to search for a starting point based on a change in the derivative of the force signal. The recommended value for triggering a start response is three standard deviations above the noise level of the derivative in the baseline. The data file must be set up to ensure that at least 10 baseline points are available. Once the data points belonging to the baseline are located by `START`, the background noise level is established using this data. This implies that the baseline data should be taken with the crosshead traveling and the test specimen slightly

Table 1.1 Group Number for the Combination of Dimensional Units Available for the Tensile Stress-Strain Data

Group No.	Length	Load	Time
1	in.	lb_f	min
2	mm	kg_f	min
3	mm	N	min

flexed. Because this procedure is impractical for stiff materials, a string of points taken after the specimen breaks, with the crosshead still moving, can be copied from the bottom of the data file. For *completely* flat baselines, the noise will be set to the square root of 1 force digit divided by the number of baseline points. In addition to START, the background noise level is used for decisions in several parts of the program. In the extreme, program execution is terminated if the signal-to-noise ratio is too low.

Using the subroutine POLYN to perform a first-order linear regression, TENSIL obtains an estimate of the modulus for a string of data corresponding to 1% strain or a tenth of the total number of points in the first acquisition section, whichever is smaller. This step provides a check against the modulus values obtained through the empirical model.

A new string of points is then selected to define the initial stress-strain response. For the first iteration, the string is guaranteed to encompass the initial 0.5% strain history beyond the starting point estimated by START. Using POLYN, the points are fitted with a quadratic function to determine the nature of the curvature. A positive second derivative indicates that the string is in the toe region at the beginning of the extension. In this case, the string is moved over one data point to encompass higher stress-strain data, whereupon the process is repeated. Another, less common circumstance will also shift the string over one point – a value less than zero for the discriminant of the quadratic function, indicating a lack of interception with the baseline.

Upon satisfying the curvature and intercept criteria, more points are added to the string at its high-strain end. The number of points added at each iteration is proportional to the number of points per percent strain. This exercise, limited to 15 iterations, is continued until the slope, at zero force, goes through a maximum. The condition of maximum slope limits the fitting of the stress-strain function to a region prior to any yield. At this juncture, the quadratic is abandoned, with the parameters serving a good initial estimates for the fitting of the nonlinear model (Equation 1.10.1) to the final string.

If the data string consists of at least nine points and the signal-to-noise is sufficient, the fit to Equation 1.10.1 is accomplished with a quasilinear Gauss-Newton least-squares routine [11]; otherwise, the results from the quadratic are used. Because strain is already dimensionless, only a linear transformation of the load data, scaled to the maximum load reading, is applied. While the error in the load does increase slightly with load, the error in the low-load data would be unduly influential if the normal assumption of a linear correlation of error with load were made.

The Gauss-Newton iterative method starts with the linearization of the nonlinear function by truncating its Taylor expansion

$$f_i(\mathbf{B} + \delta\mathbf{B}) = f_i(\mathbf{B}) + \sum_{j=1}^{N} \frac{\partial f_i}{\partial \beta_j} \delta\beta_j \tag{1.11}$$

where f_i refers to the function evaluated at data point i, and $\mathbf{B}$ denotes the vector of the function parameters β_j. Once the initial estimates to the parameters are chosen, the corrections, $\delta\mathbf{B}$, that will minimize the objective function are determined from

$$\delta\mathbf{B} = (\mathbf{A}^{\mathrm{T}}\mathbf{A})^{-1}(\mathbf{A}^{\mathrm{T}}\,\mathbf{r}) \tag{1.12}$$

where $\mathbf{r}$ is the vector of residuals and $\mathbf{A}$ is the matrix formed by taking the partial derivatives of the function with respect to the parameters; that is,

$$A_{ij} = \frac{\partial f_i}{\partial \beta_j} \tag{1.13}$$

where the subscript i denotes the evaluation at data point i.

The subroutine `GAUSSN` carries out the minimization with `NIT` iterations starting from an initial approximation of the solution vector $\mathbf{B}$ consisting of `N` variables. In `TENSIL`, this approximation is done internally by using the parameters obtained from the quadratic fit. The number of iterations is restricted to

$$\mathrm{NIT} \leq 2 + 2\mathrm{N}^2 \tag{1.14}$$

since convergence is expected to be rapid if the initial estimates of the parameters are close to the true solution. Otherwise, the method is likely to experience obvious convergence problems. Convergence is attained if

$$\mathrm{CC} > \frac{|\delta\beta_j|}{|\beta_j|} \tag{1.15}$$

for all parameters. The convergence criterion, `CC`, has a default value of 10^{-4} in the module `SPECC`. The criterion should be no smaller than the square root of the computer's floating-point precision because nothing is to be gained by using smaller values [10].

The minimization routine uses the analytical forms of the first partial derivatives; these are included in the module FSS. The matrix operations, described by Equation 1.12, are carried out in the external subroutines MATMUL and MATINV for matrix multiplication and inversion, respectively. The MATINV subroutine uses the lower-triangular and upper-triangular (*LU*) matrix decomposition routines available in *Numerical Recipes* [10]. Refer to Appendix A for the program dependencies.

Expressed in terms of 95% confidence intervals, the errors in the fitted parameters are estimated from the diagonal components of the covariance matrix,

$$\pm\; t_{95,\,\nu}\;\; s_{\varepsilon}\sqrt{C_{ii}} \tag{1.16}$$

where $t_{95,\,\nu}$ is the *t*-statistic as a function of the appropriate degrees of freedom, ν, s_{ε} is the standard error of estimate for the data set, and C_{ii} is the corresponding diagonal element for the estimated covariance matrix of the standard error [10,11]. From Equation 1.12,

$$C_{ii} \;\equiv\; \left[(\mathbf{A}^{\mathrm{T}}\mathbf{A})^{-1}\right]_{ii} \tag{1.17}$$

To summarize, the quasilinear method GAUSSN is capable of providing rapid and precise determination of the optimum values of the function parameters as well as estimates of their precision. However, reasonable initial estimates of the parameters are required or the search is likely to fail, or converge to a local rather than global minimum. Linear approximations of nonlinear functions are often used to obtain initial estimates. For the stress-strain model in Equation 1.10.1, a quadratic function is applied to extract these estimates. Briefly, the estimate for E_o is based on the first derivative of the quadratic, the solution to the quadratic at zero force gives an estimate for β_2, and the approximation $e^x \cong 1 + x$ at small strains provides an initial estimate to β_3. While these estimates may be close to the true solution, they do not provide the maximum likelihood estimates or the corresponding precision of the parameters; thus, the final nonlinear optimization step is essential.

Given the final data string, the following calculations are carried out. The tangent modulus is calculated from the zero-force value of the derivative of the stress-strain function (β_1), and the 1-s relaxation and 1% secant moduli are then calculated from Equations 1.9 and 1.10.1, respectively. If the data string does not cover at least one second of the tensile experiment time scale, the 1-s relaxation modulus is based on a fit of a quadratic function around the vicinity of 1 s. A statistical test on the quadratic term is performed at a 0.05 significance level. If the null hypothesis cannot be rejected, a straight-line fit is used. The corresponding confidence limits are obtained via error propagation analysis. The nominal stresses and corresponding residuals are calculated along with a standard error of estimate.

The standard error of estimate is compared with the standard deviation of the baseline points. This ratio, which is the square root of the F-statistic, should be close to 1.0 [10,11]. If $F > 1$, the reason may be only mechanical noise being transmitted through the extending test specimen to the load cell of the testing instrument. However, if $F >> 1$, there may be a problem with a lack of fit of the model to the experimental stress-strain data. In this case, the size of the data string used to fit the stress-strain function and the residuals should be carefully examined.

With the work on the moduli calculations finished, `TENSIL` tries to find the breakpoint. Oddly enough, the breakpoint may not be obvious for some specimens: the force rises and falls as the specimen stretches out or tears. A simple, but reliable, algorithm is used for finding the breakpoint. `TENSIL` starts off by searching for a minimum in the first derivative (maximum negative slope) which should occur as the test specimen fails. (A sharp yield could cause confusion, but this has not been found to be a problem in practice.) Once this point is found, the search reverses direction to find the first positive slope between adjacent points: the first of these points is considered the breakpoint. The elongation and tensile strength at break may then be calculated. In addition, the nominal tensile strength from the maximum force reading is provided.

A final important task is performed by `TENSIL`: the calculation of two checks to guard against gross errors in the recording of crosshead speed, data acquisition rate, or force amplification. The maximum force and the total crosshead travel are calculated, the former by using the routine `AMAXI`. These values may be compared with display readings from the testing instrument.

1.3 Execution of TENSIL

The execution of `TENSIL` requires the subroutines `AMAXI`, `GAUSSN`, `POLYN`, `SPECC`, and `STAT` found in Appendix B. Refer to Appendix A for the program dependencies. For example, the subroutine `GAUSSN` calls the subroutines `MATINV` and `MATMUL` found in Appendix B. The remaining subroutines, inherent to running `TENSIL`, are included in the code listed in Section 1.5.

Before the program `TENSIL` is executed, a data file must be set up by using an editor or a text processor that outputs ASCII code. The first record is set aside for the title of the file; include anything up to 79 characters containing identification information. The second record requires an integer entry specifying a group number representing the data's dimensional units. A menu for three combinations of dimensional units corresponding to length, load, and time is listed in Table 1.1. The subroutine `CONVSI` provides the appropriate conversion factors to the SI system. The third record contains information on crosshead speed, gauge length and an average cross-sectional area along the gauge length of the test specimen. When an extension indicator is unavailable, enter the initial grip separation as the gauge length. The fourth record is for calibration constants. The first entry is the calibrated load reading per digit from the A/D converter. The second entry is the calibrated extension reading per digit from the A/D converter; enter 0 if extensometer data is unavailable. The fifth record is for the data acquisition time intervals. Enter the time intervals, in seconds, for each acquisition section; use 0 for the second entry if only one acquisition section is used. The subsequent records are for force and extension (if available). Be sure to include at least 10 points of zero-force baseline readings. More baseline points can be used, but they are unnecessary and can be edited out. Up to 1000 points, or data pairs, can be accommodated. If a second section of data at a different time interval has been included, flag it, using 99999 for each data column. (Here, we are anticipating that nothing more precise than a 16-bit converter would be used for data acquisition.) Finish the file with at least one baseline point. Because specimen breakage within the range of the extensometer clamps will cause an abrupt increase in the extensometer output, the resulting extension data should be edited. Listed in Section 1.6.1 is a sample data file which has been abbreviated due to its length.

The execution of `TENSIL` is interactive. The first interaction concerns smoothing. We have found that, while very messy files can be digested by `TENSIL`, bad points occurring in an inopportune position in the file can produce misleading results. Thus, the file should be edited or smoothed to remove obvious outliers. Data plagued with heavy coherent noise might well benefit from low-pass filtering, but our experience is that filtering has little effect on the final results, and, if the frequency cutoff is low enough, tends to lower the calculated modulus. `TENSIL` offers an option to remove erratic data points via `MEDFIL` for load, extension, or both data columns.

The second interaction calls the routine `SPECC` to specify the convergence criterion for the Gauss-Newton minimization method. As discussed in the preceding section, the default value is 10^{-4} and has been found to be adequate in all trial cases.

The last interaction sets up the format for the output file. The user has the option of producing a detailed or a summary file; enter `D` or `S`, respectively. If all goes well, the results for the checks, mechanical properties, and the fit to the stress-strain function are displayed on the terminal screen. The summary file includes information on the specimen geometry, speed of testing, and the checks for maximum load and total crosshead travel. The data given in parentheses are the corresponding values for the original units of measurement. Next, the mechanical properties are listed along with their 95% confidence intervals. Also provided are a listing of the data string used to fit Equation 1.10.1 and the strain corresponding to that string, the maximum likelihood estimates and their precision, the standard error of estimate of the function, and the ratio corresponding to the F-statistic. A difference between the start of the data string and the estimated starting point found by `START` indicates a compensation for a toe region. A tabulation of the nominal stress-versus-strain data follows for the string of data used to fit the stress-strain function. A glance at the randomness of the residuals should give a good indication of how well the function fits the data. If the detailed (D) output option has been chosen, the output includes all pertinent information used in the calculations plus a tabulation of nominal stress and strain for all data points.

Sections 1.6.2 and 1.6.3, respectively, contain a listing of a typical terminal dialog during the TENSIL's execution and the resulting detailed output file.

1.4 Application of **TENSIL**

As illustrative examples, we picked tensile test results for a variety of materials. The first group of examples deals with specimens of uniform original cross section and includes two sets in which specimens were cut from thin sheets and a set where a thin strip was microtomed off an electrical insulation cable to analyze possible deterioration in its mechanical properties. The specimens were tested without an extensometer at an initial strain rate of 0.033 s^{-1} and exhibited an E_0 range of almost three decades. Because of the relatively high extensions of these polymers, multiple time intervals were employed. The mechanical data obtained from TENSIL are listed in Table 1.2.

The first example, a silicone elastomer, illustrates the algorithm's ability to deal with the toe region. From Figure 1.4, the toe region consists of the first three points immediately following the apparent starting point, covering 0.01 strain unit. The compensation correctly fits Equation 1.10.1 to a string of points starting from the thirteenth point and ending at a strain of 0.077.

Table 1.2 TENSIL Calculations of the Mechanical Properties from Force-Only Data Sets Using an Initial Nominal Strain Rate of 0.033 s^{-1}

Property[a]	Silicone	Viton®	XLPE
Initial modulus	4.14 ± 0.25	96.0 ± 3.5	856. ± 7
1% secant modulus	4.10 ± 0.45	73.5 ± 2.7	522. ± 38
1-s relaxation modulus	3.91 ± 0.32	29.1 ± 1.4	89.8 ± 3.2
Tensile strength	4.67	11.1	14.6
Tensile strength at break	4.67	11.1	14.6
% Elongation at break	341	234	426

[a] Reported values for modulus and tensile strength are in MPa. The reported errors for moduli are the 95% confidence intervals.

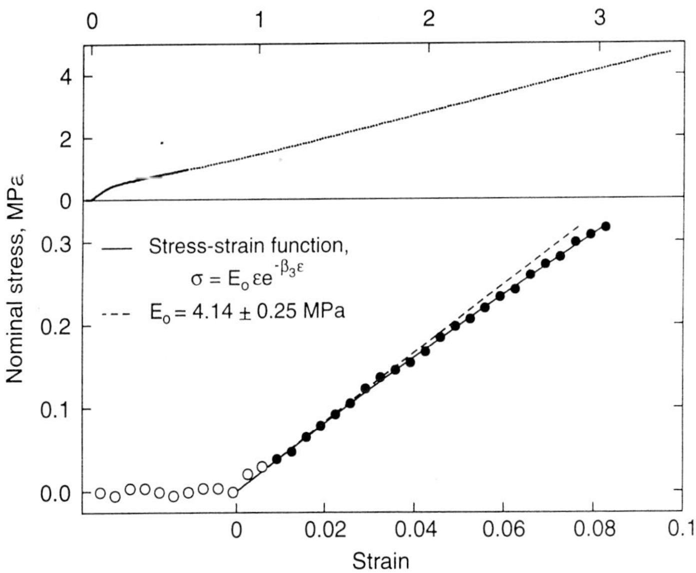

Figure 1.4 Results for a silicone elastomer exhibiting a toe region. Upper plot shows the entire stress-strain response using two data acquisition rates. Lower plot shows the data string (●) used to fit the stress-strain model (Equation 1.10.1).

The next two examples illustrate TENSIL's ability to handle data that exhibit essentially no Hookean region at low strains. The tensile responses for a fluoropolymer elastomer Viton® and a cross-linked polyethylene (XLPE) electrical cable insulator are shown in Figures 1.5 and 1.6, respectively. No toe compensation was required for the fluoroelastomer. For the polyethylene specimen, however, the initial data string had to move 0.005 strain unit to satisfy the curvature and intercept criteria.

These two examples also highlight the importance of choosing an appropriate data acquisition rate in relation to the initial strain rate. As discussed in Section 1.2, the smallest possible initial data string corresponds to 0.005 strain unit and yet must have a minimum of nine points to employ the three-parameter nonlinear model (Equation 1.10.1). While the data acquisition rate was twice as fast as that used for the silicone elastomer, the fits of Equation 1.10.1 to the fluoropolymer and polyethylene data were accomplished with strings of 14 and 10 points, respectively, both of which correspond to approximately 2% strain. The calculated F-values were 3.8 and 8.2, respectively. Clearly, a faster acquisition rate for the initial response would increase the precision by including more points for every 0.005 strain unit. However, the F-value of 8.2 does suggest that the goodness-of-fit of the data to the stress-strain model should be examined.

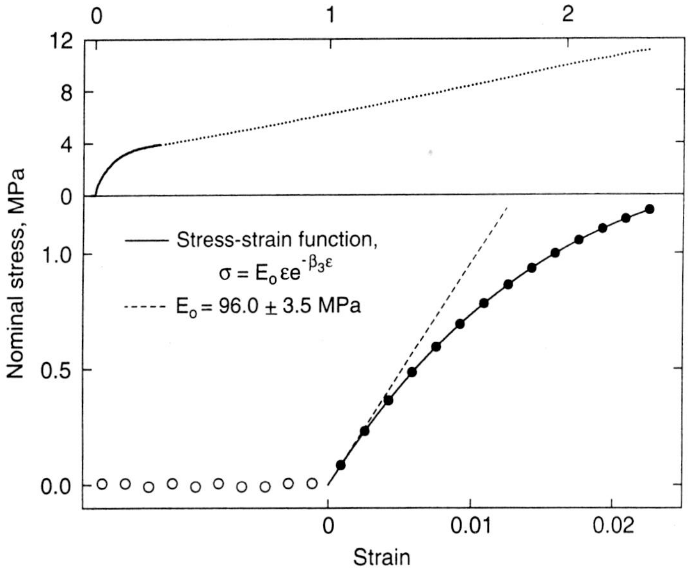

Figure 1.5 Results for a fluoroelastomer Viton®. Upper plot shows the entire stress-strain response using two data acquisition rates. Lower plot shows the data string (●) used to fit the stress-strain model (Equation 1.10.1).

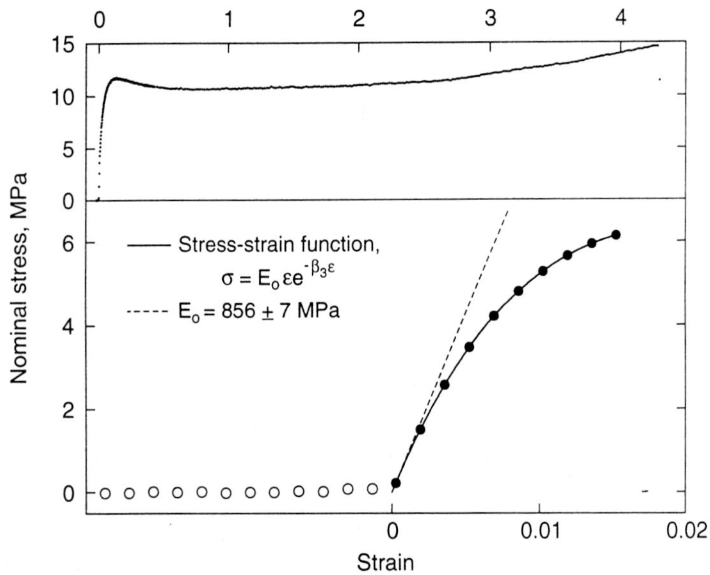

Figure 1.6 Results for a cross-linked polyethylene (XLPE). Upper plot shows the entire stress-strain response. Lower plot shows the data string (●) used to fit the stress-strain model (Equation 1.10.1).

The second group of examples concerns data collected on glassy polymers by using a contact-type extensometer and a single time interval for data acquisition. The specimens were Type V in accordance with the ASTM D 638 test method and were approximately 1 mm thick. The testing speed was 0.02 in/min. Compared to the strain rate of 3.3×10^{-4} s^{-1} based on the crosshead speed, the strain rates obtained from the extensometers were 2.0 and 2.7×10^{-4} s^{-1} for polystyrene and poly(methyl methacrylate), respectively. The mechanical data obtained from TENSIL are listed in Table 1.3.

Table 1.3 TENSIL Calculations of the Mechanical Properties from Force-Extension Data Sets Using an Initial Nominal Strain Rate of 0.0003 s^{-1} on ASTM D 638 Type V Specimens

Property	Polystyrene[a]	PMMA[b]
Initial modulus, GPa	3.78 ± 0.16[c]	3.13 ± 0.08
1% secant modulus, GPa	2.95 ± 0.21	2.76 ± 0.07
1-s relaxation modulus, GPa	3.71 ± 0.16	3.01 ± 0.08
Tensile strength, MPa	37.2	63.8
Tensile strength at break, MPa	37.2	63.8
% Elongation at break	1.3	4.3

[a] Lustrex®, M_w = 69,000. Specimen broke at gauge mark.
[b] Polysciences poly(methyl methacrylate), M_w = 350,000. Specimen within gauge marks.
[c] The reported errors for moduli are the 95% confidence intervals.

Typical results for a Monsanto Lustrex® polystyrene ($M_w = 6.9 \times 10^4$) and a Polysciences poly(methyl methacrylate) ($M_w = 3.5 \times 10^5$) are shown in Figure 1.7. For the polystyrene data, the terminal dialog for running TENSIL is listed in Section 1.6.2; the data (abridged) and output files are listed in Sections 1.6.1 and 1.6.3, respectively. The checks for maximum load and crosshead travel agree well with the instrument readings listed in the first record of the data file. Noteworthy from the plots is the noise in the extensometer data, probably due to slippage. This problem is evident across the entire tensile curve, but is most severe below 1% strain. Extensometer noise would not contribute to baseline variance. Thus, the high F-values of 12.2 and 8.6 for the polystyrene and poly(methyl methacrylate) specimens, respectively, are not surprising. For this type of noise, passing the data through the smoothing program MEDFIL will have no effect because the force is always increasing with strain, albeit in a steplike manner. Nevertheless, TENSIL performed satisfactorily.

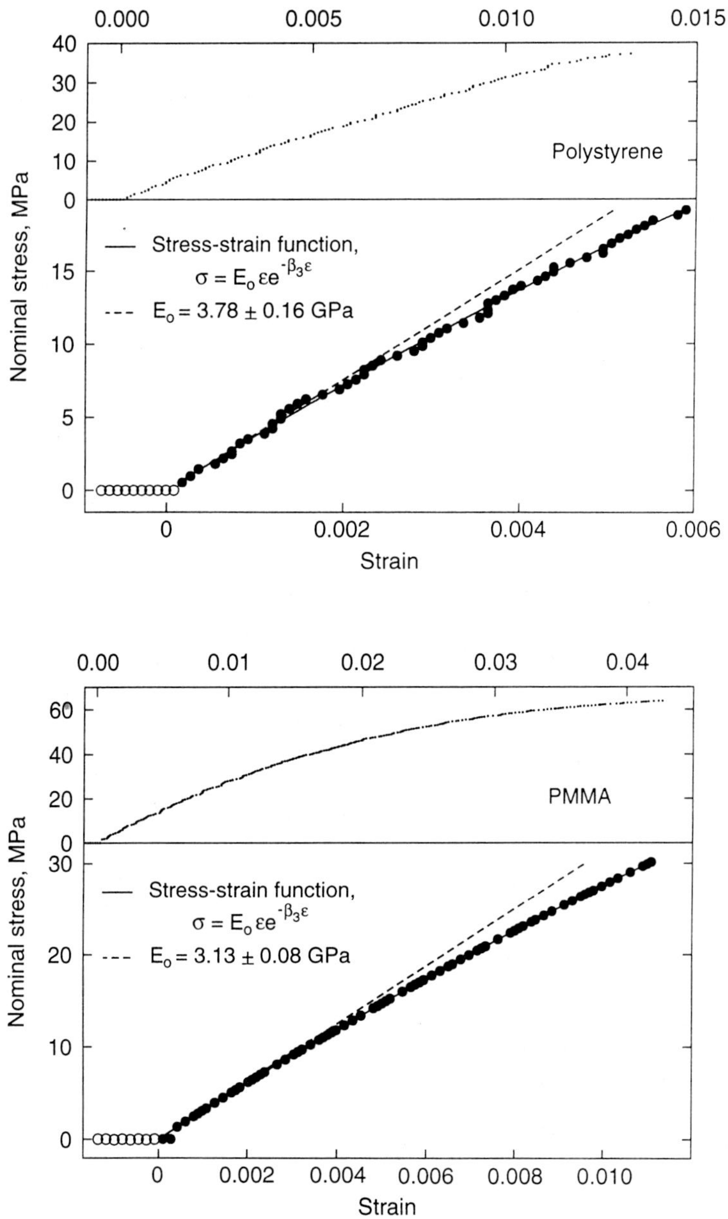

Figure 1.7 Results for ASTM D 638 Type V specimens of a Lustrex® polystyrene and a poly(methyl methacrylate), Polysciences Cat. #4552, using a contact-type extensometer. Upper plots show the entire stress-strain response. Lower plots show the data strings (●) used to fit the stress-strain model (Equation 1.10.1).

1.5 Program Listing for TENSIL

Imbedded in the code are subprograms inherent to the execution of TENSIL. These subprograms include CALTEN, CONVSI, FSS, MEDFIL (XMED and SORT), SMOOTH, and START.

```
      PROGRAM TENSIL
C Applies the stress (Y) - strain (X-A2) model
C                    Y=A1*(X-A2)*EXP(-A3*(X-A2))
C  to load-time or load-extension data. Data may be collected using two
C  different acquisition rates.
C Requires the subroutines AMAXI, GAUSSN, POLYN, SPECC, STAT, and STOPP.
C Data file format:
C        1st record - Title, up to 79 characters
C        2nd record - the data file's dimensional units group number
C                     Group No.     Length     Load    Time
C                        1            in.       lb      min
C                        2            mm        kg      min
C                        3            mm        N       min
C        3rd record - Crosshead speed, gauge length, cross-sectional area
C        4th record - Calibration constants: load/digit, extension/digit
C                     Use 0 for second entry if extensometer data is
C                     unavailable.
C        5th record - first time interval, second time interval (seconds).
C                     Use 0 for second entry if second time interval is
C                     not used.
C subsequent records - A/D reading(s) of load (and extension)
C    Include at least 10 points of baseline.
C    Start second time interval with entry of 99999 for each data column.
C subsequent records - A/D reading(s) for 2nd time interval.
      IMPLICIT DOUBLE PRECISION (A-H,O-Z)
      PARAMETER (NMAX1=1001,NMAX2=500,NPD=10,NPAR=3)
      CHARACTER ANS*1,TITLE*80
      DIMENSION A(NPAR),AERR(NPAR),C(NPAR),ERR(NPAR),IPN(NMAX1)
      DIMENSION RES(NMAX1),R(NMAX2,1),S(NMAX2,NPD),ST(NPD,NMAX2)
      DIMENSION X(NMAX2),Y(NMAX2),WT(NMAX1)
      COMMON /BLOC1/ W(NMAX1),E(NMAX1),T(NMAX1),B(NPAR)
      EXTERNAL FSS
C t-statistic for 95% confidence interval with N degrees of freedom.
      T95(N)=1.96+2.3539/N+3.1902/N**2+0.7004/N**3+4.5015/N**4
10    CALL MSHEAD
      CALL UNITS(NTER,NINT,NOUT)
      READ(NINT,'(A80)')TITLE
      READ(NINT,*)IDGRP
      READ(NINT,*)XHS,XL,AREA
      XHS=XHS/6.D1
      READ(NINT,*)WCAL,ECAL
      ITYPE=1
      IF(ECAL.GT.0.)ITYPE=2
C Convert to SI units.
      CALL CONVSI(IDGRP,WTFAC,XLFAC)
      STRFAC=WCAL*WTFAC/(AREA*XLFAC*XLFAC)/1.D6
      READ(NINT,*)TI1,TI2
      NP2=0
C Read in A/D digits.
      DO 40 I=1,NMAX1
        IF(ITYPE.EQ.1)THEN
          READ(NINT,*,END=50)W(I)
        ELSE IF(ITYPE.EQ.2)THEN
          READ(NINT,*,END=50)W(I),E(I)
        ENDIF
        IF(I.EQ.NMAX1)CALL STOPP(NTER,NMAX1-1)
        T(I)=(I-1)*TI1
        IPN(I)=I
C Check if a second time interval has been utilized.
```

```
          IF(W(I).EQ.99999)THEN
            DO 20 J=I,NMAX1
              IF(ITYPE.EQ.1)THEN
                READ(NINT,*,END=30)W(J)
              ELSE IF(ITYPE.EQ.2)THEN
                READ(NINT,*,END=30)W(J),E(J)
              ENDIF
              IF(J.EQ.NMAX1)CALL STOPP(NTER,NMAX1-1)
              T(J)=(J-I+1)*TI2+T(I-1)
              IPN(J)=J
20          CONTINUE
30          NP2=J-I
            GOTO 50
          ENDIF
40      CONTINUE
50      NP1=I-1
60      NPZ=NP1+NP2
C If desired, smooth out noise. In doing so, original data is destroyed.
        WRITE(NTER,*)' Is smoothing of force signal desired? (Y/N)'
        READ(NTER,'(A1)')ANS
        IF(ANS.EQ.'Y'.OR.ANS.EQ.'y')CALL SMOOTH(NTER,NP1,W,RES)
        IF(ITYPE.EQ.2)THEN
          WRITE(NTER,*)' Is smoothing of extension signal desired? (Y/N)'
          READ(NTER,'(A1)')ANS
          IF(ANS.EQ.'Y'.OR.ANS.EQ.'y')CALL SMOOTH(NTER,NP1,E,RES)
        ENDIF
        CALL AMAXI(NPZ,W,WMAX,IWMAX)
        IF(ITYPE.EQ.2)CALL AMAXI(NPZ,E,EMAX,IEMAX)
C Search for starting point IS based on force data.
C The parameter PAR controls the decision that extension has started.
        PAR=3.
        IS=1
        CALL START(NP1,W,IS,PAR,WT)
        IS=IS+2
        IS0=IS
        CALL STAT(IS,W,WBL,WSD)
        IF(WSD.LT.SQRT(1.D0/REAL(IS)))WSD=SQRT(1.D0/REAL(IS))
        IF(ITYPE.EQ.1)THEN
          SR=XHS/XL
        ELSE
          CALL STAT(IS,E,EBL,ESD)
          IF(ESD.LT.SQRT(1.D0/REAL(IS)))ESD=SQRT(1.D0/REAL(IS))
          J=0
          DO 70 I=IS,IS+NP1/2
            J=J+1
            X(J)=T(I)
            Y(J)=E(I)*ECAL/XL
70        CONTINUE
          CALL POLYN(2,J,X,Y,A,AERR,RES,SEE)
          SR=A(2)
        ENDIF
        PPPS=1.0D-2/SR/TI1
        NP3=MIN(NP1/10,INT(3.*PPPS))
        WSN=(W(IS+NP3)-WBL)/WSD/NP3
        IF(WSN.LT..01)THEN
          WRITE(NTER,*)' The signal/noise ratio is too low!'
          STOP
        ENDIF
C Estimate the modulus using first-order linear regression.
        J=0
        DO 80 I=IS,IS+NP3
          J=J+1
          IF(ITYPE.EQ.1)THEN
            X(J)=T(I)*SR
          ELSE IF(ITYPE.EQ.2)THEN
            X(J)=(E(I)-EBL)*ECAL/XL
          ENDIF
          Y(J)=W(I)-WBL
80      CONTINUE
```

```
      CALL POLYN(2,J,X,Y,A,AERR,RES,SEE)
      XMOD=STRFAC*A(2)
      ERR0=T95(J-2)*STRFAC*AERR(2)
C Update S/N using signal at strain PS and dividing by number of points.
      PS=1./XMOD
      ITH=INT(PS/SR/TI1)+IS+4
      WSN=(W(ITH)-W(IS))/WSD/(IPN(ITH)-IPN(IS))
      IF(ITYPE.EQ.2)ESN=(E(ITH)-E(IS))/ESD/(IPN(ITH)-IPN(IS))
      DVZT=0.D0
      FAC=0.D0
C The length of point sequence to be included in calculation.
      IP=MAX(INT(PPPS/2)+1,INT(10/WSN)+3)
      DFAC=TI1*IP*SR
C Check for reasonable DFAC.
      DFACM=(IPN(IWMAX)-IPN(IS))*SR*TI1/5.D0
      IF(DFAC.GT.DFACM)DFAC=DFACM
      NCOUNT=0
      FAC=DMIN1(DFAC,1.0D-3)
90    FAC=FAC+DFAC
100   NPT=IS+INT(FAC/SR/TI1)+1
      IF(NPT.LE.IWMAX)THEN
        NCOUNT=NCOUNT+1
        IF(NCOUNT.LE.15)THEN
          J=0
          DO 110 I=IS,NPT
            J=J+1
            IF(ITYPE.EQ.1)THEN
              X(J)=T(I)*SR
            ELSE IF(ITYPE.EQ.2)THEN
              X(J)=(E(I)-EBL)*ECAL/XL
            ENDIF
            Y(J)=(W(I)-WBL)/WMAX
110       CONTINUE
          CALL POLYN(3,J,X,Y,A,AERR,RES,SEE)
          Q=A(2)**2-4.*A(1)*A(3)
C If Q ≤ 0, there is no intersection with the baseline.
          IF(Q.LE.0.)THEN
            IS=IS+1
            GOTO 100
          ENDIF
C Calculate the initial guesses for the stress-strain function.
          B(2)=(-A(2)+SQRT(Q))/2./A(3)
          B(1)=A(2)+2.*B(2)*A(3)
          B(3)=-A(3)/B(1)
          DVZ=B(1)*WMAX*TI1*SR
C If B(3) < 0, then the curve is the wrong way (due to toe).
          IF(B(3).LT.0.)THEN
            IS=IS+1
            GOTO 100
          ENDIF
C Eliminate points below zero strain.
          IF(ITYPE.EQ.1)THEN
            IB2=INT(B(2)/SR/TI1)+1
          ELSE IF(ITYPE.EQ.2)THEN
            DO 120 I=1,J
              IF((E(I)-EBL)*ECAL/XL-B(2).GE.0)THEN
                IB2=I
                GOTO 130
              ENDIF
120         CONTINUE
          ENDIF
130       IF(IB2.GT.IS)IS=IB2
C Check for increase in slope; if slope increases, add more points.
          IF(DVZ.GT.DVZT)THEN
            DVZT=DVZ
            IST=IS
            NPE=NPT
            GOTO 90
          ENDIF
```

```
        ENDIF
      ENDIF
140   IST=IS
      NPE=NPT
      IF(J.GT.8.AND.WSN.GT..1)THEN
C Specify the convergence criterion CC.
        CALL SPECC(NTER,CC)
        CALL GAUSSN(FSS,CC,NPAR,J,X,Y,B,S,ST,R,ERR,SEE)
      ELSE
        WRITE(NTER,'(/2X,A/)')'The GAUSSN routine was not used.'
        ERR(2)=AERR(1)/SQRT(Q)+ABS((-1.+A(2)/SQRT(Q))/2./A(3))*AERR(2)+
     &         ABS(-A(1)/SQRT(Q)-B(2))*AERR(3)/A(3)
        ERR(1)=AERR(2)+2.*(B(2)*AERR(3)+A(3)*ERR(2))
        ERR(3)=AERR(3)+ERR(1)/B(1)**2
        SEE=1.
      ENDIF
C Calculate the 95% confidence intervals.
      DO 150 I=1,NPAR
        ERR(I)=T95(J-NPAR)*SEE*ERR(I)
150   CONTINUE
C Calculate the initial modulus, B(1).
      B(1)=B(1)*WMAX*STRFAC
      ERR(1)=ERR(1)*WMAX*STRFAC
C Calculate the 1% secant modulus, XMOD2.
      XMOD2=B(1)*EXP(-B(3)/1.D2)
      ERR2=XMOD2*(ERR(1)/B(1)/(WMAX*STRFAC)+ERR(3)/1.D2)
C Calculate the 1-s relaxation modulus, XMOD3.
      IF(ITYPE.EQ.1)THEN
        OSS=XHS/XL
      ELSE IF(ITYPE.EQ.2)THEN
        IF(TI1*(NP1-IST).GE.1.)THEN
          ISM=INT(1./TI1)+IST
        ELSE IF(TI1*(NP1-IST).LT.1.AND.TI2.GT.0.)THEN
          ISM=INT((1.-TI1*(NP1-IST))/TI2)+NP1
        ELSE
          ISM=0
        ENDIF
        IF(ISM.GT.0.)THEN
          OSS=(E(ISM)-EBL)*ECAL/XL-B(2)
        ELSE
          OSS=0
        ENDIF
      ENDIF
      XMOD3=B(1)*EXP(-B(3)*OSS)*(1.-B(3)*OSS)
      ERR3=XMOD3/B(1)*ERR(1)+XMOD3*OSS*ABS(2.*OSS*B(3)-1.)*ERR(3)
C If out of range of data, calculate the 1-s modulus separately.
       IF((NPE-IST+1)*TI1.LT.1.AND.TI1*(NP1-IST)+TI2*NP2.GT.1.)THEN
         NPC=MIN(IWMAX-6,IDINT(1./TI1)+IST)
         K=0
         DO 160 I=MAX(IST,NPC-5),NPC+6
           K=K+1
           X(K)=T(I)-B(2)/SR-1.D0
           IF(ITYPE.EQ.1)THEN
             Y(K)=(W(I)-WBL)*STRFAC/(T(I)*SR-B(2))
           ELSE IF(ITYPE.EQ.2)THEN
             Y(K)=(W(I)-WBL)*STRFAC/((E(I)-EBL)*ECAL/XL-B(2))
           ENDIF
160      CONTINUE
         NO=NPAR
         CALL POLYN(NO,K,X,Y,C,AERR,RES,SEE)
         IF(NO.EQ.NPAR.AND.ABS(C(NO)/ERR(NO)).LT.T95(K-NO))THEN
           NO=NO-1
           CALL POLYN(NO,K,X,Y,C,AERR,RES,SEE)
         ENDIF
         XMOD3=C(1)+C(2)
         ERR3=T95(K-NO)*(AERR(1)+AERR(2))
       ENDIF
      NIN=IST
      DWM=0.D0
```

```
      DO 170 I=IST,NPZ-1
        DW=W(I)-W(I+1)
        IF(DW.GT.DWM)THEN
          DWM=DW
          NIN=I
        ENDIF
170   CONTINUE
      DO 180 I=NIN,IST,-1
        IF(W(I-1).LE.W(I))GOTO 190
180   CONTINUE
190   IB=I
      IF(ITYPE.EQ.1)THEN
        EBRK=SR*T(IB)-B(2)
      ELSE IF(ITYPE.EQ.2)THEN
        EBRK=(E(IB)-EBL)*ECAL/XL-B(2)
      ENDIF
      STBK=(W(IB)-WBL)*STRFAC
C Generate output file.
      WRITE(NTER,*)' Generate detailed (D) or summary (S) results file?'
      READ(NTER,'(A1)')ANS
      IF(ANS.EQ.'D'.OR.ANS.EQ.'d')IRF=1
      WRITE(NTER,200)
      WRITE(NOUT,200)
200   FORMAT(/20X,'Tensile Stress-Strain Analysis'/20X,30('-')/
     &       /19X,'Y = B1*(X-B2)*exp(-B(3)*(X-B2))'/
     &       /32X,'Y = stress'/25X,'X - B(2) = strain'/)
      WRITE(NOUT,'(/A80)')TITLE
      WRITE(NOUT,210)XL*XLFAC*1.E3,XL,AREA*XLFAC*XLFAC*1.E6,AREA
210   FORMAT(/13X,'Test specimen length =',F10.2,' mm',5X,'(',F8.2,')'/
     &      13X,'Cross-sectional area =',F10.2,' mm**2  (',1P,E8.2,')')
      CS=XHS*XLFAC*6.E4
      CML=WCAL*WTFAC*(WMAX-WBL)
      CCT=XHS*XLFAC*(NP1*TI1+NP2*TI2)*1.E3
      WRITE(NTER,220)CS,XHS*60.,CML,CML/WTFAC,CCT,CCT/XLFAC/1.E3
      WRITE(NOUT,220)CS,XHS*60.,CML,CML/WTFAC,CCT,CCT/XLFAC/1.E3
220   FORMAT(       /18X,'Crosshead speed =',F10.2,' mm/min (',F8.2,')'
     &     /10X,'Calculated maximum load =',F10.2,' N',6X,'(',F8.2,')'
     &  /6X,'Calculated crosshead travel =',F10.2,' mm',5X,'(',F8.2,')')
      IF(IRF.EQ.1)THEN
        WRITE(NOUT,230)SR,TI1,PPPS,NPZ,ISO
230     FORMAT(/6X,'Initial nominal strain rate =',1P,E10.2,' 1/s'
     &        /11X,'Initial point interval =',0P,F10.3,' s'
     &        /14X,'Points per % strain =',F10.1
     &        /11X,'Total number of points =',I10
     &        /9X,'Estimated starting point =',I10)
        WRITE(NOUT,240)WSN,WMAX,IWMAX,WBL,WSD,W(NIN),NIN,W(IB),IB
240     FORMAT(/5X,'Force signal/noise per point =',F10.2
     &         /5X,'Maximum force signal (point) =',F10.1,1X,'(',I4,')'
     &        /19X,'Force baseline =',F10.1
     &         /7X,'Force baseline noise level =',F10.3
     &        /17X,'Inflection point =',F10.1,1X,'(',I4,')'
     &        /22X,'Break point =',F10.1,1X,'(',I4,')')
        IF(ITYPE.EQ.2)WRITE(NOUT,250)ESN,EMAX,IEMAX,EBL,ESD
250     FORMAT(1X,'Extension signal/noise per point =',F10.2
     &         /1X,'Maximum extension signal (point) =',F10.1,1X,'(',I4,
     &     ')'/15X,'Extension baseline =',F10.1
     &         /3X,'Extension baseline noise level =',F10.3)
      ENDIF
      WRITE(NTER,260)XMOD,ERR0,B(1),ERR(1),XMOD2,ERR2
      WRITE(NOUT,260)XMOD,ERR0,B(1),ERR(1),XMOD2,ERR2
260   FORMAT(/16X,'Estimated modulus =',1P,E10.2,' +/-',E10.2,' MPa'
     &      /26X,'Modulus =',E10.2,' +/-',E10.2,' MPa'
     &      /16X,'1% secant modulus =',E10.2,' +/-',E10.2,' MPa')
      IF(OSS.GT.0)THEN
        WRITE(NTER,270)XMOD3,ERR3
        WRITE(NOUT,270)XMOD3,ERR3
270     FORMAT(11X,'1-s relaxation modulus =',1P,E10.2,' +/-',E10.2,
     &         ' MPa')
      ENDIF
```

```
      WRITE(NTER,280)STRFAC*(WMAX-WBL),STBK,EBRK*1.0E2
      WRITE(NOUT,280)STRFAC*(WMAX-WBL),STBK,EBRK*1.0E2
280   FORMAT(/17X,'Tensile strength =',1P,E10.2,' MPa'
     &        /8X,'Tensile strength at break =',E10.2,' MPa'
     &       /14X,'Elongation at break =',0P,F10.1,' %')
      WRITE(NTER,290)IST,NPE,J*TI1*SR
      WRITE(NOUT,290)IST,NPE,J*TI1*SR
290   FORMAT(/2X,'The stress-strain function results:'/3X,
     &       'The points used in the calculation are from',I3,' to',I4
     &       /3X,'corresponds to a strain of',F6.3/)
      WRITE(NTER,300)(I,B(I),ERR(I),I=1,NPAR)
      WRITE(NOUT,300)(I,B(I),ERR(I),I=1,NPAR)
300   FORMAT(29X,'B(',I1,') =',1P,E10.2,' +/-',E10.2)
      SUM=0.D0
      DO 310 I=IST,NPE
         CALL CALTEN(I,ITYPE,XHS,XL,WBL,EBL,ECAL,EPS,WOBS,FCAL)
         SUM=SUM+(WOBS*STRFAC-FCAL)**2
310   CONTINUE
      SEE=SQRT(SUM/FLOAT(NPE-IST-2))
      WRITE(NOUT,320)SEE,SEE/STRFAC/WSD
320   FORMAT(/7X,'Standard error of estimate =',1P,E10.2
     &        /6X,'SEE(function)/SEE(baseline) =',0P,F10.1)
      WRITE(NOUT,330)
330   FORMAT(/T35,'Nominal Stress, MPa'/T6,'Point',T27,35('-')/T5,
     &       'Number',T16,'Strain',T28,'Observed',T40,'Calculated',
     &      T53,'Residual'/T5,6('-'),3(3X,(10('-'))),3X,9('-'))
      DO 360 I=1,NPE
        CALL CALTEN(I,ITYPE,XHS,XL,WBL,EBL,ECAL,EPS,WOBS,FCAL)
        WOBS=WOBS*STRFAC
        IF(EPS.LT.0.OR.EPS.LE.-B(2))THEN
          WRITE(NOUT,340)IPN(I),WOBS
340       FORMAT(T6,I4,T19,'-',T27,1P,E10.3,T45,'-',T57,'-')
        ELSE
          WRITE(NOUT,350)IPN(I),EPS,WOBS,FCAL,WOBS-FCAL
350       FORMAT(T6,I4,T11,1P,3E13.3,T54,0P,F6.3)
        ENDIF
360   CONTINUE
      IF(IRF.EQ.1)THEN
        DO 380 I=NPE+1,NIN
          CALL CALTEN(I,ITYPE,XHS,XL,WBL,EBL,ECAL,EPS,WOBS,FCAL)
          WRITE(NOUT,370)IPN(I),EPS,WOBS*STRFAC
370       FORMAT(T6,I4,T11,1P,2E13.3,T45,'-',T57,'-')
380     CONTINUE
      ENDIF
      CALL MSTAIL(*10)
      STOP
      END
C
      SUBROUTINE SMOOTH(NTER,NP,S,ST)
C Calls smoothing routine MEDFIL to smooth data NP points of S data.
      IMPLICIT DOUBLE PRECISION (A-H,O-Z)
      DIMENSION S(NP),ST(NP)
      WRITE(NTER,10)
10    FORMAT(2X,'Enter amount of smoothing, i.e.,the number of points'/
     &       2X,'(between 3 and 19) over which data is smoothed.')
      READ(NTER,*)NPP
C NS is number of smoothing stages. If NS=1, vector ST contains the
C  filtered data. If NS > 1, vector S contains the smoothed data.
      WRITE(NTER,*)' Enter number of smoothing stages.'
      WRITE(NTER,*)' TENSIL recommends 2 stages.'
      READ(NTER,*)NS
      CALL MEDFIL(NP,S,NS,NPP,ST)
      IF(NS.EQ.1)THEN
        DO 20 I=1,NP
          S(I)=ST(I)
20      CONTINUE
      ENDIF
      RETURN
      END
```

```
      SUBROUTINE MEDFIL(NP,V,NS,NPP,VT)
C Smooths data in vector V(NP) by replacing midpoint of strip of NPP odd
C  points (3<NPP<19) with median of the set. If number of passes through
C  smoothing routine (input) NS > 1, V contains smoothed data. If NS = 1,
C  VT contains smoothed data and V is unchanged. VT is a working vector.
C  End points are handled by using fewer elements in the strip.
C Requires the function subprogram XMED.
      IMPLICIT DOUBLE PRECISION (A-H,O-Z)
      DIMENSION V(NP),VT(NP),X(20),W(20)
C Check for non-odd number of points in strip.
      NPS=NPP
      IF(NPS/2*2.EQ.NPS)NPS=NPS+1
      IF(NPS.LT.3)NPS=3
      IF(NPS.GE.20)NPS=19
      NPW=NPS/2
      DO 80 N=1,NS
C Do center strips.
        DO 20 I=NPW+1,NP-NPW
          DO 10 J=1,NPS
            X(J)=V(I+J-NPW-1)
10        CONTINUE
          VT(I)=XMED(NPS,X,W)
20      CONTINUE
C Do low-end points.
        DO 40 I=NPW,2,-1
          DO 30 J=1,2*I-1
            X(J)=V(J)
30        CONTINUE
          VT(I)=XMED(J-1,X,W)
40      CONTINUE
        VT(1)=V(1)
C Do high-end points.
        DO 60 I=NP-NPW+1,NP-1
          DO 50 J=1,2*(NP-I)
            X(J)=V(2*I-NP+J-1)
50        CONTINUE
          VT(I)=XMED(J-1,X,W)
60      CONTINUE
        VT(NP)=V(NP)
        IF(NS.NE.1)THEN
          DO 70 I=1,NP
            V(I)=VT(I)
70        CONTINUE
        ENDIF
80    CONTINUE
      RETURN
      END
C
      DOUBLE PRECISION FUNCTION XMED(NP,X,W)
C Returns median value of array X containing NP elements. If NP is even,
C  XMED returns average of two values at the center.
C Requires the subroutine SORT.
      IMPLICIT DOUBLE PRECISION (A-H,O-Z)
      DIMENSION X(NP),W(NP)
      DO 10 I=1,NP
        W(I)=X(I)
10    CONTINUE
      CALL SORT(NP,W)
      MID=NP/2
C If NP is even...
      IF(MID*2.EQ.NP)THEN
        XMED=(W(MID)+W(MID+1))/2.
        RETURN
C If NP is odd...
      ELSE
        XMED=W(MID+1)
        RETURN
      ENDIF
      END
```

```
      SUBROUTINE CALTEN(I,ITYPE,XHS,XL,WBL,EBL,ECAL,EPS,WOBS,FCAL)
C Calculates strain (EPS) and nominal stress (WOBS and FCAL).
      IMPLICIT DOUBLE PRECISION (A-H,O-Z)
      PARAMETER (NMAX1=1001,NPAR=3)
      COMMON /BLOC1/ W(NMAX1),E(NMAX1),T(NMAX1),B(NPAR)
      IF(ITYPE.EQ.1)THEN
        EPS=T(I)*XHS/XL-B(2)
      ELSE IF(ITYPE.EQ.2)THEN
        EPS=(E(I)-EBL)*ECAL/XL-B(2)
      ENDIF
      WOBS=W(I)-WBL
      FCAL=B(1)*EPS*EXP(-B(3)*EPS)
      RETURN
      END
C
      SUBROUTINE CONVSI(IDGRP,WTFAC,XLFAC)
C Provides factors based on IDGRP to convert to SI units.
      IMPLICIT DOUBLE PRECISION (A-H,O-Z)
      WTFAC=1.D0
      XLFAC=1.0D-3
      IF(IDGRP.EQ.1)THEN
        WTFAC=4.448221D0
        XLFAC=2.54D-2
      ELSE IF(IDGRP.EQ.2)THEN
        WTFAC=9.80665D0
      ENDIF
      RETURN
      END
C
      SUBROUTINE FSS(N,X,B,F,DER)
C Evaluates the function F and its first partial derivatives DER(N)
C  at point X when called by GAUSSN.
      IMPLICIT DOUBLE PRECISION (A-H,O-Z)
      DIMENSION B(N),DER(N)
      EPS=X-B(2)
C Determine the function value F(X).
      F=B(1)*EPS*EXP(-B(3)*EPS)
C Determine the first derivatives of the function F with respect to:
C the first parameter B(1);
      DER(1)=F/B(1)
C the second parameter B(2); and,
      DER(2)=-F/EPS+F*B(3)
C the third parameter B(3).
      DER(3)=-F*EPS
      RETURN
      END
C
      SUBROUTINE SORT(NP,W)
C Sorts vector W(NP) into ascending numerical order.  Uses the Box-Lacey
C  Combsort routine: BYTE Magazine, McGraw-Hill, April 1991, p. 315.;
C  plus a revision for IGAP calc. by R. Hills (BYTE, August 1991, p 18).
      IMPLICIT DOUBLE PRECISION (A-H,O-Z)
      DIMENSION W(NP)
      IGAP=NP
      DO 20 K=1,NP
        IGAP=MAX(IGAP*8/11,1)
        ISWIT=0
        DO 10 I=1,NP-IGAP
          J=I+IGAP
          IF(W(I).GT.W(J))THEN
            HOLD=W(I)
            W(I)=W(J)
            W(J)=HOLD
            ISWIT=ISWIT+1
          ENDIF
10      CONTINUE
        IF(ISWIT.EQ.0.AND.IGAP.EQ.1)RETURN
20    CONTINUE
      END
```

```
      SUBROUTINE START(NP,V,IS,PAR,W)
C Given equally spaced signals V of dimension NP and an initial estimate
C  IS, the routine searches for the start IS based on the number of
C  standard deviations PAR above which signal derivative must rise
C  (suggested value is 3).  W is the working vector.
C Requires the subroutine MEDFIL.
      IMPLICIT DOUBLE PRECISION (A-H,O-Z)
      DIMENSION V(NP),W(NP),VD(5)
      IF(NP.LT.20.OR.NP-IS.LT.5)THEN
        PAUSE ' There are too few points for START.'
        STOP
      ENDIF
      IF(PAR.LT.1.)PAR=1.D0
      NS=1
      NPP=3
C Smooth for better resolution of the starting point.
      CALL MEDFIL(NP,V,NS,NPP,W)
C Do seven-point slope determination. Use working vector W so that raw
C  data is unchanged.
      DO 10 I=1,4
        K=I-1
        VD(I)=(-3.D0*W(IS+K)-2.D0*W(IS+K+1)-W(IS+K+2)+W(IS+K+4)
     &         +2.D0*W(IS+K+5)+3.D0*W(IS+K+6))/2.8D1
10    CONTINUE
C Calculate average slope AVE and standard deviation SD around trial
C  starting point.
      SUM=VD(1)+VD(2)+VD(3)
      SUM2=VD(1)**2+VD(2)**2+VD(3)**2
      AVE=SUM/3.D0
      SD=SQRT((SUM2-SUM**2/3.D0)/2.D0)
      N=3
C Get slope at one point beyond trial starting point and check against
C  average.
      DO 20 I=IS+6,NP-3
        VD(5)=(-3.D0*W(I-3)-2.D0*W(I-2)-W(I-1)+W(I+1)+2.D0*W(I+2)
     &          +3.D0*W(I+3))/2.8D1
         IF(ABS(VD(5)-AVE).GT.PAR*SD.AND.VD(4)*VD(5).GT.0.
     &      .AND.ABS(VD(5)).GT.ABS(VD(4)))THEN
           IS=I-1
           RETURN
         ELSE
           N=N+1
           VD(3)=VD(4)
           VD(4)=VD(5)
           SUM=SUM+VD(3)
           SUM2=SUM2+VD(3)**2
           SD=SQRT((SUM2-SUM**2/N)/(N-1))
           AVE=SUM/N
         ENDIF
20    CONTINUE
      PAUSE ' No start found.'
      STOP
      END
```

1.6 Sample Listings of INPUT/OUTPUT Files for TENSIL

1.6.1 Data File for TENSIL

Shown here is an abridged data file, with extension data available and one data acquisition time interval, in the format required for running TENSIL.

```
PS Monsanto Lustrex ASTM D 638 Type V; 29.97 lb; 0.01 in. travel
1
0.02   0.405   5.5811E-03
0.048877       3.785E-05
0.500  0.
     2047      2044
     2047      2045
     2047      2044
     2047      2044
     2047      2045
     2047      2044
     2047      2044
     2047      2044
     2047      2044
     2047      2044
     2056      2047
     2063      2048
     2071      2049
     2077      2051
     2083      2052
     2088      2053
     2091      2053
     2100      2054
     2105      2055
     2111      2057
     2117      2058
     2122      2058
     2128      2059
     2133      2059
     2139      2060
     2145      2061
     2150      2062
     2155      2064
     2161      2066
     2167      2067
     2172      2068
     2178      2069
     2183      2069
     2188      2070
     2194      2071
     2199      2073
     2204      2075
     2210      2076
     2214      2076
     2219      2077
     2225      2078
     2230      2079
     2236      2081
     2242      2083
     2247      2084
     2251      2084
     2258      2084
     2262      2085
     2267      2086
     2274      2087
     2278      2088
     2284      2090
     2289      2091
     2294      2092
     2299      2092
     2304      2094
     2310      2096
       .         .
       .         .
       .         .
     2657      2181
     2660      2184
     2663      2187
     2047      2045
```

1.6.2 Terminal Dialog with **TENSIL**

Reproduced here is the terminal dialog for running TENSIL with the data file TENSIL.DAT listed in Section 1.6.1.

```
>TENSIL
 Enter name of data file (N, if unavailable).
TENSIL.DAT
 Results to be written to NEW file: TENSIL.RES
 OK (Y/N)?
Y
 Is smoothing of force signal desired? (Y/N)
N
 Is smoothing of extension signal desired? (Y/N)
N
 Use the default convergence criterion, 1.E-4? (Y/N)
Y
 Generate a detailed (D) of summary (S) results file?
D

                  Crosshead speed =        .51 mm/min (       .02)
          Calculated maximum load =     133.93 N       (    30.11)
      Calculated crosshead travel =        .54 mm      (       .02)

                Estimated modulus =  3.49E+03 +/-  2.85E+02 MPa
                          Modulus =  3.78E+03 +/-  1.57E+02 MPa
                1% secant modulus =  2.95E+03 +/-  2.06E+02 MPa
            1-s relaxation modulus =  3.71E+03 +/-  1.63E+02 MPa

                 Tensile strength =  3.72E+01 MPa
        Tensile strength at break =  3.72E+01 MPa
              Elongation at break =       1.3 %

  The stress-strain function results:
   The points used in the calculation are from 11 to  68
   corresponds to a strain of  .006

                             B(1) =  3.78E+03 +/-  1.57E+02
                             B(2) =  8.23E-05 +/-  5.43E-05
                             B(3) =  2.47E+01 +/-  6.95E+00

 The results are in TENSIL.RES
 Run another set of data? (Y/N)
N
Stop - Program terminated.
```

1.6.3 Output File from **TENSIL**

Reproduced here is the detailed version of an output file from TENSIL using the data and terminal dialog listed in Sections 1.6.1 and 1.6.2, respectively. The results are plotted in Figure 1.7.

```
                    Tensile Stress-Strain Analysis
                    ------------------------------

                  Y = B1*(X-B2)*exp(-B(3)*(X-B2))

                               Y = stress
                        X - B(2) = strain


PS Monsanto Lustrex ASTM D 638 Type V; 29.97 lb; 0.01 in. travel

             Test specimen length =     10.29 mm       (       .41)
             Cross-sectional area =      3.60 mm**2    (5.58E-03)

                  Crosshead speed =       .51 mm/min (       .02)
         Calculated maximum load =    133.93 N        (     30.11)
      Calculated crosshead travel =       .54 mm      (       .02)

     Initial nominal strain rate =  2.04E-04 1/s
          Initial point interval =      .500 s
               Points per % strain =      98.2
           Total number of points =       128
         Estimated starting point =        10

      Force signal/noise per point =     21.61
      Maximum force signal (point) =    2663.0 ( 127)
                    Force baseline =    2047.0
         Force baseline noise level =      .316
                  Inflection point =    2663.0 ( 127)
                       Break point =    2663.0 ( 127)
 Extension signal/noise per point =      3.56
 Maximum extension signal (point) =    2187.0 ( 127)
                Extension baseline =    2044.2
     Extension baseline noise level =      .422

               Estimated modulus =  3.49E+03 +/-   2.85E+02 MPa
                         Modulus =  3.78E+03 +/-   1.57E+02 MPa
               1% secant modulus =  2.95E+03 +/-   2.06E+02 MPa
          1-s relaxation modulus =  3.71E+03 +/-   1.63E+02 MPa

                Tensile strength =  3.72E+01 MPa
       Tensile strength at break =  3.72E+01 MPa
             Elongation at break =       1.3 %

 The stress-strain function results:
  The points used in the calculation are from 11 to  68
  corresponds to a strain of  .006

                            B(1) =  3.78E+03 +/-   1.57E+02
                            B(2) =  8.23E-05 +/-   5.43E-05
                            B(3) =  2.47E+01 +/-   6.95E+00

      Standard error of estimate =  2.34E-01
     SEE(function)/SEE(baseline) =      12.2
```

Point Number	Strain	Nominal Stress, MPa Observed	Calculated	Residual
1	-	0.000E+00	-	-
2	-	0.000E+00	-	-
3	-	0.000E+00	-	-
4	-	0.000E+00	-	-
5	-	0.000E+00	-	-
6	-	0.000E+00	-	-
7	-	0.000E+00	-	-
8	-	0.000E+00	-	-
9	-	0.000E+00	-	-
10	-	0.000E+00	-	-
11	1.794E-04	5.434E-01	6.751E-01	-.132
12	2.728E-04	9.661E-01	1.024E+00	-.058
13	3.663E-04	1.449E+00	1.372E+00	.077
14	5.532E-04	1.811E+00	2.063E+00	-.251
15	6.467E-04	2.174E+00	2.406E+00	-.232
16	7.401E-04	2.476E+00	2.747E+00	-.271
17	7.401E-04	2.657E+00	2.747E+00	-.090
18	8.336E-04	3.200E+00	3.087E+00	.114
19	9.270E-04	3.502E+00	3.425E+00	.077
20	1.114E-03	3.864E+00	4.096E+00	-.232
21	1.207E-03	4.227E+00	4.430E+00	-.203
22	1.207E-03	4.529E+00	4.430E+00	.099
23	1.301E-03	4.891E+00	4.762E+00	.129
24	1.301E-03	5.193E+00	4.762E+00	.431
25	1.394E-03	5.555E+00	5.092E+00	.463
26	1.488E-03	5.917E+00	5.421E+00	.497
27	1.581E-03	6.219E+00	5.748E+00	.471
28	1.768E-03	6.521E+00	6.398E+00	.123
29	1.955E-03	6.883E+00	7.042E+00	-.158
30	2.049E-03	7.246E+00	7.361E+00	.116
31	2.142E-03	7.548E+00	7.680E+00	-.132
32	2.235E-03	7.910E+00	7.996E+00	-.086
33	2.235E-03	8.212E+00	7.996E+00	.216
34	2.329E-03	8.514E+00	8.311E+00	.203
35	2.422E-03	8.876E+00	8.625E+00	.251
36	2.609E-03	9.178E+00	9.248E+00	-.070
37	2.796E-03	9.480E+00	9.865E+00	-.385
38	2.890E-03	9.842E+00	1.017E+01	-.329
39	2.890E-03	1.008E+01	1.017E+01	-.087
40	2.983E-03	1.039E+01	1.048E+01	-.090
41	3.077E-03	1.075E+01	1.078E+01	-.031
42	3.170E-03	1.105E+01	1.108E+01	-.031
43	3.357E-03	1.141E+01	1.168E+01	-.268
44	3.544E-03	1.177E+01	1.227E+01	-.499
45	3.637E-03	1.208E+01	1.257E+01	-.492
46	3.637E-03	1.232E+01	1.257E+01	-.250
47	3.637E-03	1.274E+01	1.257E+01	.172
48	3.731E-03	1.298E+01	1.286E+01	.121
49	3.824E-03	1.328E+01	1.315E+01	.131
50	3.918E-03	1.371E+01	1.344E+01	.263
51	4.011E-03	1.395E+01	1.373E+01	.215
52	4.198E-03	1.431E+01	1.431E+01	.004
53	4.291E-03	1.461E+01	1.459E+01	.021
54	4.385E-03	1.491E+01	1.487E+01	.039
55	4.385E-03	1.522E+01	1.487E+01	.341
56	4.572E-03	1.552E+01	1.544E+01	.081
57	4.759E-03	1.588E+01	1.599E+01	-.114
58	4.946E-03	1.618E+01	1.655E+01	-.364
59	4.946E-03	1.648E+01	1.655E+01	-.062
60	5.039E-03	1.685E+01	1.682E+01	.026
61	5.133E-03	1.721E+01	1.709E+01	.116
62	5.226E-03	1.745E+01	1.736E+01	.086
63	5.320E-03	1.781E+01	1.763E+01	.179
64	5.413E-03	1.805E+01	1.790E+01	.152
65	5.506E-03	1.842E+01	1.817E+01	.247

66	5.787E-03	1.878E+01	1.896E+01	-.184
67	5.880E-03	1.914E+01	1.922E+01	-.084
68	5.974E-03	1.944E+01	1.949E+01	-.042
69	5.974E-03	1.974E+01	-	-
70	6.161E-03	1.999E+01	-	-
71	6.348E-03	2.035E+01	-	-
72	6.534E-03	2.071E+01	-	-
73	6.628E-03	2.101E+01	-	-
74	6.628E-03	2.131E+01	-	-
75	6.628E-03	2.168E+01	-	-
76	6.815E-03	2.198E+01	-	-
77	7.002E-03	2.228E+01	-	-
78	7.189E-03	2.264E+01	-	-
79	7.189E-03	2.301E+01	-	-
80	7.282E-03	2.331E+01	-	-
81	7.376E-03	2.361E+01	-	-
82	7.469E-03	2.391E+01	-	-
83	7.562E-03	2.433E+01	-	-
84	7.656E-03	2.439E+01	-	-
85	7.749E-03	2.494E+01	-	-
86	7.843E-03	2.524E+01	-	-
87	8.030E-03	2.554E+01	-	-
88	8.123E-03	2.590E+01	-	-
89	8.310E-03	2.621E+01	-	-
90	8.404E-03	2.651E+01	-	-
91	8.404E-03	2.687E+01	-	-
92	8.591E-03	2.723E+01	-	-
93	8.777E-03	2.747E+01	-	-
94	8.964E-03	2.778E+01	-	-
95	9.058E-03	2.814E+01	-	-
96	9.058E-03	2.826E+01	-	-
97	9.151E-03	2.880E+01	-	-
98	9.151E-03	2.904E+01	-	-
99	9.338E-03	2.941E+01	-	-
100	9.432E-03	2.971E+01	-	-
101	9.619E-03	3.001E+01	-	-
102	9.712E-03	3.031E+01	-	-
103	9.805E-03	3.073E+01	-	-
104	9.899E-03	3.092E+01	-	-
105	9.992E-03	3.122E+01	-	-
106	1.018E-02	3.152E+01	-	-
107	1.027E-02	3.182E+01	-	-
108	1.037E-02	3.212E+01	-	-
109	1.055E-02	3.236E+01	-	-
110	1.074E-02	3.267E+01	-	-
111	1.093E-02	3.303E+01	-	-
112	1.102E-02	3.309E+01	-	-
113	1.111E-02	3.363E+01	-	-
114	1.111E-02	3.387E+01	-	-
115	1.111E-02	3.418E+01	-	-
116	1.121E-02	3.460E+01	-	-
117	1.139E-02	3.472E+01	-	-
118	1.167E-02	3.502E+01	-	-
119	1.186E-02	3.532E+01	-	-
120	1.195E-02	3.563E+01	-	-
121	1.214E-02	3.605E+01	-	-
122	1.233E-02	3.617E+01	-	-
123	1.252E-02	3.641E+01	-	-
124	1.261E-02	3.647E+01	-	-
125	1.270E-02	3.683E+01	-	-
126	1.298E-02	3.701E+01	-	-
127	1.326E-02	3.719E+01	-	-

References

1. ASTM Standard Test Method D 638-91, "Standard Test Method for Tensile Properties of Plastics," 1993 *Annual Book of ASTM Standards.* ASTM, Philadelphia, Vol. 08.01, 1993, pp. 161-173.
2. ASTM Standard Test Method D 882-91, "Standard Test Methods for Tensile Properties of Thin Plastic Sheeting," 1993 *Annual Book of ASTM Standards.* ASTM, Philadelphia, Vol. 08.01, 1993, pp. 316-324.
3. Birley, A. W.; Haworth, B.; Batchelor, J.: *Physics of Plastics: Processing, Properties and Materials Engineering.* Hanser, New York, 1992.
4. Smith, T. L.: "Nonlinear Viscoelastic Response of Amorphous Elastomers to Constant Strain Rates," *Trans. Soc. Rheol.* **6**, 61-80 (1962).
5. Smith, T. L.: "Linear Viscoelastic Response to a Deformation at Constant Strain Rate: Derivation of Physical Properties of a Densely Crosslinked Elastomer," *Trans. Soc. Rheol.* **20**, 103-117 (1976).
6. Smith, T. L.: *Treatise on Materials Science and Technology.* Academic Press, New York, 1977, Vol. 10, Part A, p. 369.
7. Matsuoka, S.: "Free Volume, Excess Entropy and Mechanical Behavior of Polymeric Glasses," *Polym. Eng. Sci.* **21**, 907-921 (1981).
8. Matsuoka, S.; Ryan, J. T.; Ghoneim, H.: "Computerized Prediction of Mechanical Behavior over a Wide Temperature-Time Range," *SPE Tech. Papers* **28**, 16-18 (1982).
9. Staats-Westover, R. F.; Matsuoka, S.: "A Critical Review of ASTM D-638 Tensile Test Method for Plastics," *SPE Tech. Papers* **28**, 25-28 (1982).
10. Press, W. H.; Flannery, B. P.; Teukolsky, S. A.; Vetterling, W. T.: *Numerical Recipes in* FORTRAN: *The Art of Scientific Computing.* Cambridge University Press, Cambridge, 1986.
11. Himmelblau, D. M.: *Process Analysis by Statistical Methods.* Wiley, New York, 1970.

Chapter 2

Tensile Modulus of Elastomers Constant-Force Method

2.1 Introduction

The equilibrium tensile or Young's modulus, E_e, of an elastomer is of critical importance to its application. Young's modulus is also a useful parameter for characterizing the covalent network making up the material, and is used to follow or describe the state of cure of the material. A commonly used expression for modulus is

$$E_e = 3f\,\frac{\rho RT}{M_c}\left(1 - \frac{2M_c}{M_n}\right) \tag{2.1}$$

where f is a front factor close to unity with a complex dependence on cross-link density, ρ is the density of the material, M_c is the molecular weight between cross-links, and M_n is the molecular weight of the main chains if the cross-links were to be removed. The parenthetical factor corrects for the fact that the dangling ends of the network do not contribute to the equilibrium modulus [1].

The modulus is measured by examining the equilibrium stress-strain properties of the elastomer. As described in Chapter 3, the experimental procedure may be time consuming. To obviate the problems connected with drift of transducers and tying up expensive instrumentation, the modulus can be easily measured in a constant-force, uniaxial extension experiment. The procedure involves hanging weights on a thin strip of the elastomer and waiting for a suitable length of time before the extension is measured. The amount of time required should be determined by an analysis similar to that described in Chapter 3. The length of the elastomer between gauge marks L can be measured by several methods, including a cathetometer accurate to 0.01 mm. Both ascending and descending stresses should be used to check for equilibrium or mechanical hysteresis effects.

The analysis of the stress-strain curve is often done in terms of the Mooney-Rivlin phenomenological equation [2-5] by considering the elastomer as an incompressible continuum capable of storing energy during deformation. The Mooney-Rivlin equation is

$$\frac{F}{A_o} = 2\left(C_1 + \frac{C_2}{\lambda}\right)\left(\lambda - \frac{1}{\lambda^2}\right) \tag{2.2}$$

where F is the force applied uniaxially to the elastomeric specimen, A_o is the undeformed cross-sectional area, the uniaxial extension $\lambda = L/L_o$ is the specimen length L normalized by the undeformed initial length L_o, and C_1 and C_2 are the Mooney-Rivlin constants independent of λ. Comprehensive reviews of the phenomenological theory are available elsewhere [6-8].

Modulus is defined as the derivative of the stress with strain at zero strain. Thus,

$$E_e = \lim_{\varepsilon \to 0} \frac{\partial(F/A_o)}{\partial \varepsilon} \tag{2.3}$$

From the relationships

$$\varepsilon = \lambda - 1 \tag{2.4.1}$$

$$A = A_o \lambda^{-1} \tag{2.4.2}$$

$$\frac{\partial}{\partial \varepsilon} = \frac{\partial}{\partial \lambda} \tag{2.4.3}$$

the derivative of the stress with strain can be expressed as

$$\frac{\partial \sigma}{\partial \varepsilon} = \frac{\partial}{\partial \lambda}\left[2\left(C_1 + \frac{C_2}{\lambda}\right)\left(\lambda^2 - \frac{1}{\lambda}\right)\right] \tag{2.5}$$

where σ is the true stress, $(F/A_o)\lambda$. The equilibrium tensile modulus is then

$$E_e = 6(C_1 + C_2) \tag{2.6}$$

Although the molecular origin of C_2 remains a subject of controversy, some investigators [9,10] feel that the connection between network structure and limiting equilibrium behavior should not involve C_2, although C_2 may not be negligible under apparently equilibrium conditions [11]. Nonaffine equilibrium deformations also give rise to C_2 [12].

The traditional analysis of Equation 2.2 is with the aid of the Mooney-Rivlin plot, where the reduced stress, $F/[A_o(\lambda - \lambda^{-2})]$, is plotted versus $1/\lambda$. This plot should give a straight line, up to the limit of Gaussian behavior or finite extensibility, with a slope of $2C_2$ and an intercept of $2C_1$. However, the method introduces a pair of annoying problems. It is sensitive to data having high scatter at low strains. This sensitivity was best illustrated by van der Hoff and Buckler, who introduced a $\pm$ 1% error in the initial length [13]. In Figure 2.1, we adapt this illustration on real data for a rubber band at 27 °C. It is clear that any deviation from the true value of L_o would lead to disastrous results. For example, a negative C_1 may result if the experimentally measured initial length is shorter than the actual L_o. In the final

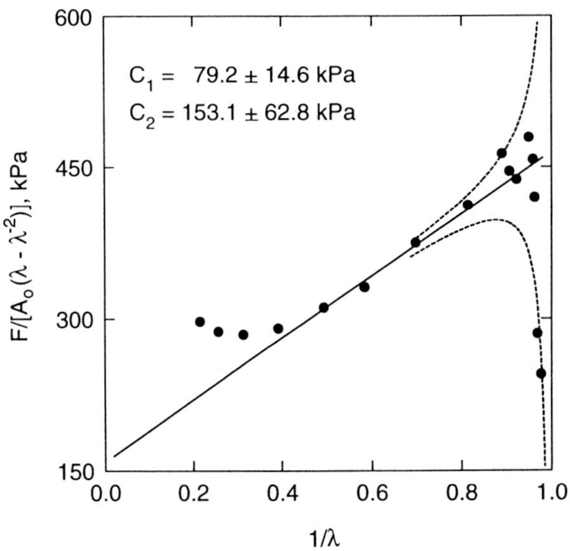

Figure 2.1 Adapting the illustrative method of van der Hoff and Buckler [13], the sensitivity of the Mooney-Rivlin plot to systematic errors at low strains is described by the 1% error bands. The constants were calculated from STRECH. The deviation from theory at high extensions was due to stress-induced crystallization. Adapted from Ref. 13, p. 756, by courtesy of Marcel Dekker, Inc.

analysis, the result of the application of a Mooney-Rivlin plot is determined by an arbitrary decision concerning which data, at low extensions, should be rejected.

The second problem is statistical in nature. The applied weight in this experiment is the error-free independent variable, whereas the length is the observed dependent variable subject to random error. When the length is determined by using a cathetometer, the variance in the extension ratio will be roughly independent of λ. Therefore, the proper way to handle the data is to fit the Mooney-Rivlin equation, solved for λ, to experimentally observed values of λ. The resulting optimization for the true extension ratio λ_c introduces a third parameter, a calculated zero length $L_{o,c}$. With this modification, Equation 2.2 can be rewritten as

$$\frac{F}{A_o} = \beta_1\left(1 + \frac{\beta_2\beta_3}{\lambda}\right)\left[\frac{\lambda}{\beta_3} - \left(\frac{\beta_3}{\lambda}\right)^2\right] \tag{2.7}$$

where $\beta_1 = 2C_1$, $\beta_2 = C_2/C_1$, and $\beta_3 = \lambda/\lambda_c = L_{o,c}/L_o$. The ratio defined by β_2 has been related to the extent to which deformation becomes nonaffine as elongation increases [12] and to the looseness with which the cross-links are embedded within the network [14]. The use of a third parameter to estimate the undeformed length is not unrealistic or new [9,15]. It circumvents the difficult task of measuring the length of a flexible specimen without applying any straightening force. Equation 2.7 is the basis for the program STRECH to be described in the next section.

2.2 Description of the Algorithm STRECH

The program STRECH is semi-interactive. Assuming the undeformed initial length as a parameter, the Mooney-Rivlin equation becomes a three-parameter nonlinear equation, described in Equation 2.7, and is included in the function FMR. The solution of this equation for λ_c is found by the Newton-Raphson iterative method in the external subroutine EQSOLN. The optimization of these solutions with the corresponding experimentally observed values, λ, is accomplished in the IMSL subroutine DBCONF which uses a quasi-Newton method and a finite-difference gradient with simple bounds on the parameters [16]. To ensure an accurate search, the routine redefines the first parameter as $\log \beta_1$. The objective function to be minimized, included in the subroutine FSTR, is expressed as

$$\sum_{i=1}^{N} \left[\frac{\lambda_{i,c} - \lambda_i}{\lambda_i} \right]^2 \tag{2.8}$$

where N is the number of observations. The denominator leads to heavier weighting of data at low values of the extension ratio in concert with the realization that the phenomenological equation is likely to provide a better description of the data at smaller strains.

The initial estimates for the first two parameters in the optimization routine are obtained through the traditional Mooney-Rivlin plot using the subroutine ALINR. ALINR fits a line using the method of least squares. STRECH allows the user the option of limiting the analysis of the Mooney-Rivlin plot within a bracketed extension range. As written, the restriction will be $1.1 < \lambda < 2.5$. The lower bound appears to eliminate most problems associated with noise and curvature for the data sets we have examined. For comparison purposes, the results from the Mooney-Rivlin plot are included in the output file together with the corresponding 95% confidence intervals and the standard error of estimate for the model. If the estimate of C_1 from the Mooney-Rivlin plot is negative, a result of underestimating L_0, the initial estimate for C_1 is based on the secant modulus at the longest extension. The secant modulus is the ratio of nominal stress to the corresponding specified strain. One source of large errors in the parameters found from the Mooney-Rivlin plot is the imprecision in the measured value of the undeformed length.

In the optimization of the objective function shown in Equation 2.8, the parameters β_i are constrained to account for the susceptibility to large errors in the low extension region. The first parameter in Equation 2.7, β_1, is given a bound of three decades in each direction. The range of the second parameter, β_2, is set between 10^{-4} and 3. The third parameter, β_3, is given an initial guess of 1.0 with a 10% tolerance. These estimates and the corresponding bounds provide the algorithm with the ability to handle data containing significant systematic deviations at low strains, as evidenced by the set in Figure 2.1.

The error in each parameter in Equation 2.7 is estimated by the expression

$$\pm\, t_{95,\nu}\; s_\varepsilon \sqrt{C_{ii}} \tag{2.9}$$

where $t_{95,\nu}$ is the t-statistic as a function of the degrees of freedom, ν, s_ε is the standard error of estimate for the data set, and C_{ii} is the diagonal element for the estimated covariance matrix of the standard errors in the fitted parameters [17]. The covariance matrix elements are determined in the subroutine `ERROR` from the inverse of the curvature matrix. The curvature matrix is one-half the Hessian matrix. The Hessian, comprising all second partial derivatives of the objective function with respect to the parameters, is approximated by using forward differences by the IMSL subroutine `DFDHES` [16].

Finally, the 95% confidence intervals associated with the Mooney-Rivlin constants, the calculated initial length, and the equilibrium tensile modulus are obtained from error propagation analysis. The quality of the fit to the data by the three-parameter model can be assessed by scrutinizing the standard error of estimate (it should be small) and examining the residuals for the variable λ in the output file; a good fit will show up as several sign changes in the residuals.

2.3 Execution of STRECH

The execution of `STRECH` requires the subroutines `ALINR`, `EQSOLN`, `ERROR`, and `SORT2` found in Appendix B. Refer to Appendix A for the program dependencies. For example, `ERROR` calls the matrix inversion routine `MATINV` also found in Appendix B. In turn, `MATINV` uses the *LU* matrix decomposition routines available from *Numerical Recipes* [17]. In addition, the optimization and error analysis procedures use the IMSL routines `DBCONF` and `DFDHES`, respectively [16].

Before using `STRECH`, you must set up a data file by using an editor or text processor that outputs ASCII code. The first record is the title for the file; insert anything up to 79 characters that could serve as a reminder of the contents or the experimental details of the file. The second record is for the cross-sectional area A_o in square centimeters (cm^2), and the best average estimate for the undeformed initial length L_o in centimeters. The succeeding records are for the pairs of raw data obtained from the experiment. The force, in grams, and the corresponding gauge length L (cm) are entered. The data pairs may be entered in any sequence; they are subsequently sorted by force with the subroutine `SORT2`. The tabulation in the output file is also in the order of increasing force to facilitate an examination for nonrandomness in the residuals. A sample listing of a data file is provided in Section 2.6.1.

As discussed in the preceding section, the program is partially interactive. A sample listing of the terminal dialog involved in running STRECH is provided in Section 2.6.2. It is the analyst's prerogative to edit the source code to specify different bounds for either the Mooney-Rivlin plot or the optimization routine, or to alter the convergence criterion in the Newton-Raphson algorithm. However, modification will not necessarily lead to convergence for difficult data sets. The listing of the source code for STRECH is in Section 2.5.

The contents of the output file include results from the Mooney-Rivlin plot and the three-parameter model plus associated errors. No direct comparison should be made between the standard error estimates from each model, the former being dimensional in units of reduced stress and the latter being dimensionless. The equilibrium tensile modulus, calculated from Equation 2.6, is given. Finally, a tabulation is provided for the raw data, the extension λ based on the calculated zero length from the optimization $L_{o,c}$, the calculated extension λ_c from Equation 2.7, the residual term $\lambda - \lambda_c$, and the true stress σ based on the calculated extension

$$\sigma = \frac{F}{A_o} \lambda_c \tag{2.10}$$

A sample listing of an output file from STRECH is provided in Section 2.6.3.

2.4 Application of STRECH

The application of STRECH can be seen by examining its performance with a published set of data for a polyisoprene elastomer at three different temperatures [18]. To set up the data files, we converted the numerical data, tabulated as stress and extension ratio, to the required format, assuming a 0.2-cm^2 cross section and a 1.0-cm gauge length. Given the limits imposed by non-Gaussian behavior and crystallization, data above $\lambda \approx 3$ were discarded. The listings of the input/output files for polyisoprene at 70 °C are provided in Section 2.6. The plots of σ versus λ, along with the calculated zero lengths $L_{o,c}$, are shown in Figure 2.2. The representative initial gauge length L_o was exceeded by the calculated undeformed length by an average of 0.1%, well within the lower confidence interval for all cases. The positive deviation of $L_{o,c}$ relative to L_o is an illustration of hysteresis and may be attributed to the viscoelastic nature of the material and stress softening, often described as the Mullins effect [19-21].

The empirical constants obtained from STRECH and from the Mooney-Rivlin plot are listed in Table 2.1. The corresponding values were not listed in the original publication [18] thereby thwarting a direct comparison. The resulting equilibrium tensile moduli from the three-parameter model are larger by an average of 1.5%, while the values of C_1 are smaller by 1.7%.

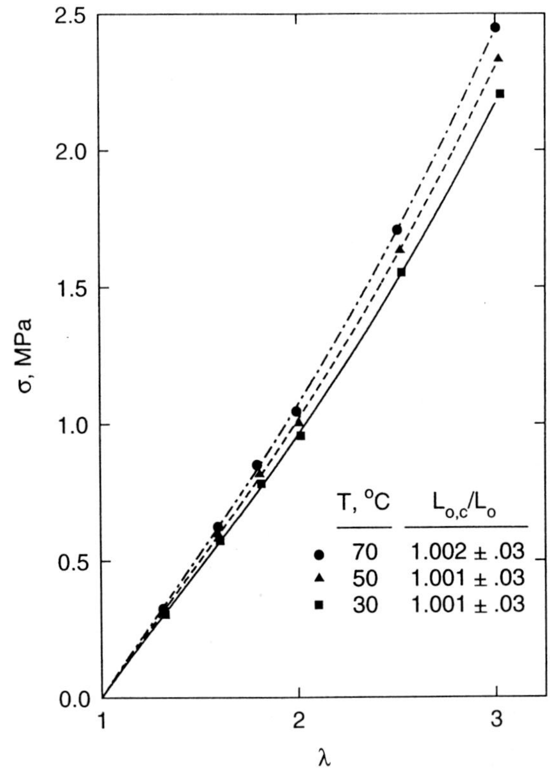

Figure 2.2 STRECH algorithm results for a polyisoprene elastomer [18].

Table 2.1 Comparison of Mooney-Rivlin Constants (in kPa) for a Polyisoprene Elastomer [18]

	STRECH		Mooney-Rivlin Plot	
T, °C	C_1	C_2	C_1	C_2
30	102 ± 8[a]	70 ± 27	103 ± 9	66 ± 17
50	109 ± 9	71 ± 28	111 ± 9	68 ± 17
70	117 ± 10	71 ± 31	120 ± 11	64 ± 20

[a] 95% confidence interval.

The second illustration is for five different bimodal networks of polydimethylsiloxane (PDMS) at 25 °C [22]. The data sets, reproduced in Figure 2.3, were obtained through the digitization of Figure 2 in Reference 22 and using the cited values $L_o \approx 2.0$ cm and $A_o \approx 0.03$ cm^2. Table 2.2 compares the constants from the

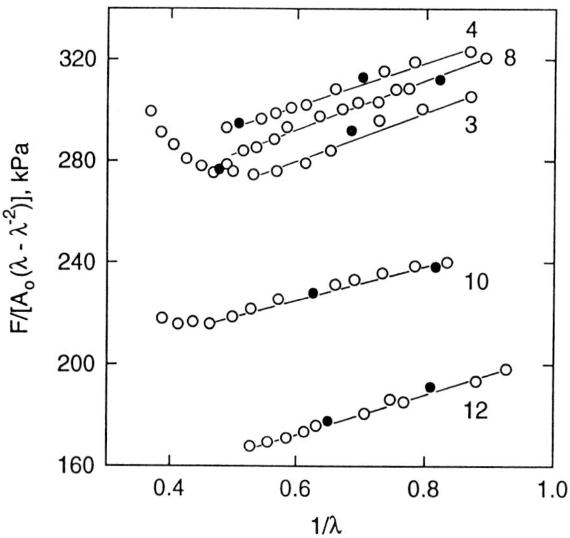

Figure 2.3 Mooney-Rivlin plots for PDMS bimodal networks at 25 °C [22]. Unfilled circles obtained using series of increasing force; filled circles obtained out of sequence as a test for hysteresis. Adapted from Ref. 22. Copyright © 1981 by Rubber Division, American Chemical Society, Inc. Reprinted by permission.

Table 2.2 Mooney-Rivlin Constants (in kPa) for PDMS Bimodal Networks at 25 °C

	Literature [22]		STRECH			
	Mooney-Rivlin Plot		Mooney-Rivlin Plot		Equation 2.7	
Network	C_1	C_2	C_1	C_2	C_1	C_2
3	110.	50.5	109.8 ± 6.5[a]	50.6 ± 9.5	109.2 ± 8.8	51.5 ± 21.
4	126.	43.1	125.3 ± 2.6	43.1 ± 3.9	124.8 ± 2.7	43.9 ± 6.6
8	116.5	50.0	116.0 ± 2.9	50.2 ± 4.2	113.2 ± 4.2	55.8 ± 11.
10	95.0	30.7	94.4 ± 1.7	31.3 ± 2.7	94.1 ± 2.3	31.7 ± 6.5
12	63.0	39.0	63.4 ± 2.0	38.8 ± 2.8	62.8 ± 2.4	39.6 ± 6.1

[a] 95% confidence interval.

Mooney-Rivlin plot obtained from the algorithm with the published values. With the exception of Network 8, the results for the three-parameter model (Equation 2.7) show excellent agreement with the published Mooney-Rivlin constants. Aside from any error occurring from the digitization procedure, the result for Network 8 is attributed to a discrepancy between $L_{o,c}$ found by STRECH and L_o. No appreciable difference was observed for the remaining data sets.

Unfortunately, the use of published data sets does not highlight the advantages of using STRECH, because data at low extensions are avoided for the reasons previously noted. At a specific extension range, usually $1.1 < \lambda < 2.5$, the traditional Mooney-Rivlin plot may lead to smaller errors in the equation parameters, implying a breakdown of the parabolic and uniform variance assumptions inherent in the numerical methods used in STRECH. To illustrate the advantages of STRECH, we numerically simulated data and introduced random error for $1.01 < \lambda < 2.5$. The results of the parameter estimation are shown in Figure 2.4.

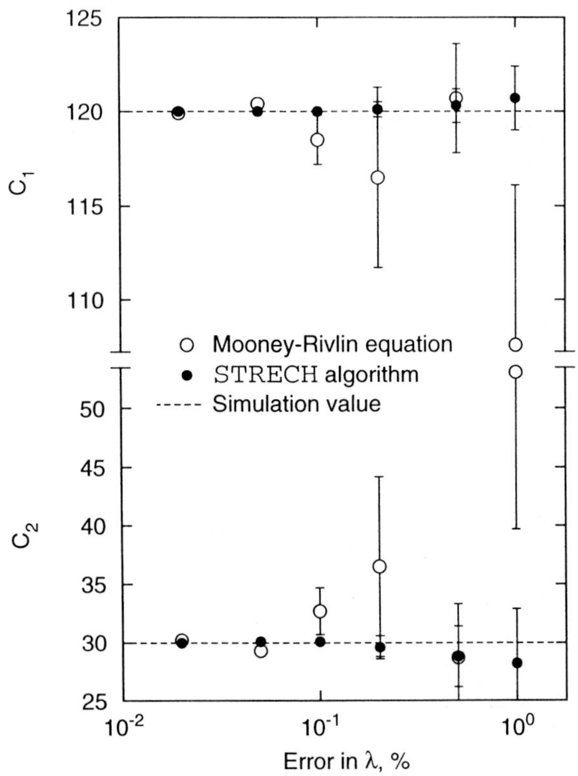

Figure 2.4 Results using numerically simulated data with random error for the extension range $1.01 < \lambda < 2.5$ show an overall better performance from program STRECH in calculating the Mooney-Rivlin constants.

As expected, the traditional Mooney-Rivlin plot performed poorly with the data sets containing more than 0.1% error; the 95% confidence intervals of C_1 and especially C_2 did not always overlap the corresponding real values of the Mooney-Rivlin constants, a serious deficiency. Such was not the case for the results from STRECH. In the limit of very small errors, the two methods agreed.

While the motivation behind the development of STRECH was to provide a more precise determination of both C_1 and C_2 from constant-force experiments, the program could be applied to stress-strain data obtained through other experimental techniques. However, the statistical aspects would not be as favorable.

2.5 Program Listing for STRECH

Imbedded into the code are the function subprograms FSTR and FMR inherent to the execution of STRECH.

```
      PROGRAM STRECH
C Calculates the Mooney-Rivlin equation coefficients (C1, C2), the zero
C  length parameter, and equilibrium modulus Ee from constant-force
C  tensile tests.
C          F/Ao = 2.*(C1 + C2/lambda)*(lambda - 1/lambda**2)
C   where         lambda=L/Lo, the strain ratio
C Requires the subroutines ALINR, EQSOLN, ERROR, SORT2, and STOPP
C  and the IMSL routine DBCONF.
C Data file format:
C          1st record - title, up to 79 characters.
C          2nd record - cross-section, Ao (cm**2), gauge length, Lo (cm).
C subsequent records - force, F (g), gauge length, L (cm).
      IMPLICIT DOUBLE PRECISION (A-H,O-Z)
      PARAMETER (NMAX=102,NPAR=3)
      CHARACTER ANS*1,TITLE*80
      DIMENSION B(NPAR),BGUESS(NPAR),BSCALE(NPAR),BLB(NPAR),BUB(NPAR)
      DIMENSION C(11),CC(NPAR),COV(NPAR),H(NPAR,NPAR),HI(NPAR,NPAR)
      DIMENSION IPARAM(7),RPARAM(7),X(NMAX),Y(NMAX)
      COMMON /BLOC1/ F(NMAX),XL(NMAX),XLC(NMAX),NP
      EXTERNAL FSTR
C t-statistic for 95% confidence interval with N degrees of freedom.
      T95(N)=1.96+2.3539/N+3.1902/N**2+0.7004/N**3+4.5015/N**4
10    CALL MSHEAD
      CALL UNITS(NTER,NINT,NOUT)
      WRITE(NOUT,20)
20    FORMAT(27X,'Analysis of the'/,21X,'Equilibrium Tensile Modulus'/
     &       21X,27('-')/)
      READ(NINT,'(A80)')TITLE
      WRITE(NOUT,'(/1X,A80/)')TITLE
C Read in cross-sectional area (AO, cm**2) and gauge length (XLO, cm).
      READ(NINT,*)AO,XLO
      WRITE(NOUT,30)AO,AO*1.D2,XLO,XLO*1.D1
30    FORMAT(1X,'Cross-sectional area =',F7.3,' cm**2',F7.2,' mm**2'/
     &     15X,'Length =',F7.3,' cm    ',F7.2,' mm',/)
C Read in force (F, gm) and gauge lengths (XL, cm).
      DO 40 I=2,NMAX
        READ(NINT,*,END=50)F(I),XL(I)
        IF(I.EQ.NMAX)CALL STOPP(NTER,NMAX-2)
C Convert force into stress (kPa) and length to strain ratio.
        F(I)=F(I)*.098066/Ao
        XL(I)=XL(I)/XLO
40    CONTINUE
50    NP=I-2
```

```
C Estimate the Mooney-Rivlin coefficients from the Mooney-Rivlin plot.
C Analysis may be restricted to a specific range of extension, IFLAG=2.
      IFLAG=1
      WRITE(NTER,60)
60    FORMAT(2X,'Restrict extension range to estimate C1 and C2 from',
     &      ' the Mooney-Rivlin plot'/2X,'(1.1 < lambda < 2.5) ? (Y/N)')
      READ(NTER,'(A1)')ANS
      IF(ANS.EQ.'Y'.OR.ANS.EQ.'y')IFLAG=2
      NMR=0
      DO 70 I=2,NP+1
        IF(IFLAG.EQ.1.OR.XL(I).GE.1.1.AND.XL(I).LE.2.5)THEN
          NMR=NMR+1
          X(NMR)=1.D0/XL(I)
          Y(NMR)=F(I)/(XL(I)-1.D0/XL(I)**2)
        ENDIF
70    CONTINUE
      WRITE(NTER,'(2X,A)')'Computing!'
      CALL ALINR(NMR,X,Y,C)
      WRITE(NOUT,80)C(1)/2.D0,C(9)/2.D0*T95(NP-2),C(2)/2.D0,
     &              C(10)/2.D0*T95(NP-2),C(8)
80    FORMAT(/1X,'Coefficients from the Mooney-Rivlin plot'/
     &        /19X,'C1 = ',F6.1,2X,'+/-',F6.1,' kPa'
     &        /19X,'C2 = ',F6.1,2X,'+/-',F6.1,' kPa'/
     &        /3X,'Standard error of estimate = ',G9.3,' kPa')
C Include zero stress and zero extension into data set.
      NP=NP+1
      F(1)=0.D0
      XL(1)=1.D0
      CALL SORT2(NP,F,XL,F,XL)
C The three parameters are:
C     B(1)=Log of twice the Mooney-Rivlin coefficient C1 (kPa);
C     B(2)=Ratio of Mooney-Rivlin coefficients C2/C1; and,
C     B(3)=XLO(true)/XLO(apparent) or lambda(apparent)/lambda(true).
C If C1 < 0, get B(1) estimate from secant modulus at longest extension.
      IF(C(1).LE.0.)C(1)=F(NP)/(XL(NP)-1.D0)/3.D0
C Set up call to IMSL routine DBCONF.
      BGUESS(1)=LOG10(C(1))
      BGUESS(2)=C(2)/C(1)
      BGUESS(3)=1.
      IBTYPE=0
C Set the bounds for each of the three parameters (IBTYPE=0).
      BLB(1)=BGUESS(1)-3.
      BUB(1)=BGUESS(1)+3.
      BLB(2)=1.D-4
      BUB(2)=3.
      BLB(3)=0.9*BGUESS(3)
      BUB(3)=1.1*BGUESS(3)
      BSCALE(1)=1.D0
      BSCALE(2)=1.D0
      BSCALE(3)=1.D0
      FSCALE=1.D0
      IPARAM(1)=0
      CALL DBCONF(FSTR,NPAR,BGUESS,IBTYPE,BLB,BUB,BSCALE,FSCALE,IPARAM,
     &     RPARAM,B,FX)
      SEE=SQRT(FX/FLOAT(NP-NPAR))
      CALL ERROR(FSTR,NPAR,B,BSCALE,FX,H,HI,COV)
      WRITE(NOUT,'(/1X,A/)')'Results for the 3-parameter model'
      WRITE(NTER,'(/1X,A/)')'Results for the 3-parameter model'
      WRITE(NTER,90)(I,B(I),T95(NP-NPAR)*SEE*SQRT(COV(I)),I=1,NPAR)
      WRITE(NOUT,90)(I,B(I),T95(NP-NPAR)*SEE*SQRT(COV(I)),I=1,NPAR)
90    FORMAT(17X,'B(',I1,') =',F7.3,2X,'+/-',F6.3)
      WRITE(NTER,100)SEE
      WRITE(NOUT,100)SEE
100   FORMAT(/3X,'Standard error of estimate = ',G9.3)
C Calculate Mooney-Rivlin coefficients and errors.
      C(1)=1.D1**B(1)/2.D0
      C(2)=C(1)*B(2)
      C(3)=XLO*B(3)
```

```
      CC(1)=(LOG(1.D1)/2.D0)**2*1.D1**(2.D0*B(1))*COV(1)
      CC(2)=CC(1)*B(2)**2+C(1)**2*COV(2)
      CC(3)=XLO**2*COV(3)
C Output Mooney-Rivlin coefficients and true zero length.
      WRITE(NTER,110)(C(I),T95(NP-NPAR)*SEE*SQRT(CC(I)),I=1,NPAR)
      WRITE(NOUT,110)(C(I),T95(NP-NPAR)*SEE*SQRT(CC(I)),I=1,NPAR)
110   FORMAT(/19X,'C1 =',F7.1,2X,'+/-',F6.1,' kPa'
     &       /19X,'C2 =',F7.1,2X,'+/-',F6.1,' kPa'/
     &       /5X,'True zero length =',F7.3,2X,'+/-',F6.3,' cm')
C Output the equilibrium tensile modulus in MPa.
      WRITE(NTER,120)6.D0*(C(1)+C(2))/1.D3,
     &                6.D0*T95(NP-NPAR)*SEE*SQRT(CC(1)+CC(2))/1.D3
      WRITE(NOUT,120)6.D0*(C(1)+C(2))/1.D3,
     &                6.D0*T95(NP-NPAR)*SEE*SQRT(CC(1)+CC(2))/1.D3
120   FORMAT(/19X,'Ee =',F7.3,2X,'+/-',F6.3,' MPa')
      STRFAC=1.D3*AO/98.066
      WRITE(NOUT,130)
130   FORMAT(//T34,'Calc.',T53,'True'/T10,'F, g',T18,'L, cm',
     &       T25,'lambda',T33,'lambda',T41,'Residual',T50,'Stress, MPa'
     &       /T9,6('-'),3(2X,(6('-'))),T41,8('-'),T50,11('-')/)
      DO 150 I=2,NP
        XLAM=XL(I)/B(3)
        XLAMC=XLC(I)/B(3)
        R=XLAM-XLAMC
        WRITE(NOUT,140)F(I)*STRFAC,XL(I)*XLO,XLAM,XLAMC,R,F(I)*XLAM/1.D3
140     FORMAT(6X,F8.1,3(F8.3),F9.4,F10.3)
150   CONTINUE
      WRITE(NOUT,160)
160   FORMAT(/T9,52('-')/T9,'Note: lambda based on true zero length; cal
     &c. lambda'/T15,'obtained from optimization of solutions to the'/
     &T15,'3-parameter model.')
      CALL MSTAIL(*10)
      STOP
      END
C
      SUBROUTINE FSTR(NPAR,B,FX)
C Evaluates the function FX of NPAR variables to be minimized when
C  called by the IMSL routine DBCONF.
C Requires the subroutine EQSOLN.
      IMPLICIT DOUBLE PRECISION (A-H,O-Z)
      PARAMETER (NMAX-102)
      DIMENSION B(NPAR)
      COMMON /BLOC1/ F(NMAX),XL(NMAX),XLC(NMAX),NP
      EXTERNAL FMR
      CC=1.D-8
      FX-0.D0
      DO 10 I=1,NP
         CALL EQSOLN(FMR,NPAR,B,F(I),XL(I),CC,XLC(I))
         FX=FX+(XLC(I)-XL(I))**2/XL(I)**2
10    CONTINUE
      RETURN
      END
C
      DOUBLE PRECISION FUNCTION FMR(NPAR,B,X)
C Evaluates X given FMR and B(NPAR) via the subroutine EQSOLN.
      IMPLICIT DOUBLE PRECISION (A-H,O-Z)
      DIMENSION B(NPAR)
      B1=1.D1**B(1)
      FMR=B1*(1.D0+B(2)*B(3)/X)*(X/B(3)-(B(3)/X)**2)
      RETURN
      END
```

2.6 Sample Listings of INPUT/OUTPUT Files for STRECH

2.6.1 Data File for STRECH

Shown is a data file in the format required for running STRECH. The force and extension values were derived from Table I of Reference 18 and based on a 0.20-cm^2 cross section and a 1.0-cm gauge length.

```
Polyisoprene at 70 C: Smith & Puett, J. Appl. Phys., 37, 346-354 (1966).
 0.20  1.00
  506.     1.32
  798.     1.60
  968.     1.80
 1070.     2.00
 1392.     2.51
 1662.     3.01
```

2.6.2 Terminal Dialog with STRECH

Reproduced here is the terminal dialog for running STRECH with the data file STRECH.DAT listed in Section 2.6.1.

```
>STRECH
 Enter name of data file (N, if not applicable).
STRECH.DAT
 Results to be written to NEW file: STRECH.RES
 OK? (Y/N)
Y
 Restrict extension range to estimate C1 and C2 from the Mooney-Rivlin plot
 (1.1 < lambda < 2.5) ? (Y/N)
N
 Computing!

 Results for the 3-parameter model

                B(1) =  2.370  +/-  .035
                B(2) =   .608  +/-  .259
                B(3) =  1.002  +/-  .027

   Standard Error of Estimate =  .103E-01

                 C1 =  117.1  +/-   9.5 kPa
                 C2 =   71.1  +/-  30.9 kPa

    True zero length =  1.002  +/-  .027 cm

                 Ee =  1.130  +/-  .194 MPa

 The results are in STRECH.RES
 Run another set of data? (Y/N)
N
Stop - Program terminated.
```

2.6.3 Output File from **STRECH**

Reproduced here is the output file from `STRECH` using the data and terminal dialog listed in Sections 2.6.1 and 2.6.2, respectively. The results from the Mooney-Rivlin plot did not avail of the restriction $1.1 < \lambda < 2.5$.

```
                    Analysis of the
              Equilibrium Tensile Modulus
              ---------------------------

Polyisoprene at 70 C: Smith & Puett, J. Appl. Phys., 37, 346-354 (1966).

Cross-sectional area =    .200 cm**2  20.00 mm**2
              Length =   1.000 cm     10.00 mm


Coefficients from the Mooney-Rivlin plot

                  C1 =   119.9  +/-   11.2 kPa
                  C2 =    64.1  +/-   20.5 kPa

  Standard Error of Estimate =  5.09      kPa

Results for the 3-parameter model

                B(1) =   2.370  +/-   .035
                B(2) =    .608  +/-   .259
                B(3) =   1.002  +/-   .027

  Standard Error of Estimate =  .103E-01

                  C1 =   117.1  +/-    9.5 kPa
                  C2 =    71.1  +/-   30.9 kPa

    True zero length =   1.002  +/-   .027 cm

                  Ee =   1.130  +/-   .194 MPa


                               Calc.              True
       F, g     L, cm  lambda  lambda  Residual Stress, MPa
      ------   ------  ------  ------  -------- -----------

       506.0    1.320   1.317   1.308    .0092        .327
       798.0    1.600   1.596   1.601   -.0046        .625
       968.0    1.800   1.796   1.821   -.0250        .852
      1070.0    2.000   1.996   1.970    .0259       1.047
      1392.0    2.510   2.504   2.505   -.0001       1.709
      1662.0    3.010   3.003   3.005   -.0015       2.448

      -----------------------------------------------------
      Note: lambda based on true zero length; calc. lambda
            obtained from optimization of solutions to the
            3-parameter model.
```

References

1. Flory, P. J.: "Network Structure and the Elastic Properties of Vulcanized Rubber" *Chem. Rev.* **35**, 51-75 (1944).
2. Mooney, M.: "A Theory of Large Elastic Deformations" *J. Appl. Phys.* **11**, 582-592 (1940).
3. Mooney, M.: "The Thermodynamics of a Strained Elastomer. I. General Analysis" *J. Appl. Phys.* **19**, 434-444 (1948).
4. Rivlin, R. S.: "Large Elastic Deformations of Isotropic Materials. I, II, III," *Phil. Trans. Roy. Soc. (Lond.)* A **240**, 459-490, 491-508, 509-525 (1948).
5. Rivlin, R. S.: "Large Elastic Deformations of Isotropic Materials. IV. Further Developments of the General Theory" *Phil. Trans. Roy. Soc. (Lond.)* A **241**, 379-397 (1948).
6. Treloar, L. R. G.: *The Physics of Rubber Elasticity.* Clarendon Press, Oxford, 1975.
7. Ogden, R. W.: "Recent Advances in the Phenomenological Theory of Rubber Elasticity," *Rubber Chem. Technol.* **59**, 361-383 (1986).
8. Ogden, R. W.: "Aspects of the Phenomenological Theory of Rubber Thermoelasticity," *Polymer* **28**, 379-385 (1987).
9. Ciferri A.; Flory, P. J.: "Stress-Strain Isotherm for Polymer Networks," *J. Appl. Phys.* **30**, 1498-1507 (1959).
10. Ciferri A.; Hermans, J. J.: "Non-Equilibrium Effects in the Stress-Strain Behavior of Rubbers for the Explanation of Deviations from Rubber Elasticity Theory," *J. Polym. Sci., Polym. Lett.* **2**, 1089-1094 (1964).
11. Roland, C. M.: "Network Recovery from Uniaxial Extension: II. The Origin of the Mullins Effect," *Rubber Chem. Technol.* **62**, 880-895 (1989).
12. Mark, J. E.; Erman, B.: *Rubberlike Elasticity. A Molecular Primer.* Wiley-Interscience, New York, 1988.
13. van der Hoff, B. M. E.; Buckler, E. J.: "Transient Changes in Topology and Energy on Extension of Polybutadiene Networks," *J. Macromol. Sci. (Chem.)* **A1**, 747-788 (1967).
14. Flory, P. J.: "Statistical Thermodynamics of Random Networks," *Proc. Roy. Soc. (Lond.)* A **351** 351-380 (1976).
15. Hasa J.; van der Hoff, B. M. E.: "Stress-Strain Behavior in Extension of Elastomer Networks with Crosslinks of Different Lengths," *J. Polym. Sci., Polym. Phys.* **11**, 297-311 (1973).
16. IMSL MATH/LIBRARY User's Manual. IMSL, Houston, 1991, Vol. 3.
17. Press, W. H.; Flannery, B. P.; Teukolsky, S. A.; Vetterling, W. T.: *Numerical Recipes in* FORTRAN: *The Art of Scientific Computing.* Cambridge University Press, Cambridge, 1986, Chapter 2.
18. Smith, K. J., Jr.; Puett, D.: "Stress-Optical Behavior of Rubber Networks at Large Deformations," *J. Appl. Phys.* **37**, 346-354 (1966).

19. Mullins, L.: "Effect of Stretching on the Properties of Rubber," *J. Rubber Res.* **16**, 275-289 (1947).
20. Mullins, L.: "Thixotropic Behavior of Carbon Black in Rubber," *J. Phys. Colloid Chem.* **54**, 239-251 (1950).
21. Mullins, L.; Tobin, N. R.: "Theoretical Model for the Elastic Behavior of Filler-Reinforced Vulcanized Rubber," *Rubber Chem. Technol.* **30**, 555-571 (1957).
22. Mark, J. E.; Andrady, A. L.: "Model Networks of End-Linked Polydimethylsiloxane Chains. X. Bimodal Networks Prepared in Two-Stage Reactions Designed to Give High Spatial Heterogeneity," *Rubber Chem. Technol.* **54**, 366-373 (1981).

Chapter 3
Physical Relaxation in Elastomers

3.1 Introduction

Elastomers exhibit varying amounts of nonequilibrium behavior usually categorized as nonrecoverable chemical processes or reversible physical relaxation. Whereas the former is mainly due to chain scission, the latter involves, among other possibilities, the path retracement of dangling structures attached to the network, the result of which is disentanglement [1]. Under certain assumptions, Curro and Salazar have described a method to distinguish the two processes by using stress relaxation experiments as a function of temperature [2]. The assumptions required a thermorheologically simple material, a temperature range above the glass transition wherein chemical effects are negligible over the experimental time scale, a correspondence between cross-link density and time, and a power-law relaxation spectrum in the terminal zone. The algorithms introduced in this chapter deal exclusively with the physical, isothermal relaxation process. Refer to Chapter 5 for methods for handling the superposition correspondence between time and cross-link density [3-8], time and concentration [3-5], or time and temperature [3-5,7].

The physical relaxation process in carefully designed networks can provide an ideal test of the concepts of reptation, because the network provides a *tube* which does not expand with time. Studies of the motions of unattached linear and star-branched molecules with the same chemical structure as the network have shown that reptation is far slower when the topological constraints are trapped entanglements rather than untrapped entanglements on dangling chain ends [9]. Because stress couples linearly to the length of the unrenewed tube, transient or dynamic rheological techniques can directly quantify these processes.

The long-term viscoelastic behavior of cross-linked networks has been attributed to the existence of dangling structures which relax to an unperturbed conformation while the network remains at a constant state of strain [10]. McKenna and Gaylord have examined the theoretical models for the long-term viscoelastic response of cross-linked dangling structures [11]. Based on the reptation mechanism or the continuous-time random-walk model, the theories all predict an apparent power-law relation of the relaxation process with time but differ in predictions for the dependence of the power-law exponent on cross-link density [1,12-17]. Note, the assumption that dangling ends are responsible for the longest relaxation times is not universally accepted. Using computer simulations, Neuburger and Eichinger [18] showed that Rouse-like modes of motion in elastomers relax more slowly than retracement of dangling ends and are responsible for the slow approach to the equilibrium stress.

An important practical result of the proper treatment of relaxation data in elastomers is the recovery of the equilibrium modulus along with its error. One of the earlier, more notable analyses of long-term viscoelastic response in elastomers was performed by Chasset and Thirion [3-6,19,20]. Expressed in terms of the isothermal relaxation modulus $E(t)$, the response at times t greater than the segmental Rouse relaxation times was expressed as

$$E(t) = E_e\left[1 + \left(\frac{t}{\tau}\right)^{-m}\right] \tag{3.1}$$

where E_e is the equilibrium modulus, τ is the characteristic time constant, and m is a material-dependent breadth parameter referred to as the Chasset-Thirion exponent with typical values of 0.1 - 0.3 [13]. To describe creep data, an expression analogous to Equation 3.1 was developed:

$$J(t) = J_e\left[1 + \left(\frac{t}{\tau}\right)^{-m}\right]^{-1} \tag{3.2}$$

where $J(t)$ is the creep compliance and J_e is the equilibrium compliance [21]. All three variables from each equation have been shown to depend on cross-link density [21]. However, deviations from the power-law time behavior at sufficiently long times occurred with increasing cross-link densities [3-5]. Because of the correspondence between time and cross-link density, McKenna and Gaylord [11] argued against a power-law relaxation function dependent on cross-link density and concluded that the apparent dependence of the power-law exponent on cross-link density was a consequence of taking data isothermally.

Lightly cross-linked elastomeric systems exhibit two relaxation mechanisms. As depicted in Figure 3.1, the first mechanism, associated with the Rouse processes, involves segmental motion between entanglements and concludes with the modulus leveling off to a plateau. This plateau corresponds to the entanglement network modulus, E_N, which has been associated with the average spacing between all entanglement points – trapped, untrapped, and chemical cross-links [10]. The second mechanism involves a slower approach to an elastic equilibrium value E_e as the dangling structures relax to an unperturbed state. As the cross-link density increases, the lengths of these structures decrease and the retracement process merges with the Rouse processes on the time scale.

The appearance of two mechanisms renders the original Chasset-Thirion equation inappropriate for describing the relaxation process of lightly cross-linked systems. Within the context of Equation 3.1, an empiricism was introduced [22] to account for the modulus corresponding to the entanglement network:

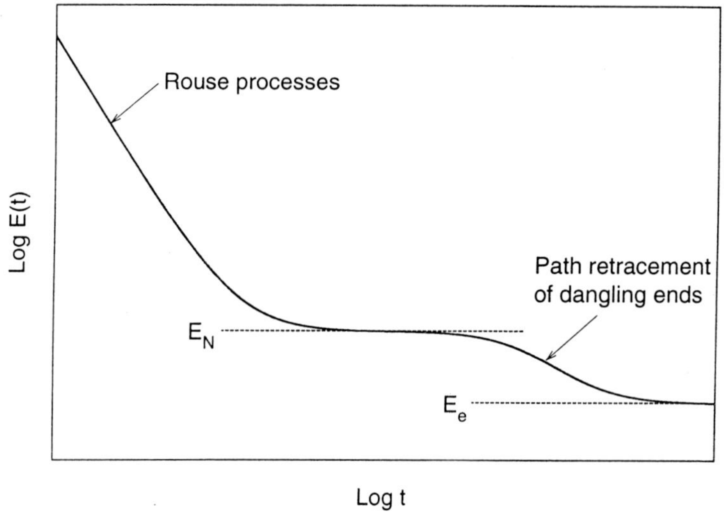

Figure 3.1 Long-term physical relaxation mechanisms for lightly cross-linked elastomers. E_N is the entanglement network modulus associated with the average spacing between all entanglement points and E_e is the equilibrium modulus.

$$\frac{E_N}{E_e} = \frac{1 + u}{u} = \frac{\nu_N}{\nu} \tag{3.3}$$

where a fourth parameter, u, has been defined as an entanglement network index. Empirically, the index u relates the entanglement network modulus to the equilibrium modulus, and the total entanglement network cross-link density ν_N to the cross-link density at mechanical equilibrium ν. The resulting modification to the Chasset-Thirion equation is then

$$E(t) = E_e \frac{1 + (1+u)(t/\tau)^{-m}}{1 + u(t/\tau)^{-m}} \tag{3.4}$$

The influence of the breadth parameter m and the entanglement network index u in describing the viscoelastic response is illustrated in Figures 3.2 and 3.3, respectively. In these calculations, we have set $u = 0.1$ in Figure 3.2 and $m = 0.2$ in Figure 3.3. In summary, the features of Equation 3.4 include a direct determination of the entanglement network modulus as an alternative to the integration of the retardation spectrum [21,23] and a reduction to the original Chasset-Thirion equation as the entanglement network index approaches zero (see Figure 3.3) – that is, as the contribution of dangling structures decreases.

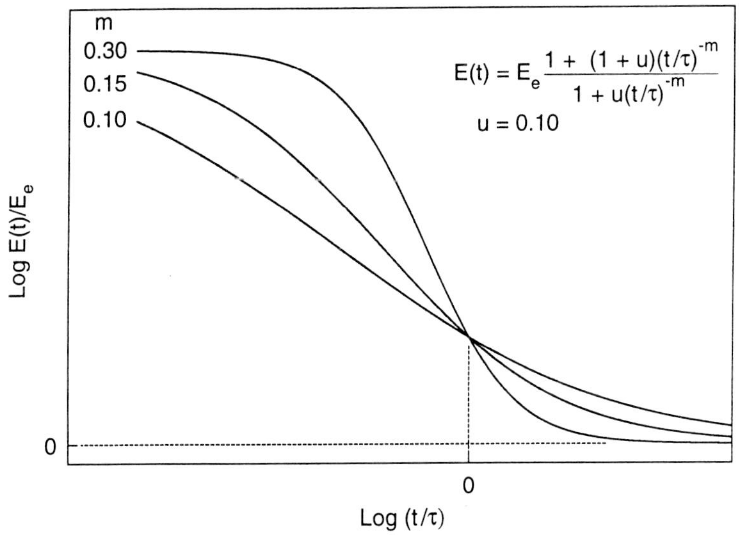

Figure 3.2 Influence of the breadth parameter *m* on the viscoelastic response of lightly cross-linked elastomers.

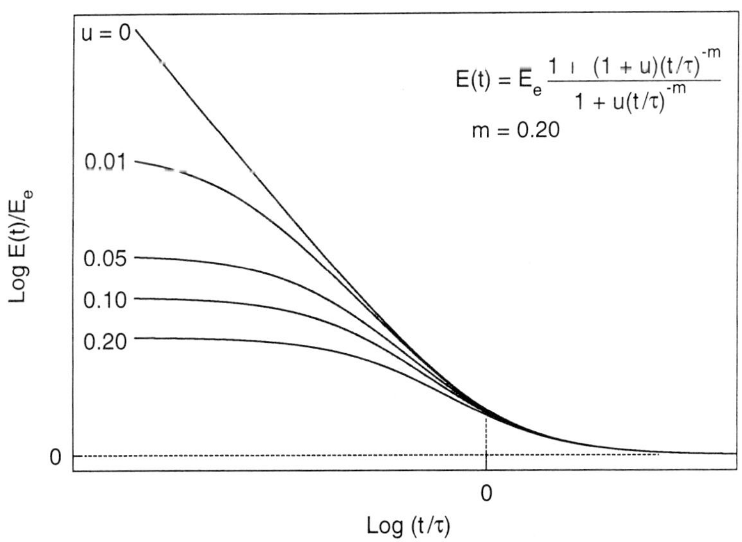

Figure 3.3 Influence of the entanglement network index *u* on the viscoelastic response of lightly cross-linked elastomers. As $u \rightarrow 0$, the Chasset-Thirion equation is recovered.

3.2 Description of the Algorithms

The program CHATH3 is for the direct application of the three-parameter Chasset-Thirion stress-relaxation model defined in Equation 3.2. The four-parameter Equation 3.4 applied in CHATH4 is recommended for lightly cross-linked systems that exhibit two relaxation mechanisms. The objective function in each algorithm is minimized by using the quasi-linear Gauss-Newton method [24] described in Chapter 1. The minimization of the objective function is carried out in the subroutine GAUSSN. To use GAUSSN, you need the analytical forms of the partial first derivatives of the function with respect to each parameter and the initial estimates of the parameters. The evaluation of the function and its derivatives is accomplished in the subroutines FCT3 and FCT4 for CHATH3 and CHATH4, respectively.

Expressed in terms of 95% confidence intervals, the errors in the fitted parameters are estimated from the diagonal components of the covariance matrix,

$$\pm\, t_{95,\,\nu}\; s_{\varepsilon} \sqrt{C_{ii}} \tag{1.16}$$

where $t_{95,\,\nu}$ is the t-statistic, a function of the degrees of freedom, ν, s_{ε} is the standard error of estimate for the data set, and C_{ii} is the corresponding diagonal element for the estimated covariance matrix of the standard error [24,25]. In CHATH4, the corresponding error in the entanglement network modulus E_N is obtained from error propagation analysis.

The goal is for the iterative search to terminate at a global (not a local) minimum. To help achieve global convergence, a couple of operator interaction steps are inserted into the algorithm. The first interaction requires that the initial estimates to the function parameters be supplied. The input of initial estimates is the most critical aspect of any minimization routine and is magnified for this method, where convergence is quadratic and therefore less forgiving. The second operator interaction is specifying the convergence criterion via the subroutine SPECC. As discussed in Chapter 1, it is recommended that the criterion be no smaller than the square root of the computer's floating-point precision, because nothing is to be gained by using smaller values [25]. The default value for the criterion is 10^{-4}.

To summarize, the quasilinear method GAUSSN is capable of providing rapid and precise determination of the optimum values of the function parameters as well as estimates of their precision. However, reasonable initial estimates of the parameters are required or the search is likely to fail. Unreasonable estimates usually result in computational problems, such as exponential overflow. Therefore, some mathematical transformations are required to get the search started on the right track. The following sections describe how the parameter estimates in each algorithm are obtained and subsequently recommended to the user. Even if the procedures outlined appear extensive, and indeed they account for most of the algorithm, they do not provide the maximum likelihood estimate or the corresponding precision of the parameters; thus, the final nonlinear optimization is always required.

3.2.1 CHATH3

If a direct experimental measurement of the equilibrium modulus existed, the Chasset-Thirion equation would be reduced to a linear problem. Prior to the use of numerical methods, the procedure for obtaining the parameters to the Chasset-Thirion equation, or its analog for creep compliance data, was to estimate, by trial and error, the value of the equilibrium modulus that produced the straightest line in a double logarithmic plot of reduced modulus versus time [19,21]. Alternatively, Chasset and Thirion [3] suggested that a value for the equilibrium modulus could be obtained by differential analysis of the relaxation response (see Equation 3.5). In the case of Equation 3.2, suggested alternative methods include plotting $J^{-1}\, d(\log J)/d(\log t)$ versus J^{-1} to obtain $-m/J_e$ as the intercept and m as the slope, and $d(\log J)/d(\log t)$ versus J to obtain m as the intercept and $-m/J_e$ as the slope [26].

In CHATH3, the procedure for obtaining the initial parameter estimates for the Chasset-Thirion equation is a two-step linearization procedure. The first step extracts the Chasset-Thirion exponent m. Equation 3.1 is differentiated with respect to time, the result of which is linearized by taking the logarithm on both sides to obtain

$$\ln\left(-\frac{dE(t)}{dt}\right) = \ln\,(mE_e\tau^m) + (-m - 1)\ln t \tag{3.5}$$

To extract reasonable values for the derivative in Equation 3.5, CHATH3 uses the external subroutine SLOPE to fit a second-order polynomial to seven data points of modulus versus the logarithm of time. The transformation of the independent variable eliminates much of the curvature in the relaxation data and, in effect, makes the degree of the polynomial chosen for the algorithm sufficient for obtaining a rough approximation. The output from SLOPE is $dE(\ln t)/d\ln t$, and a simple mathematical manipulation yields

$$-\frac{dE(t)}{dt} = \frac{1}{t}\left(-\frac{dE(\ln t)}{d\ln t}\right) \tag{3.6}$$

The slope in Equation 3.5, $-m - 1$, is provided by the linear regression subroutine ALINR. Once the parameter m has been estimated, the Chasset-Thirion equation reduces to the two-parameter linear expression

$$E(t) = E_e + (E_e\tau^m)\, t^{-m} \tag{3.7}$$

By recalling ALINR, CHATH3 obtains the estimates to the parameters E_e and τ from the intercept and slope, respectively.

3.2.2 CHATH4

A somewhat more complicated iterative procedure is required for obtaining the initial parameter estimates of the modified Chasset-Thirion equation. Because Equation 3.4 cannot be linearized by transformation, a starting value of 0.15 for the breadth parameter m is initially assigned. This value is roughly the mean of those cited in the literature. The approximation for the characteristic time constant τ is limited between the shortest (t_{min}) and longest (t_{max}) experimental times. The function subprograms AMIN and AMAX search for the bounds t_{min} and t_{max}, respectively. A better starting value for the time constant is the inflection point of a cubic polynomial fitted to ln $E(t)$ versus ln t. However, if the resulting value for τ is outside the range of t_{min} to t_{max}, the initial estimate is set to the geometric mean of these limits, $(t_{min}\, t_{max})^{1/2}$. While τ is likely to be within the experimental time window, it need not be, and the limits are provided as a precaution against completely unrealistic values.

To estimate the equilibrium modulus, Equation 3.4 is expanded in the following manner:

$$E(t) = E_e \left[1 + \sum_{i=1}^{\infty} \frac{[u(t/\tau)^{-m}]^i}{u} \right] \tag{3.8}$$

A data set of $E(t)$ and t^{-m} for times greater than τ is set up, and the truncation to a second-order polynomial, $i = 2$, allows the subroutine POLYN to determine the intercept. The resulting value is checked using the conditions $E(t_{max}) < E_e < 0.6\, E(t_{max})$, where $E(t_{max})$ should correspond to the smallest value of $E(t)$. If the condition remains out of bounds or less than three data points are in the region of $t > \tau$, the initial estimate for the equilibrium modulus is temporarily set to 90% of the smallest value recorded. Note, a value of the characteristic time may also be obtained from any coefficient of the higher-order terms. Once an approximation to the equilibrium modulus E_e is found, more realistic values for τ and m are found by rearranging Equation 3.4 as

$$\frac{E_e}{E(t) - E_e} = u + \left(\frac{t}{\tau}\right)^m \tag{3.9}$$

First, the parameter u is eliminated from this expression by taking the derivative with respect to time:

$$\frac{d\{E_e/[E(t)-E_e]\}}{dt} = \frac{m}{\tau}\left(\frac{t}{\tau}\right)^{m-1} = m\tau^{-m}t^{m-1} \tag{3.10}$$

A logarithmic transformation on Equation 3.10 results in

$$\ln \frac{d\{E_e/[E(t)-E_e]\}}{dt} = \ln(m\tau^{-m}) + (m-1)\ln t \tag{3.11}$$

and allows the use of the linear regression subroutine ALINR to extract the parameter estimates for m and τ from the slope and intercept, respectively. Three additional iterations are used to improve the approximations of the three parameters. At the conclusion of this procedure, the initial approximation for τ is based on the geometric mean of the estimates obtained from Equations 3.8 and 3.11.

The substitution of the initial estimates for the first three parameters into Equation 3.9 will provide the parameter u from the intercept of the linear regression analysis. The initial estimate for u is subjected to limits because the uncertainties in the first three parameter estimates are expected to compound the error in the value of u. The lower limit is set to 10^{-3}; anything smaller implies that the original Chasset-Thirion equation is adequate. The upper limit is set to $E_e/[E(t_{min}) - E_e]$, where $E(t_{min})$ corresponds to the earliest modulus value recorded. If the intercept of Equation 3.9 is less than the lower limit, the initial estimate for u is reset to the upper 95% confidence interval of the intercept. If the new estimate of u is still less than 10^{-3}, the user is alerted to the apparent inappropriateness of the modified Chasset-Thirion equation. If the user wishes to proceed with the search or if the approximation exceeds the upper limit, the initial estimate is set to one-half the upper limit.

3.3 Execution of the Programs

The algorithms require the function subprogram AMIN and the subroutines ALINR, CTIN, CTOUT GAUSSN, POLYN, SLOPE, and SPECC. In addition, CHATH4 requires the function subprogram AMAX. The program listings for CHATH3, CHATH4, CTIN, and CTOUT are provided in Section 3.5. Inherent only to the execution to both Chapter 3 programs, CTIN and CTOUT are subroutines that read the contents of the data file and write the results to the output file, respectively. The listings for the other codes are found in Appendix B. Refer to Appendix A for the program dependencies. For example, the subroutine GAUSSN calls the subroutines MATINV and MATMUL also found in Appendix B.

The programs were set up to require identical data file formats and terminal interaction. The data file must be set up by using an editor or a text processor that outputs ASCII code. The first record is set aside for the title of the file; insert anything up to 79 characters containing identification information. The second and succeeding records are for the time and modulus data pairs in chronological order so as to avoid problems with the derivative-finding routine SLOPE. An alternative would be to insert the subroutine SORT2 available from Appendix B. The proper statement to call SORT2 is included in CTIN as a comment line. Exclude data at early times that clearly correspond to the tail end of the glass transition (see Figure 3.5) and, at long times, show the effect of chemical processes. Refer to Section 3.6.1 for a listing of a sample data file.

The execution of the programs requires a couple of operator decisions. As detailed in the previous section, a recommendation of the initial parameter estimates and the default convergence criterion are made available to the operator. It is the prerogative of the operator whether to continue or reenter different values. In the case of CHATH4, the operator may be flagged as to the inappropriateness of Equation 3.4. While the program will be allowed to proceed with no guarantee of convergence, the data set should be treated with CHATH3. A listing of a typical terminal dialog is provided in Section 3.6.2.

The contents of the output file include a tabulation of the parameters defined by Equations 3.1 and 3.4, and associated 95% confidence intervals. Next, the observed relaxation data are tabulated along with the calculated modulus, in dimensional and logarithmic form, and the corresponding residual. An examination of the residuals for patterns may yield important information (see Figure 3.7). For example, a string of positive residuals at early times is a good indicator of the presence of glass transition data whereas a string of negative residuals at the longest times is a sign that chemical relaxation has started. The final tabulation in the output consists of reduced variables based on the empirical parameters. Refer to Section 3.6.3 for a listing of a sample output file.

3.4 Application of the Programs

3.4.1 CHATH3

The example to illustrate the capabilities of CHATH3 uses the stress relaxation results of Chasset and Thirion for natural rubber cross-linked with 3.5% dicumyl peroxide [3-6,19,20]. They reported data for five networks, coded F to J, with cross-link densities ranging from 0.46 to 1.69 $\times$ 10^{-4} mol cm^{-3}. The experiments were carried out to approximately 10^5 s in uniaxial extension at 50% elongation and 30 °C. These data sets were also examined by McKenna and Gaylord [11], who digitized the curves from the original version of Figure 3.4. A copy of the digitized data was generously supplied for testing CHATH3 [27]. Except for the characteristic time of Sample J, the results in Table 3.1 show excellent agreement with those obtained by McKenna and Gaylord, who also used a nonlinear least-squares regression analysis.

Table 3.1 Analysis of the Stress-Relaxation Data of Thirion and Chasset [4,5] for Natural Rubber Cross-Linked with Dicumyl Peroxide

	CHATH3 Program			McKenna and Gaylord[a]		
Sample	$\hat{E}_e$[b]	τ, s	m	$\hat{E}_e$	τ, s	m
F	$0.937 \pm 2.7 \times 10^{-3}$[c]	$7.31 \times 10^{-3} \pm 1.3 \times 10^{-4}$	$0.119 \pm 1.3 \times 10^{-3}$	$0.937 \pm 2.7 \times 10^{-3}$	$7.13 \times 10^{-3} \pm 1.3 \times 10^{-4}$	$0.119 \pm 1.2 \times 10^{-3}$
G	$0.964 \pm 2.3 \times 10^{-3}$	$1.89 \times 10^{-4} \pm 1.6 \times 10^{-5}$	$0.132 \pm 2.1 \times 10^{-3}$	$0.964 \pm 2.3 \times 10^{-3}$	$1.89 \times 10^{-4} \pm 1.6 \times 10^{-5}$	$0.132 \pm 2.0 \times 10^{-3}$
H	$0.999 \pm 9.7 \times 10^{-4}$	$8.91 \times 10^{-5} \pm 1.2 \times 10^{-5}$	$0.185 \pm 2.8 \times 10^{-3}$	$0.999 \pm 9.6 \times 10^{-4}$	$8.91 \times 10^{-5} \pm 1.2 \times 10^{-5}$	$0.185 \pm 2.8 \times 10^{-3}$
I	$0.995 \pm 5.9 \times 10^{-4}$	$1.90 \times 10^{-6} \pm 3.9 \times 10^{-7}$	$0.179 \pm 3.0 \times 10^{-3}$	$0.995 \pm 5.8 \times 10^{-4}$	$1.92 \times 10^{-6} \pm 4.0 \times 10^{-7}$	$0.179 \pm 3.0 \times 10^{-3}$
J	$1.001 \pm 4.6 \times 10^{-4}$	$3.23 \times 10^{-6} \pm 1.1 \times 10^{-6}$	$0.232 \pm 6.1 \times 10^{-3}$	$1.001 \pm 5.0 \times 10^{-3}$	$1.02 \times 10^{-5} \pm 1.1 \times 10^{-6}$	$0.230 \pm 6.6 \times 10^{-3}$

[a] Adapted from Ref. 11. Reprinted by permission of the publishers, Butterworth Heinemann Ltd. © 1988.
[b] Normalized equilibrium modulus. See Ref. 11.
[c] 95% confidence interval.

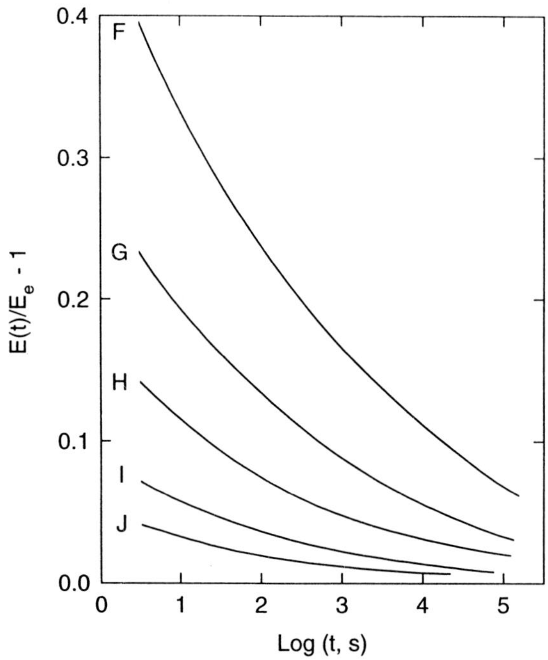

Figure 3.4 The transient response of natural rubber cross-linked with 3.5% dicumyl peroxide at 50% elongation and 30 °C [4,5]. The corresponding cross-link densities (10^{-4} mol cm^{-3}) are 0.46 (F), 0.62 (G), 0.80 (H), 1.20 (I), and 1.69 (J). Adapted from Ref. 11. Reprinted by permission of the publishers, Butterworth Heinemann Ltd. © 1988.

3.4.2 CHATH4

The first example to illustrate the capabilities of CHATH4 uses dynamic shear and creep data, measured between -30 and 55 °C, from a very lightly cross-linked styrene-butadiene copolymer [10,26]. Characteristics of this network include a cross-link density of 0.41×10^{-4} mol cm^{-3}, a number-average molecular weight of 10^5, a molecular weight between cross-links of 23,000, and an entanglement molecular weight of 3000. The numerical data, tabulated in Appendix D of Reference 9, are reduced to 25 °C. The complete data set is available elsewhere [28]. The conversion of the dynamic function to the transient function $G(t)$ was accomplished by using available numerical formulae [29]. The data, in Figure 3.5, show two dispersion regions characteristic of lightly cross-linked networks. Excluding the data for $t < 10^{-4}$ s, CHATH4 estimates the entanglement network and equilibrium moduli to be 1470 ± 317 and 187 ± 10 kPa, respectively. Mancke and Ferry reported an entanglement network modulus of 900 kPa from direct integration

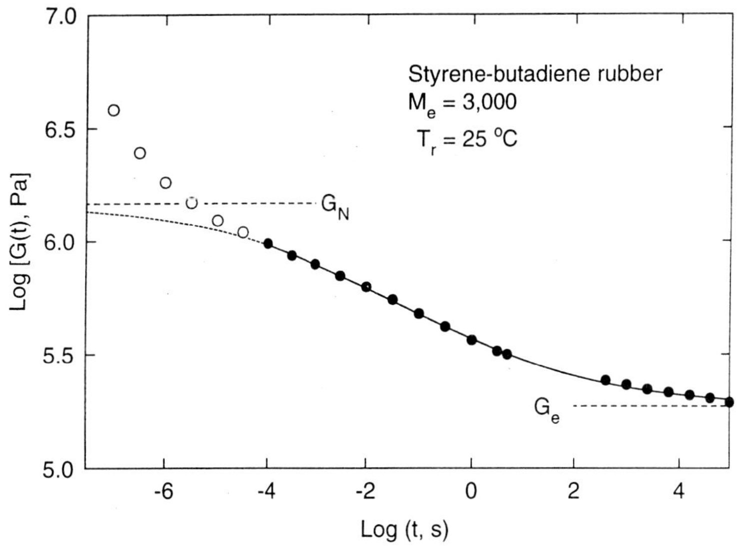

Figure 3.5 Results from the CHATH4 algorithm (–) describing the relaxation mechanism for a very lightly cross-linked styrene-butadiene rubber [10,26]. Note, the hollow circles are excluded from the actual analysis since Equation 3.4 does not account for the glass transition. Adapted from Ref. 10. Copyright © 1980 by John Wiley & Sons, Inc. Reprinted by permission.

of loss compliance (J'') data [26, Equation 5] and, by applying the alternative methods discussed in Section 3.2.1, an equilibrium modulus ($1/J_e$) of 155 kPa from Equation 3.2. Mancke and Ferry also cite an equilibrium modulus of 176 kPa [26, Table I] derived by Thirion and Chasset using their stress-relaxation measurements.

The last example is for polyisoprene (Shell 307 60713) lightly cross-linked with dicumyl peroxide (0.3 wt%). The data were obtained from dynamic torsion measurements over a frequency range of 0.01 to 100 rad/s at 35 °C and converted to the transient function $G(t)$. From Figure 3.6, the presence of an inflection around 0.26 s implies a dispersion maximum – that is, a relaxation mechanism distinct from the glassy modes. Using the three-parameter Chasset-Thirion equation, CHATH3 underestimates the equilibrium modulus (G_e = 122 ± 19 kPa) relative to the modified four-parameter function (G_e = 193 ± 2 kPa) used by CHATH4. Figure 3.7 provides the plots of the residuals as a percentage of G_e. The plots show that the four-parameter function (s_ε = 0.22 kPa) effectively removes the nonrandomness of the residuals produced by the fit from the three-parameter Chasset-Thirion equation (s_ε = 1.72 kPa).

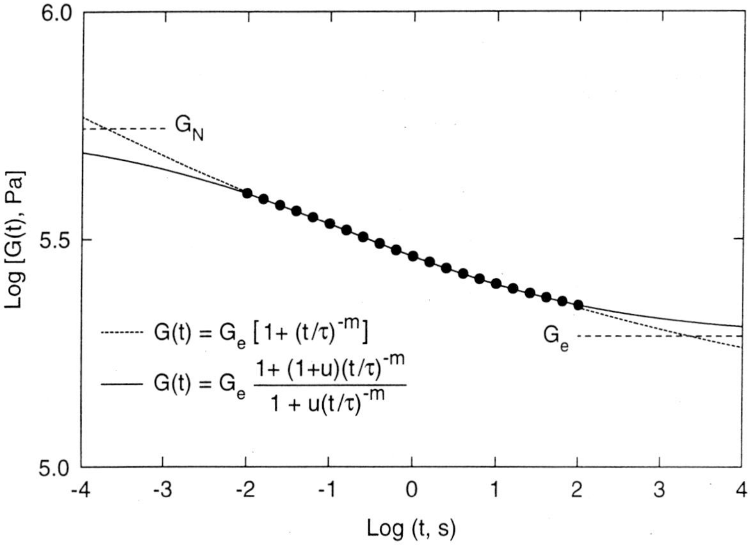

Figure 3.6 Results describing the relaxation mechanism for a very lightly cross-linked polyisoprene. The presence of an inflection around 0.2 s indicates that the four-parameter function is better suited for describing the long-term relaxation: G_e = 193 ± 2 kPa and G_N = 555 ± 22 kPa.

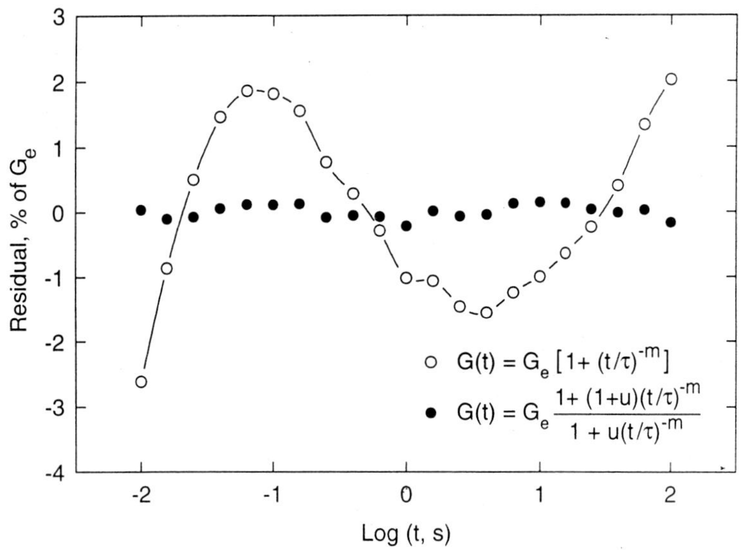

Figure 3.7 The residual plot, with respect to the equilibrium modulus G_e of the polyisoprene in Figure 3.6, shows a much better fit for the four-parameter function (s_ε = 0.22 kPa) as compared to the Chasset-Thirion equation (s_ε = 1.72 kPa).

3.5 Program Listings

3.5.1 CHATH3

Imbedded into the code is the subroutine FCT3 inherent to the execution of CHATH3.

```
      PROGRAM CHATH3
C Calculates parameters in the Chasset-Thirion equation to describe
C  long-term physical relaxation properties in elastomers.
C                    E(t) = Ee (1 + (tau/time)**m)
C Requires the subroutines ALINR, AMIN, CTIN, CTOUT, GAUSSN, POLYN,
C  SLOPE; and, SPECC.
C Data file format:
C        1st record - Title, up to 79 characters
C subsequent records - Time, Modulus
C If data are not in chronological order, activate call to SORT2 in
C   subroutine CTIN.
      IMPLICIT DOUBLE PRECISION (A-H,O-Z)
      PARAMETER (NMAX=201,NPAR=3,NPD=10)
      CHARACTER AC(NPAR)*35,ANS*1,TITLE*80
      DIMENSION A(NMAX,NPD),AT(NPD,NMAX),B(NPAR),ERR(NPAR),R(NMAX,1)
      DIMENSION RES(NMAX),U(NMAX),V(NMAX),W(NMAX),X(NMAX),Y(NMAX),Z(11)
      EXTERNAL FCT3
      DATA (AC(I), I=1,NPAR) /'Equilibrium Relaxation Modulus,  Ee',
     &                       '               Characteristic time, tau',
     &                       '          Chasset-Thirion exponent,    m'/
C t-statistic for 95% confidence interval with N degrees of freedom.
      T95(N)=1.96+2.3539/N+3.1902/N**2+0.7004/N**3+4.5015/N**4
10    CALL MSHEAD
      CALL UNITS(NTER,NINT,NOUT)
C Read data file.
      CALL CTIN(NTER,NINT,NMAX,NP,TITLE,X,Y)
C Estimate parameters by the linearization of the Chasset-Thirion equation.
      DO 20 I=1,NP
        U(I)=LOG(X(I))
20    CONTINUE
C Estimate dE(t)/d(ln t) using a second-order polynomial fit to 7 points.
      CALL SLOPE(NP,U,Y,7,3,V,RES)
      DO 30 I=1,NP
        V(I)=V(I)/X(I)
30    CONTINUE
      IF(-V(1).LE.0.)V(1)=(Y(2)-Y(1))/(X(2)-X(1))
      IF(-V(1).LE.0.)THEN
        WRITE(NTER,*)' Data are faulty; check data file.'
        V(1)=AMIN(NP,V)
      ENDIF
      V(NP+1)=0.D0
      DO 40 I=2,NP
        IF(-V(I).LE.0.)V(I)=(V(I-1)+V(I)+V(I+1))/3.D0
        IF(-V(I).LE.0.)V(I)=V(I-1)
        W(I)=LOG(-V(I))
40    CONTINUE
      U(1)=LOG(X(1))
      W(1)=LOG(-V(1))
C Obtain an initial estimate of the parameter m, B(3).
      CALL ALINR(NP,U,W,Z)
      B(3)=-Z(2)-1.D0
      DO 50 I=1,NP
        U(I)=X(I)**(-B(3))
50    CONTINUE
```

```
C Obtain intial estimates of the parameters Ee, B(1), and tau, B(2).
      CALL ALINR(NP,U,Y,Z)
      B(1)=Z(1)
      B(2)=(Z(2)/Z(1))**(1.D0/B(3))
60    WRITE(NTER,70)
70    FORMAT(2X,'Input initial guesses for parameters: Ee, tau, m.')
      WRITE(NTER,80)(B(I),I=1,NPAR)
80    FORMAT(2X,'The suggested values are: ',2E11.3,F6.3/
     &       2X,'Proceed with the proposed values? (Y/N)')
      READ(NTER,'(A1)')ANS
      IF(ANS.EQ.'N'.OR.ANS.EQ.'n')THEN
        WRITE(NTER,70)
        READ(NTER,*)(B(I),I=1,NPAR)
      ENDIF
C Specify convergence criterion.
      CALL SPECC(NTER,CC)
C Search on log of time constant.
      B(2)=LOG(B(2))
C Begin nonlinear fitting.
      WRITE(NTER,'(2X,A/)')'Computing!'
      CALL GAUSSN(FCT3,CC,NPAR,NP,X,Y,B,A,AT,R,ERR,SEE)
      B(2)=EXP(B(2))
C Calculate the 95% confidence intervals for each parameter.
      DO 100 I=1,NPAR
        ERR(I)=T95(NP-NPAR)*SEE*ERR(I)
        IF(I.EQ.2)ERR(I)=B(2)*ERR(I)
        WRITE(NTER,90)AC(I),B(I),ERR(I)
90      FORMAT(2X,A35,' =',1P,E10.3,' +/- ',E9.3)
100   CONTINUE
      WRITE(NTER,110)SEE
110   FORMAT(/11X,'Standard error of estimate = ',1P,E9.3/)
      WRITE(NTER,'(2X,A)')'Another initial guess? (Y/N)'
      READ(NTER,'(A1)')ANS
      IF(ANS.EQ.'Y'.OR.ANS.EQ.'y')GOTO 60
C Prepare output file.
      WRITE(NOUT,120)
120   FORMAT(/19X,'Physical Relaxation in Elastomers'
     &       /23X,'Chasset-Thirion Analysis'/19X,33('-'))
      WRITE(NOUT,130)TITLE
130   FORMAT(/21X,'E(t) = Ee (1 + (tau/time)**m)'///A80//)
      WRITE(NOUT,90)(AC(I),B(I),ERR(I),I=1,NPAR)
      CALL CTOUT(NOUT,NP,NPAR,B,R,SEE,X,Y)
      CALL MSTAIL(*10)
      STOP
      END
C
      SUBROUTINE FCT3(N,X,B,F,DER)
C Evaluates the function F and the its first partial derivatives DER(N)
C  at point X when called by GAUSSN.
      IMPLICIT DOUBLE PRECISION (A-H,O-Z)
      DIMENSION B(N),DER(N)
C Calculate the transient term of the Chasset-Thirion equation.
      RT=(EXP(B(2))/X)**B(3)
C Determine the function value F(X).
      F=B(1)*(1.D0+RT)
C Determine the first derivatives of the function F with respect to the
C  first parameter, B(1)
      DER(1)=1.D0+RT
C  second parameter, B(2)
      DER(2)=RT*B(1)*B(3)
C  third parameter, B(3).
      DER(3)=B(1)*RT*(B(2)-LOG(X))
      RETURN
      END
```

3.5.2 CHATH4

Imbedded into the code is the subroutine FCT4 inherent to the execution of CHATH4.

```
      PROGRAM CHATH4
C Calculates parameters in the modified Chasset-Thirion equation to describe
C  long-term physical relaxation properties in lightly cross-linked networks
C  exhibiting an entanglement network modulus, En.
C                           1 + (1+u)(tau/time)**m
C                 E(t) = Ee ----------------------
C                            1 + u*(tau/time)**m
C                   En = Ee (1+u)/u
C Requires the subroutines ALINR, AMAX, AMIN, CTIN, CTOUT, GAUSSN, POLYN,
C  SLOPE, and SPECC.
C Data file format:
C         1st record - Title, up to 79 characters
C subsequent records - Time, Modulus
C If data are not in chronological order, activate call to SORT2 in
C   subroutine CTIN.
      IMPLICIT DOUBLE PRECISION (A-H,O-Z)
      PARAMETER (NMAX=201,NPAR=4,NPD=10)
      CHARACTER AC(NPAR+1)*35,ANS*1,TITLE*80
      DIMENSION A(NMAX,NPD),AT(NPD,NMAX),B(NPAR),ERR(NPAR),R(NMAX,1)
      DIMENSION RES(NMAX),U(NMAX),V(NMAX),X(NMAX),XL(NMAX),Y(NMAX)
      DIMENSION YL(NMAX),Z(11)
      EXTERNAL FCT4
      DATA (AC(I), I=1,NPAR+1) /'Equilibrium Relaxation Modulus,  Ee',
     &                          '           Characteristic time, tau',
     &                          '        Chasset-Thirion exponent,   m',
     &                          '   Untrapped Entanglement Index,   u',
     &                          '   Entanglement Network Modulus,  En'/
C t-statistic for 95% confidence interval with N degrees of freedom.
      T95(N)=1.96+2.3539/N+3.1902/N**2+0.7004/N**3+4.5015/N**4
10    CALL MSHEAD
      CALL UNITS(NTER,NINT,NOUT)
C Read data file.
      CALL CTIN(NTER,NINT,NMAX,NP,TITLE,X,Y)
      DO 20 I=1,NP
        XL(I)=LOG(X(I))
        YL(I)=LOG(Y(I))
20    CONTINUE
C Estimate parameters.  Use an initial guess for the Chasset-Thirion
C  exponent m, and improves on this thrice.
      B(3)=0.15
C Search for inflection to estimate the characteristic time, B(2).
      CALL POLYN(NPAR,NP,YL,XL,Z,ERR,RES,SEE)
      XC=-Z(3)/Z(4)/3.
      YC=Z(1)
      DO 30 I=1,NPAR-1
        YC=YC+Z(I+1)*XC**I
30    CONTINUE
      B(2)=EXP(YC)
      B2=SQRT(AMAX(NP,X)*AMIN(NP,X))
      IF(B(2).LT.AMIN(NP,X).OR.B(2).GT.AMAX(NP,X)/100.)B(2)=B2
      DO 70 K=1,NPAR
        NCO=3
        NPQ=0
        DO 40 J=1,NP
          IF(X(J).GT.B(2))THEN
            NPQ=NPQ+1
            U(NPQ)=1./X(J)**B(3)
            V(NPQ)=Y(J)
          ENDIF
40      CONTINUE
```

```
         B(1)=AMIN(NP,Y)
         IF(NPQ.GT.2)THEN
           IF(NPQ.LT.5)NCO=2
C Obtain initial estimate of parameter Ee, B(1), restricted between a
C  lower limit BL and an upper limit BU.
           BL=0.6*AMIN(NP,Y)
           BU=B(1)
           CALL POLYN(NCO,NPQ,U,V,Z,ERR,RES,SEE)
           B(1)=Z(1)
           IF(B(1).GE.BU.OR.B(1).LT.BL)B(1)=0.9*BU
           IF(K.GT.1)B(2)=(Z(2)/Z(1))**(1./B(3))
             DO 50 I=1,NP
               YL(I)=B(1)/(Y(I)-B(1))
50           CONTINUE
C Estimate d(Ee/(E(t)-Ee))/dt using a second-order polynomial fit to
C  seven points.
           CALL SLOPE(NP,X,YL,7,3,V,RES)
           IF(V(1).LE.0.)THEN
             WRITE(NTER,*)' Data are faulty; check data file.'
             V(1)=AMAX(NP,V)
           ENDIF
           V(NP+1)=0.D0
           DO 60 I=2,NP
             IF(V(I).LE.0.)V(I)=(V(I-1)+V(I)+V(I+1))/3.
             IF(V(I).LE.0.)V(I)=V(I-1)
             YL(I)=LOG(V(I))
60         CONTINUE
           YL(1)=LOG(V(1))
C Obtain initial estimates of parameters tau, B(2), and m, B(3).
            CALL ALINR(NP,XL,YL,Z)
            B(3)=1.+Z(2)
            IF(K.EQ.NPAR)B(2)=SQRT(B(2)*(EXP(Z(1))/B(3))**(-1./B(3)))
          ENDIF
70     CONTINUE
       DO 80 I=1,NP
         U(I)=(X(I)/B(2))**B(3)
         V(I)=B(1)/(Y(I)-B(1))
80     CONTINUE
C Obtain initial estimate of parameter u, B(4), restricted between a
C  lower limit BL and an upper limit BU.
       BL=1.D-3
       BU=B(1)/(AMAX(NP,Y)-B(1))
       CALL ALINR(NP,U,V,Z)
       B(4)=Z(1)
C Check if four-parameter model is appropriate.
       IF(B(4).LT.BL)B(4)=Z(1)+T95(NP-2)*Z(9)
       IF(B(4).LT.BL)THEN
         WRITE(NTER,*)' The four-parameter model may be inappropriate.'
         WRITE(NTER,*)' Proceed? (Y/N)'
         READ(NTER,'(A1)')ANS
         IF(ANS.EQ.'Y'.OR.ANS.EQ.'y')THEN
           B(4)=0.5*BU
         ELSE
           WRITE(NTER,*)' Terminating the execution of CHATH4.'
           STOP
         ENDIF
       ENDIF
       IF(B(4).GE.BU)B(4)=0.5*BU
90     WRITE(NTER,100)
100    FORMAT(2X,'Input initial guesses for parameters: Ee, tau, m, u.')
       WRITE(NTER,110)(B(I),I=1,NPAR)
110    FORMAT(2X,'The proposed values are: ',2E11.3,2F8.4/
      &       2X,'Proceed with the proposed values? (Y/N)')
       READ(NTER,'(A1)')ANS
       IF(ANS.EQ.'N'.OR.ANS.EQ.'n')THEN
         WRITE(NTER,100)
         READ(NTER,*)(B(I),I=1,NPAR)
       ENDIF
```

```
C Specify convergence criterion.
      CALL SPECC(NTER,CC)
C Search on log of time constant.
      B(2)=LOG(B(2))
C Begin nonlinear fitting.
      WRITE(NTER,'(2X,A/)')'Computing!'
      CALL GAUSSN(FCT4,CC,NPAR,NP,X,Y,B,A,AT,R,ERR,SEE)
      B(2)=EXP(B(2))
C Calculate the 95% confidence intervals for each parameter.
      DO 130 I=1,NPAR
        ERR(I)=T95(NP-NPAR)*SEE*ERR(I)
        IF(I.EQ.2)ERR(I)=B(2)*ERR(I)
        WRITE(NTER,120)AC(I),B(I),ERR(I)
120     FORMAT(2X,A35,' =',1P,E10.3,' +/- ',E9.3)
130   CONTINUE
      WRITE(NTER,140)SEE
140   FORMAT(/11X,'Standard error of estimate =',1P,E10.3/)
      WRITE(NTER,'(2X,A)')'Another initial guess? (Y/N)'
      READ(NTER,'(A1)')ANS
      IF(ANS.EQ.'Y'.OR.ANS.EQ.'y')GOTO 90
C Prepare output file.
      WRITE(NOUT,150)
150   FORMAT(19X,'Physical Relaxation in Elastomers'
     &       /19X,'Modified Chasset-Thirion Analysis'/19X,33('-'))
      WRITE(NOUT,160)TITLE
160   FORMAT(/29X,'1 + (1+u)(tau/time)**m'/19X,'E(t) = Ee',1X,22('-')/
     &         31X,'1 + u(tau/time)**m'//21X,'En = Ee (1+u)/u'///A80//)
      WRITE(NOUT,120)(AC(I),B(I),ERR(I),I=1,NPAR)
      EN=B(1)*(1.D0+B(4))/B(4)
      WRITE(NOUT,120)AC(5),EN,EN*ERR(1)/B(1)+B(1)*ERR(4)/B(4)**2
      CALL CTOUT(NOUT,NP,NPAR,B,R,SEE,X,Y)
      CALL MSTAIL(*10)
      STOP
      END
C
      SUBROUTINE FCT4(N,X,B,F,DER)
C Evaluates the function F and its first partial derivatives DER(N)
C  at point X when called by GAUSSN.
      IMPLICIT DOUBLE PRECISION (A-H,O-Z)
      DIMENSION B(N),DER(N)
C Calculate the transient term of the Chasset-Thirion equation.
      RT=(EXP(B(2))/X)**B(3)
C Determine the function value F(X).
      F=B(1)*(1.D0+RT/(1.D0+B(4)*RT))
C Determine the first derivatives of the function F with respect to the
C  first parameter, B(1)
      DER(1)=1.D0+RT/(1.D0+B(4)*RT)
C  second parameter, B(2)
      DER(2)=B(1)*B(3)*RT/(1.D0+B(4)*RT)**2
C  third parameter, B(3)
      DER(3)=RT*B(1)*(B(2)-LOG(X))/(1.D0+B(4)*RT)**2
C  fourth parameter, B(4)
      DER(4)=-B(1)*(RT/(1.D0+B(4)*RT))**2
      RETURN
      END
```

3.5.3 CTIN

```
      SUBROUTINE CTIN(NTER,NINT,NMAX,NP,TITLE,X,Y)
C Reads in data file for CHATH3 and CHATH4 consisting of NP pairs of time X
C  and relaxation modulus Y data.  If data are not in chronological order,
C  remove comment label from statement calling the subroutine SORT2.  NTER
C  and NINT are the units specified for the terminal device and input file,
C  respectively.
C Requires the subroutine STOPP.
      CHARACTER TITLE*80
      DOUBLE PRECISION X(NMAX),Y(NMAX)
      READ(NINT,'(A80)')TITLE
C Read in time (X) and modulus (Y) data pairs.
      DO 10 I=1,NMAX
        READ(NINT,*,END=20)X(I),Y(I)
        IF(I.EQ.NMAX)CALL STOPP(NTER,NMAX-1)
10    CONTINUE
20    NP=I-1
C     CALL SORT2(NP,X,Y,X,Y)
      RETURN
      END
```

3.5.4 CTOUT

```
      SUBROUTINE CTOUT(NOUT,NP,NPAR,B,R,SEE,X,Y)
C Creates partial output file for CHATH3 and CHATH4.
C NOUT is the unit specified for the the output file.
      IMPLICIT DOUBLE PRECISION (A-H,O-Z)
      DIMENSION B(NPAR),R(NP,1),X(NP),Y(NP)
      WRITE(NOUT,10)
10    FORMAT(/T27,'Observed',T45,'Calculated'/T22,17('-'),T41,17('-')/
     &        T16,'Log',T34,'Log',T53,'Log'/T6,'Time',T15,'Time',T20,
     &        2(3X,'Modulus',2X,'Modulus'),T62,'Residual'/T2,9('-'),
     &        T13,7('-'),T20,2(2X,9('-'),1X,7('-')),T61,10('-')/)
      DO 30 I=1,NP
         WRITE(NOUT,20)X(I),LOG10(X(I)),Y(I),LOG10(Y(I)),Y(I)-R(I,1),
     &                 LOG10(Y(I)-R(I,1)),R(I,1)
20       FORMAT(1P,E10.3,2X,0P,F7.3,2(1P,E11.3,2X,0P,F6.3),T61,1P,E10.3)
30    CONTINUE
      WRITE(NOUT,40)SEE
40    FORMAT(/1X,'Standard error of estimate =',1P,E10.3/)
      WRITE(NOUT,*)'Variable units are identical to those of data'
      WRITE(NOUT,50)
50    FORMAT(/1X,'Calculated quantities are:'/
     &       /3X,' Tr = Time/tau',              29X,' L(Tr) = Log10(Tr)'
     &       /3X,' Er = Modulus/Ee',            27X,' L(Er) = Log10(Er)'
     &       /3X,'Erc = Calculated modulus/Ee',16X,'L(Erc) = Log10(Erc)'
     &       /3X,'Res = (Residual/Ee)*100'/)
      WRITE(NOUT,60)
60    FORMAT( T8,'Tr',T19,'Er',T30,'Erc',T39,'Res',T47,'L(Tr)',
     &       T56,'L(Er)',T64,'L(Erc)'/T2,3(2X,9('-')),T38,5('-'),
     &       T43,3(3X,6('-'))/)
      DO 80 I=1,NP
         WRITE(NOUT,70)X(I)/B(2),Y(I)/B(1),(Y(I)-R(I,1))/B(1),
     &                R(I,1)*1.0D2/B(1),LOG10(X(I)/B(2)),
     &                LOG10(Y(I)/B(1)),LOG10((Y(I)-R(I,1))/B(1))
70       FORMAT(1X,1P,3E11.3,0P,F8.2,3F9.3)
80    CONTINUE
      RETURN
      END
```

3.6 Sample Listings of INPUT/OUTPUT Files

3.6.1 Data File

Shown is a data file in the format required for running either CHATH3 or CHATH4.

```
Polyisoprene with 0.3% dicup in air at 35 C. (G (t, s) = Pa)
    .10000E-01    .39953E+06
    .15848E-01    .38771E+06
    .25119E-01    .37611E+06
    .39809E-01    .36467E+06
    .63091E-01    .35318E+06
    .10000E+00    .34174E+06
    .15848E+00    .33060E+06
    .25119E+00    .31936E+06
    .39809E+00    .30900E+06
    .63091E+00    .29902E+06
    .10000E+01    .28931E+06
    .15848E+01    .28088E+06
    .25119E+01    .27243E+06
    .39809E+01    .26475E+06
    .63091E+01    .25794E+06
    .10000E+02    .25141E+06
    .15848E+02    .24536E+06
    .25119E+02    .23969E+06
    .39809E+02    .23461E+06
    .63091E+02    .23018E+06
    .10000E+03    .22573E+06
```

3.6.2 Terminal Dialog with CHATH4

Reproduced here is the terminal dialog for running CHATH4 with the data file CHATH.DAT listed in Section 3.6.1. Except for the parameter *u*, the dialog for running CHATH3 would be identical.

```
>CHATH4
 Enter name of data file (N, if not applicable).
CHATH.DAT
 Results to be written to NEW file: CHATH.RES
 OK (Y/N)?
Y
 Input initial guesses for parameters: Ee, tau, m, u.
 The proposed values are:    .188E+06   .230E+00   .2533   .4446
 Proceed with the proposed values? (Y/N)
Y
 Use the default convergence criterion, 1.E-4? (Y/N)
Y
 Computing!

 Equilibrium Relaxation Modulus,  Ee = 1.933E+05 +/- 1.910E+03
            Characteristic time, tau = 2.549E-01 +/- 1.571E-02
       Chasset-Thirion exponent,   m = 2.811E-01 +/- 8.239E-03
   Untrapped Entanglement Index,   u = 5.349E-01 +/- 2.404E-02

          Standard error of estimate = 2.191E+02

 Another initial guess? (Y/N)
N
```

3.6.3 Output File from CHATH4

Reproduced here is the output file from CHATH4 using the data and terminal dialog listed in Sections 3.6.1 and 3.6.2, respectively.

```
                    Physical Relaxation in Elastomers
                    Modified Chasset-Thirion Analysis
                    ---------------------------------

                              1 + (1+u)(tau/time)**m
                    E(t) = Ee ----------------------
                                1 + u(tau/time)**m

                      En = Ee (1+u)/u


Polyisoprene with 0.3% dicup in air at 35 C. (G (t, s) = Pa)


 Equilibrium Relaxation Modulus,  Ee = 1.933E+05 +/- 1.910E+03
            Characteristic time, tau = 2.549E-01 +/- 1.571E-02
       Chasset-Thirion exponent,   m = 2.811E-01 +/- 8.239E-03
   Untrapped Entanglement Index,   u = 5.349E-01 +/- 2.404E-02
   Entanglement Network Modulus,  En = 5.546E+05 +/- 2.171E+04

                          Observed           Calculated
                     -----------------  -----------------
             Log              Log                Log
   Time      Time    Modulus  Modulus   Modulus  Modulus      Residual
---------  -------  --------- -------  --------- -------     ----------

1.000E-02   -2.000  3.995E+05   5.602  3.995E+05   5.601      6.470E+01
1.585E-02   -1.800  3.877E+05   5.589  3.879E+05   5.589     -2.073E+02
2.512E-02   -1.600  3.761E+05   5.575  3.762E+05   5.575     -1.399E+02
3.981E-02   -1.400  3.647E+05   5.562  3.646E+05   5.562      1.041E+02
6.309E-02   -1.200  3.532E+05   5.548  3.530E+05   5.548      2.205E+02
1.000E-01   -1.000  3.417E+05   5.534  3.415E+05   5.533      2.168E+02
1.585E-01    -.800  3.306E+05   5.519  3.304E+05   5.519      2.477E+02
2.512E-01    -.600  3.194E+05   5.504  3.195E+05   5.504     -1.614E+02
3.981E-01    -.400  3.090E+05   5.490  3.091E+05   5.490     -1.067E+02
6.309E-01    -.200  2.990E+05   5.476  2.992E+05   5.476     -1.437E+02
1.000E+00     .000  2.893E+05   5.461  2.897E+05   5.462     -4.267E+02
1.585E+00     .200  2.809E+05   5.449  2.809E+05   5.448      1.644E+01
2.512E+00     .400  2.724E+05   5.435  2.726E+05   5.435     -1.301E+02
3.981E+00     .600  2.648E+05   5.423  2.648E+05   5.423     -8.982E+01
6.309E+00     .800  2.579E+05   5.412  2.577E+05   5.411      2.407E+02
1.000E+01    1.000  2.514E+05   5.400  2.511E+05   5.400      2.826E+02
1.585E+01    1.200  2.454E+05   5.390  2.451E+05   5.389      2.495E+02
2.512E+01    1.400  2.397E+05   5.380  2.396E+05   5.380      6.760E+01
3.981E+01    1.600  2.346E+05   5.370  2.346E+05   5.370     -2.880E+01
6.309E+01    1.800  2.302E+05   5.362  2.301E+05   5.362      5.182E+01
1.000E+02    2.000  2.257E+05   5.354  2.261E+05   5.354     -3.281E+02

Standard error of estimate = 2.191E+02

Variable units are identical to those of data

Calculated quantities are:

   Tr = Time/tau                                  L(Tr)  = Log10(Tr)
   Er = Modulus/Ee                                L(Er)  = Log10(Er)
  Erc = Calculated modulus/Ee                     L(Erc) = Log10(Erc)
  Res = (Residual/Ee)*100
```

Tr	Er	Erc	Res	L(Tr)	L(Er)	L(Erc)
3.924E-02	2.067E+00	2.067E+00	.03	-1.406	.315	.315
6.218E-02	2.006E+00	2.007E+00	-.11	-1.206	.302	.303
9.856E-02	1.946E+00	1.947E+00	-.07	-1.006	.289	.289
1.562E-01	1.887E+00	1.886E+00	.05	-.806	.276	.276
2.476E-01	1.827E+00	1.826E+00	.11	-.606	.262	.262
3.924E-01	1.768E+00	1.767E+00	.11	-.406	.248	.247
6.218E-01	1.711E+00	1.709E+00	.13	-.206	.233	.233
9.856E-01	1.652E+00	1.653E+00	-.08	-.006	.218	.218
1.562E+00	1.599E+00	1.599E+00	-.06	.194	.204	.204
2.476E+00	1.547E+00	1.548E+00	-.07	.394	.190	.190
3.924E+00	1.497E+00	1.499E+00	-.22	.594	.175	.176
6.218E+00	1.453E+00	1.453E+00	.01	.794	.162	.162
9.856E+00	1.410E+00	1.410E+00	-.07	.994	.149	.149
1.562E+01	1.370E+00	1.370E+00	-.05	1.194	.137	.137
2.476E+01	1.335E+00	1.333E+00	.12	1.394	.125	.125
3.924E+01	1.301E+00	1.299E+00	.15	1.594	.114	.114
6.218E+01	1.270E+00	1.268E+00	.13	1.794	.104	.103
9.856E+01	1.240E+00	1.240E+00	.03	1.994	.093	.093
1.562E+02	1.214E+00	1.214E+00	-.01	2.194	.084	.084
2.476E+02	1.191E+00	1.191E+00	.03	2.394	.076	.076
3.924E+02	1.168E+00	1.170E+00	-.17	2.594	.067	.068

References

1. de Gennes, P. G.: "Reptation of Stars," *J. Physique* **36**, 1199-1203 (1975).
2. Curro, J. G.; Salazar, E. A.: "Physical and Chemical Stress Relaxation of Elastomers," *J. Appl. Polym. Sci.* **19**, 2571-2581 (1975).
3. Chasset, R.; Thirion, P.: "Viscoelastic Relaxation of Rubber Vulcanizates between the Glass Transition and Equilibrium," *Proceedings of the International Conference on the Physics of Non-Crystalline Solids.* (J. A. Prins, ed.), American Elsevier, New York, 1965, pp. 345-359.
4. Thirion, P.; Chasset, R.: "Relaxation Viscoélastique des Vulcanisats de Caoutchouc en Extension," *Chim. Ind. Gén. Chim.* **97**, 617-626 (1967).
5. Thirion, P.; Chasset, R.: "Les Rapports entre l'Élasticité et la Structure des Réseaux Macromoléculaires: Esquisse d'une Nouvelle Approche," *Rev. Gén. Cauot. Plast.* **45**, 859-866 (1968).
6. Thirion, P.; Chasset, R.: "Recent Observations on the Mechanical Properties of Polymer Networks," *Pure Appl. Chem.,* **23**, 183-199 (1970).
7. Plazek, D. J.: "Effect of Crosslink Density on the Creep Behavior of Natural Rubber Vulcanizates," *J. Polym. Sci. A-2* **4**, 745-763 (1966).
8. Arenz, R. J.: "Effect of Crosslinking Density on Biaxial Relaxation of SBR by Using Reduced Variables," *J. Polym. Sci., Phys. Ed.* **12**, 131-143 (1974).
9. Kan, H-C.; Ferry, J. D.; Fetters, L. J.: "Rubber Networks Containing Unattached Macromolecules. 5. Stress Relaxation in Styrene-Butadiene-Styrene Block Copolymer with Unattached Linear and Star Polybutadienes," *Macromol.,* **13**, 1571-1577 (1980).

10. Ferry, J. D.: *Viscoelastic Properties of Polymers.* Wiley, New York, 1980.
11. McKenna, G. B.; Gaylord, R. J.: "Relaxation of Crosslinked Networks: Theoretical Models and Apparent Power Law Behaviour," *Polymer* **29**, 2027-2032 (1988).
12. Curro, J. G.; Pincus, P.: "A Theoretical Basis for Viscoelastic Relaxation of Elastomers in the Long-Time Limit," *Macromol.* **16**, 559-562 (1983).
13. Curro, J. G.; Pearson, D. S.; Helfand, E.: "Viscoelasticity of Randomly Cross-Linked Polymer Networks. Relaxation of Dangling Chains," *Macromol.,* **18**, 1157-1162 (1985).
14. Thirion, P.; Monnerie, L.: "Double-Step Retraction of a Polyisoprene Vulcanizate: Comparison with Molecular Models," *J. Polym. Sci., Phys. Ed.* **24**, 2307-2318 (1986).
15. Thirion, P.; Monnerie, L.: "Response of Polyisoprene and Polybutadiene Vulcanisates to Double-Step Retractions: Analysis by a Disengagement Process of Pendant Chains," *J. Polym. Sci., Phys. Ed.* **25**, 1033-1041 (1987).
16. Helfand, E.; Pearson, D. S.: "Statistics of the Entanglement of Polymers: Unentangled Loops and Primitive Paths," *J. Chem. Phys.* **79**, 2054-2059 (1983).
17. Gaylord, R. J.; Weiss, G. H.; DiMarzio, E. A.: "Nonequilibrium Mechanical Response of a Cross-Linked Network," *Macromol.* **19**, 927-929 (1986).
18. Neuburger, N. A.; Eichinger, B. E.: "Computer Simulation of the Dynamics of Trifunctional and Tetrafunctional End-linked Random Elastic Networks," *J. Chem. Phys.* **83**, 884-891 (1985).
19. Thirion, P.; Chasset, R.: "Contributions Relatives de la Viscoélasticité et du Vieillissement a la Relaxation des Vulcanisats de Caoutchouc," *Rev. Gén. Cauot.* **41**, 271-276 (1964); "Relative Contributions of Viscoelasticity and Aging to the Relaxation of Cured Rubber," *Rubber Chem. Technol.* **37**, 617-626 (1964).
20. Chasset, R.; Thirion, P.: *Proceedings of the Natural Rubber Conference.* Kuala Lumpur, 1968.
21. Dickie, R. A.; Ferry, J. D.: "Dynamic Mechanical Properties of Cross-Linked Rubbers. III. Dicumyl Peroxide Vulcanizates of Natural Rubber," *J. Phys. Chem.* **70**, 2594-2600 (1966).
22. Lee, H. H.; Shaw, M. T.: "Relaxation Processes in Elastomers," presented at the 131st Meeting of the Rubber Division of the American Chemical Society, Montreal, 1987, poster XIV.
23. Sanders, J. F.: "Viscoelastic Properties of 1,2, and *cis*-1,4 Polybutadienes and Butyl Rubber," Ph.D. thesis, University of Wisconsin, 1968.
24. Himmelblau, D. M.: *Process Analysis by Statistical Methods.* Wiley, New York, 1970.
25. Press, W. H.; Flannery, B. P.; Teukolsky, S. A.; Vetterling, W. T.: *Numerical Recipes in* FORTRAN: *The Art of Scientific Computing* Cambridge University Press, Cambridge, 1986.

26. Mancke, R. G.; Ferry, J. D.: "Dynamic Mechanical Properties of Cross-Linked Rubbers. IV. Dicumyl Peroxide Vulcanizates of Styrene-Butadiene Rubber," *Trans. Soc. Rheol.* **12**, 335-350 (1968).
27. McKenna, G. B.: National Institute of Standards and Technology, private communication, 1991. (A misprint in the reported value of τ for Sample G in Reference 10 was verified; instead of -2, the exponent should read -4.)
28. Mancke, R. G.: "Linear Viscoelastic Properties of Cross-Linked Polymers – Natural Rubber, Polybutadiene, and Styrene-Butadiene Copolymer," Ph.D. thesis, University of Wisconsin, 1967.
29. Schwarzl, F. R.: "Numerical Calculation of Stress Relaxation Modulus from Dynamic Data for Linear Viscoelastic Materials," *Rheol. Acta* **14**, 581-590 (1975).

Chapter 4

Molecular Weight Distribution From the Viscosity Function

4.1 Introduction

While the molecular weight of a polymer is critical to the overall performance of the material, the molecular weight distribution plays an especially strong role in its processing and fabrication. Usually, an analytical laboratory determines the molecular weight distribution by dissolving the polymer and running a size-exclusion chromatogram (SEC). For some polymers, however, solvents are not readily available. Solution methods also take time, not so much for the SEC run itself but often for dissolving the polymer. Modern rheological methods allow four decades of frequency to be gathered in about 20 minutes by using the melt from a process stream. No solvent is involved and no solids have to filtered. Thus, rheological methods are ideal for process control [1].

There is no question that the molecular weight distribution of a polymer influences the viscosity function of its melt. By way of illustration, Figure 4.1 shows the shear viscosities for two polystyrene melts: one having a narrow distribution ($M_w/M_n < 1.1$), the other a broad distribution ($M_w/M_n = 2.5$) [2]. The narrow distribution melt goes from Newtonian behavior at low shear rates to power-law behavior over a very narrow shear-rate range. The broader distribution melt requires many decades of rate before the power-law region is entered. Thus, a qualitative indicator of the breadth of the molecular weight distribution is the shear-rate range over which there is a finite second derivative of the viscosity curve. All we need is a method to translate the shear-rate scale into a molecular-weight scale.

The quantitative translation of the viscosity function into a molecular weight distribution has received much attention [3-10]; this subject and other similar transforms have been recently and thoroughly reviewed by Tuminello [10]. The better formulated approaches tend to result in transforms which are very difficult to use. There is also some question of whether the nuances of a sophisticated approach will be lost during its application to real (i.e., noisy) data. The simpler transforms rely on assumed bulk mixing rules which are probably valid only for linear polymers. The easiest to illustrate is a logarithmic mixing rule combined with the observed behavior of very narrow fractions. The viscosity function for such fractions can be idealized by a truncated power law as illustrated in Figure 4.2. The higher molecular weight fractions have a higher viscosity but enter the power-law region at lower shear rates, roughly on a common line with a slope of $-\nu$. The range for ν is

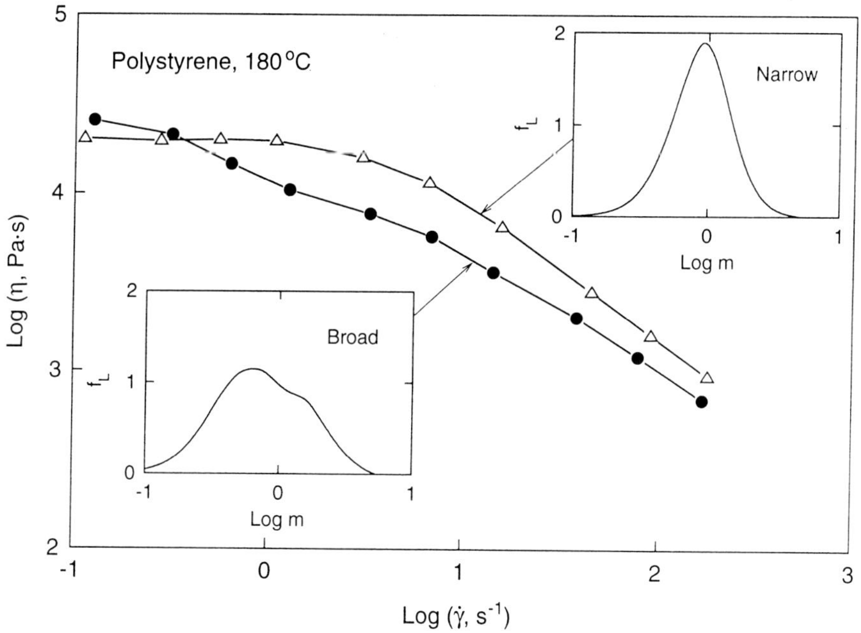

Figure 4.1 Viscosity data for two polystyrenes, one having a broad and the other a narrow molecular weight distribution [2]. The insets show the inversion of these data by program MWD. Adapted from Ref. 2. Copyright © 1970 by The Society of Rheology, Inc. Reprinted by permission of the American Institute of Physics.

generally 0.5 to 1.0. If we mix equal amounts of these two fractions and assume the logarithmic mixing rule

$$\log \eta_m = w_1 \log \eta_1 + w_2 \log \eta_2 \tag{4.1}$$

the dashed line shown in Figure 4.2 should result. The corresponding cumulative molecular weight distributions, $F(M)$, are shown in the inset. For this mixing rule, it can be seen that the value of $F(M)$ is equal to 1 + slope/ν; that is,

$$F(M) = 1 + \frac{1}{\nu} \frac{d \ln \eta}{d \ln \dot{\gamma}} \bigg|_{\dot{\gamma} = g(M)} \tag{4.2}$$

if a correspondence between $\dot{\gamma}$ and M can be made. If this mixing rule were to hold, the viscosity function would have to be void of inflections because $F(M)$ must be a monotonically increasing function with M. However, inflections have been observed [8].

If the more general mixing rule

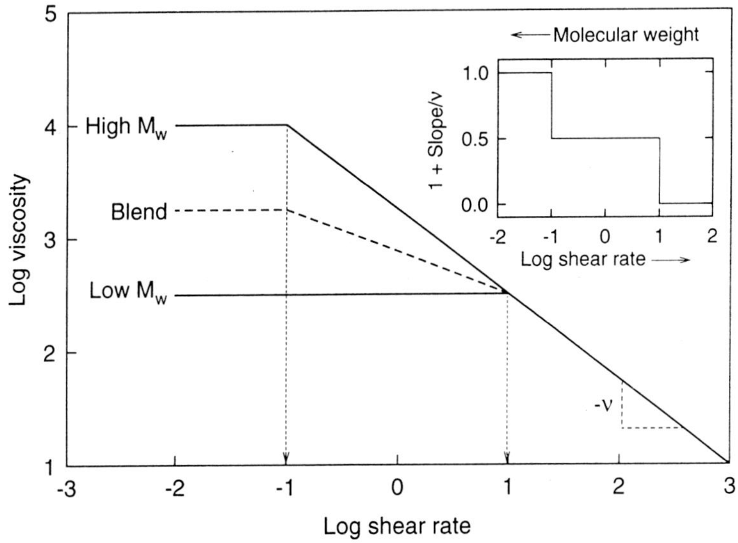

Figure 4.2 Explanation of the inversion technique using a log-normal mixing rule as an illustration. The inset shows the start of derived cumulative distribution based on the slopes of the segments of the blend's viscosity function (dashed line).

$$\eta^{1/\alpha} = w_1 \eta_1^{1/\alpha} + w_2 \eta_2^{1/\alpha} \tag{4.3}$$

were assumed, it can be shown that the transform is

$$F(m) = 1 + \frac{1}{\nu} \left(\frac{\eta}{\eta_o} \right)^{1/\alpha} \left(\frac{\dot{\gamma}}{\dot{\gamma}_c} \right)^{\nu/\alpha} \frac{d \ln \eta}{d \ln \dot{\gamma}} \Bigg|_{\dot{\gamma} = \dot{\gamma}_c m^{-\alpha/\nu}} \tag{4.4}$$

where m is a reduced molecular weight M/M_w, η_o is the zero-shear-rate viscosity, and $\dot{\gamma}_c$ is the critical shear rate. If one imagines a truncated, power-law model enveloping the viscosity curve (refer to Figure 4.6 for a real example), the resulting critical shear rate is at the break in this model.

If $\alpha = 3.4$, the mixing rule is consistent with the exclusive dependence of η_o on M_w, which is known to be empirically valid. However, little is known about the mixing rules that are applicable at high shear rates, and it is conceivable that the mixing rule will be a function of shear stress and other variables. Note that $\alpha = \infty$ corresponds to the logarithmic mixing rule.

The application of the transform expressed by Equation 4.4, and its limitations, can be illustrated with an example proposed by Malkin and Teishev [9]. Starting with a double-box distribution shown in the top inset of Figure 4.3, the application of the mixing rule in Equation 4.3 gives the viscosity function shown in this figure.

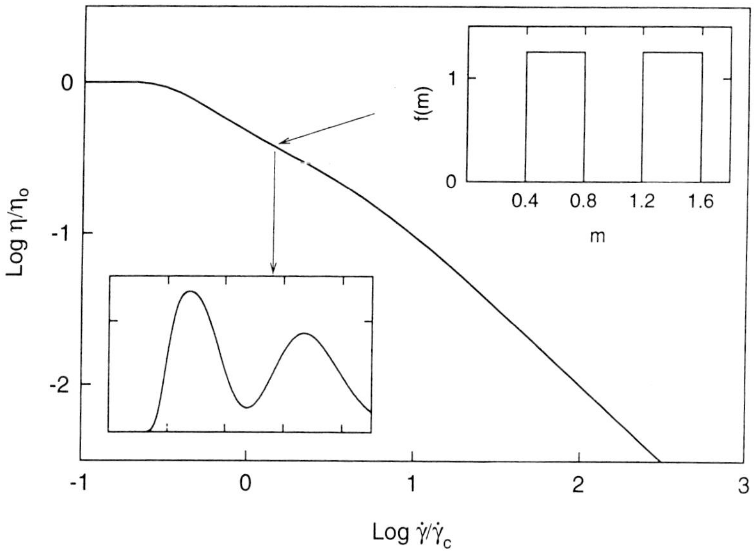

Figure 4.3 Conversion of the example molecular weight distribution of Malkin and Teishev to a viscosity function and back again to the distribution. Rounding of the corners is evident (scales of the insets are identical).

Before the power-law region is completely reached, there is a slight bump, because of the box at low m. On application of the transform of Equation 4.4, as handled by program MWD in this chapter, the distribution in the lower inset results. As expected, there is a significant amount of rounding in applying the forward and inverse transforms, but the qualitative aspect of bimodality is preserved.

The two messages emerging from this exercise are that the viscosity data must be especially precise and there must be a sufficient shear rate range covered. Furthermore, any low-parameter function applied to fit the data may cause the important details (such as bimodality) to disappear [8]. This was demonstrated by Malkin and Teishev [9]. The requirement for viscosity data over a very broad range will probably mean that superposition of data from different temperatures will be needed; see Chapter 5 for details.

A note of caution should be given at this point concerning the plotting of $F(M)$ and its differentiation to yield a differential distribution. If we plot F versus $\log M$, we should probably write the redundant $F(\log M)$ to remind ourselves that the visible slope will be the common $dF(\log M)/d\log M$. This will have quite a different shape than the function $f(M) = dF(M)/dM$. The latter may be only weakly bimodal in cases where the former is strongly bimodal; refer to work by Friedman for specific examples [11]. The output from size-exclusion chromatography is close to the function $dF(\log M)/d\log M$ because the elution time goes roughly as $-\log M$.

Be advised that appearing in the literature are differential distribution functions labeled "amount," "fraction," "relative amount," or with no label. Often there is little indication of whether these functions refer to dF/dM, $dF/d\log M$, or something else. We find only a confusion of nomenclature and thus have chosen to use the general terms $F(x)$ and $f(x) = dF/dx$.

4.2 Description of the Algorithm MWD

The program MWD has no fancy features, but there are a few tricks. The difficult step in the application of Equation 4.4 is taking a numerical derivative; recall that it cannot be done by fitting a low-parameter model to the data. Rheological data accumulated by time-temperature superposition will be dense in the areas of overlap and sparse near the ends (see Chapter 5). The overlap region can lead to problems with many derivative techniques because the points will be very closely spaced on the log $\dot{\gamma}$ axis (even nonunique), but not all that close in viscosity. A remedy is to spread the points out along the abscissa. After MWD sorts the data set by using SORT2, the subroutine EVEN is called to spread out the points using interpolation only. If the points are unique, a minimum-curvature Akima interpolation [12] is used; otherwise, a linear interpolation is used. The effect of applying EVEN is illustrated in Figure 4.4. The data [8] were not unique so linear interpolation was invoked. As can be seen from the viscosity curves at the top of the figure, the interpolation has only a slight smoothing effect, but it makes the set much more differentiable. Interpolation carries another small bonus; smoothing and differentiation on evenly spaced abscissa values are much faster.

The influence of the spreading process on the derivatives is illustrated in the bottom curves of Figure 4.4 where the derivatives are plotted. The derivatives are based on a second-order polynomial function fitted to five consecutive points. Upon careful examination, one can spot a few errant points from the original set thereby illustrating the principal advantage of the spreading step: elimination of the most deviant derivatives. Note that the derivatives do not hit zero at low shear rates, nor do they approach a constant value at high rates; these problems must also be addressed.

Are there penalties from the spreading step? There is a distortion due to overweighting the influence of any problems near the end of the set. At low shear rates, errors attributable to poor signal-to-noise ratio are expected, while at high rates errors due to viscous heating, slip, and flow instabilities need to be handled. If dynamic data are used, nonlinearities and viscous heating can influence the results at high frequencies. In both steady and dynamic experiments, the influence of the glass transition may start to be felt [8]. Ameliorating these is the fact that $F(M)$ must go to 0 and 1 in the limits of high and low shear rates, respectively. However, a weak high or low molecular weight tail is very apt to be lost.

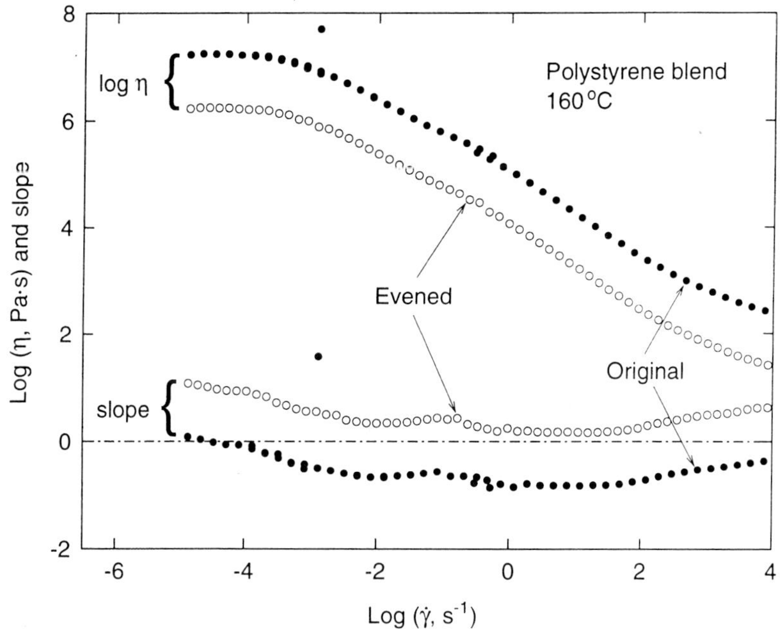

Figure 4.4 Illustration of the spreading of data along the abscissa to aid numerical differentiation. The *evened* curves have been shifted vertically one decade for clarity. Adapted from Ref. 8. Copyright © 1991 by Society of Plastics Engineers. Reprinted by permission.

As mentioned, something must be done about data at frequencies high enough to be entering the $G' \propto \omega^{1/2}$ region (glass transition). This results in an inflection in the viscosity at high shear rates which has nothing to do with molecular weight. The program MWD handles this by searching for an inflection at high frequencies. If one is found, all the data above the inflection (along with a few below) are replaced with extrapolated points. The extrapolation is done by using a Vinogradov-type viscosity function [13]

$$\eta = \frac{\eta_o}{\sum_{i=0}^{N} a_i \dot{\gamma}^{i\nu/N}} \tag{4.5}$$

where N is the number of terms. This function has some attractive features: it is linear when inverted (the inverse of viscosity is fluidity), and it is guaranteed to approach a logarithmic slope of -ν eventually. The advantage of fitting the fluidity rather than the viscosity is that the high fluidity values at high shear rates are fitted with low relative error. The performance of this function ($N = 3$) is illustrated in Figure 4.5. The fit to the function is poor at high viscosities but excellent at low viscosities (high fluidities). Decidedly, this would be the worst possible function to differentiate at low shear rates.

A word about the slope -ν. If the molecular-weight-dependent part of the relaxation spectrum extends only to a finite frequency, the viscosity should fall with a -1 slope corresponding to a flat rubbery plateau in the relaxation modulus. Empirical slopes tend to range from -0.66 to -0.85 (see Chapter 9); Ferry cites -0.82 as a frequently found value [14, p. 258]. Theory gives various results which depend on the treatment of the high-frequency end of the relaxation process. The program `MWD` allows the operator to choose; guidance is provided by listing the highest slope magnitude found in the data set.

Equation 4.4 requires the zero-shear-rate viscosity and the critical shear rate. The former is obtained in `MWD` by fitting up to the first decade of data to a linearized Bueche-Harding model [15,16]. Starting with four points, more points are added until the estimated error for the zero-shear-rate viscosity bottoms out. The critical shear rate is taken as the intersection of the zero-shear-rate viscosity and the power-law slope (-ν) found at high shear rates by using the fluidity fit (Equation 4.5). This exercise guarantees that the calculated cumulative weight fractions will lie between 0 and 1 and avoids a renormalization at a later stage. By doing so, we are claiming

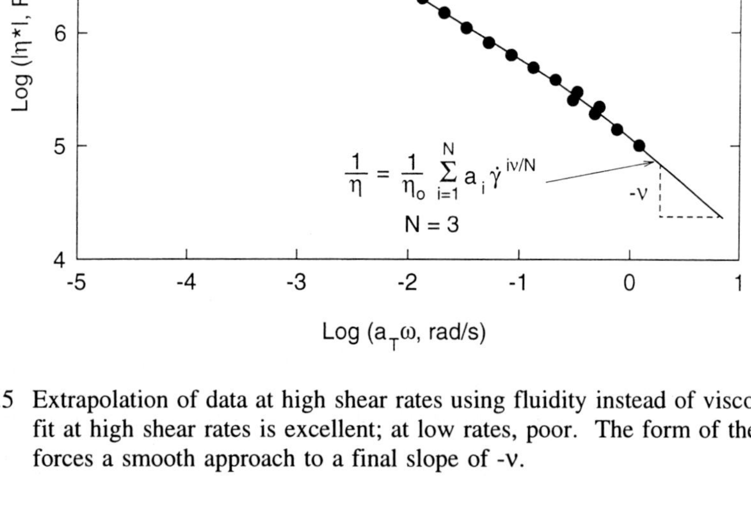

Figure 4.5 Extrapolation of data at high shear rates using fluidity instead of viscosity. The fit at high shear rates is excellent; at low rates, poor. The form of the equation forces a smooth approach to a final slope of -ν.

that we know (or think we know) much more about the behavior of the viscosity function than we do about the molecular weight distribution function. There is also an element here of wanting to take care of problems early.

With known well-behaved functions at the head and the tail of the data set, points are added at both ends until the second derivatives fall below a certain level, again to guarantee normalization. The process is illustrated with real data for a polystyrene melt [17] shown in Figure 4.6. Because normalization is most difficult at the low molecular-weight end (high shear rates), the number of points added there can be large. While the molecular weight distribution could be calculated analytically in these areas from the fitted expressions, the additional points help with the numerical derivatives at both ends of the original data set. Figure 4.6 also shows the truncated power-law envelope for this data set.

The next step is the derivative; it can be put off only so long. We chose a smoothing derivative using a second-order moving strip covering about a decade. Because the *knee* in the viscosity function of a narrowly distributed polymer covers about two decades of frequency, our strip should not miss even the smallest possible bump in the viscosity function. With evenly spaced points, the slope will have optimal statistical properties except at the ends of the data set. However, the ends

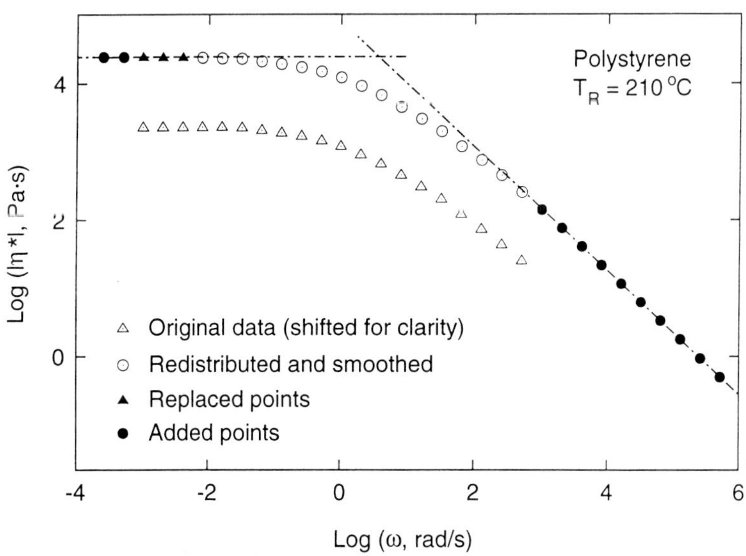

Figure 4.6 Example showing points added at both ends of set and the replacement of three points at low shear rates. The original data are shifted one decade down for clarity. Adapted from Ref. 17. Copyright © 1985 by Springer-Verlag. Reprinted by permission.

have been padded with perfectly smooth points, so we do not have too much to worry about.

Before the second differentiation a slight smoothing is applied by using a moving strip of only three points. Smoothing is an ill-defined problem [18, pp. 495-497]. We know what we want to do – remove the error in the points – but the problem is that we do not have any idea on what curve the error-free points should lie, particularly for this problem. It should be a least-squares curve, if the errors are normally distributed, but there is an infinite set of least-squares curves. Finally, we feel that the *true* curve should have a very limited number of inflections and quite low local curvature. The subroutine DCURVE has, as its merit function, a sum of the squares of the distances of the new points away from the old points and the sum of the squares of the curvatures at each point. The user gets to choose which is more important via the input variable EMP. If EMP is set to 1.0, almost any curvature (including the *real* curvature) will disappear. If EMP is set to 0.0, the smoothed points will be the same as the original data because the first term of the merit function is emphasized. We set the variable to 0.2 to give a slight bit of smoothing yet to leave any low-frequency curvature intact. DCURVE depends on uniform spacing of the points.

Because the viscosity function itself does not contain any information about the absolute molecular weight (unless one is willing to accept a molecular theory), the primary result of the application of Equation 4.4 is a weight fraction at a particular shear rate. The shear rate is converted to molecular weight by the equation

$$m = \frac{M}{M_w} = \left(\frac{\dot{\gamma}}{\dot{\gamma}_c}\right)^{-\nu/\alpha} \tag{4.6}$$

While we do not know M_w, the ratio of the number-average molecular weight to the weight-average molecular weight can be found. This important ratio is given by

$$\frac{M_w}{M_n} = \int_0^\infty \frac{f(m)}{m}\, dm \tag{4.7}$$

By using the trapezoidal rule, Equation 4.7 reduces to

$$\frac{M_w}{M_n} = \sum_{i=1}^{N-1} \frac{f_i + f_{i+1}}{m_i + m_{i+1}} (m_i - m_{i+1}) \tag{4.8}$$

where $f(m)$ is calculated from the differentiated form of Equation 4.4; that is,

$$f(m) = -\frac{1}{m\nu^2}\left(\frac{\eta}{\eta_o}\right)^{1/\alpha}\left(\frac{\dot{\gamma}}{\dot{\gamma}_c}\right)^{\nu/\alpha}\left[\alpha\frac{d^2\ln\eta}{d\ln\dot{\gamma}^2} + \nu\frac{d\ln\eta}{d\ln\dot{\gamma}} + \left(\frac{d\ln\eta}{d\ln\dot{\gamma}}\right)^2\right] \tag{4.9}$$

The normalization of $f(m)$ is checked, again by trapezoidal integration, and reported in the results file.

Finally, MWD calculates the distribution function

$$f_L(\log m) = \frac{dF}{d\log m} \tag{4.10}$$

4.3 Execution of MWD

The execution of MWD requires the subroutines ALINR, AMAXI, AMINI, DCURVE, POLYN, SLOPE, and SORT2 found in Appendix B. Inherent to the execution of MWD, the subprograms EVEN, VALUE, and LNTER are included in the code listed in Section 4.5. Refer to Appendix A for the program dependencies. For example, POLYN calls the routines LUBKSB, LUDCMP, and MPROVE from *Numerical Recipes* [18].

The program requires data for viscosity versus shear rate for the material of interest. The data should extend from the Newtonian region into the power-law region. The broader the molecular weight distribution, the wider must be the shear-rate range. For some melts, the viscosity can be approximated by the Cox-Merz rule from dynamic experiments [19]:

$$\eta(\dot{\gamma}) = \left|\eta^*(\omega)\right|\Big|_{\omega = \dot{\gamma}} \tag{4.11}$$

The format of the ASCII data file is identical to that required by the Chapter 9 program VISFIT to fit functions to viscosity data. The first record is set aside for the title of the file; include anything up to 79 characters containing identification information. The second record specifies the variable representing the ordinate. Enter V or S for viscosity or shear stress, respectively. Data may be entered as dimensional numbers or in the corresponding logarithmic (base 10) values. The format should be specified in the third record by entering N for dimensional numbers or L for logarithmic values. The succeeding records are either pairs of shear-rate and viscosity data or shear-rate and shear-stress data. The data pairs can be in any sequence. A sample listing of a data file is provided in Section 4.6.1.

The operation of MWD is semi-interactive; refer to Section 4.6.2 for a sample terminal dialog. After the name of the data file is entered, the program reports the number of points, the number of points per decade, the maximum reported viscosity, and the estimate of the zero-shear-rate viscosity. The program will provide the option of accepting the estimate or entering a new value for η_o. After establishing η_o, MWD will print the maximum slope found in the power-law region and ask for the desired value of ν. Entering a low value for ν will tend to narrow the distribution, while a steeper slope will add extrapolated points to the data and broaden the distribution. This effect is shown in Table 4.1, where the magnitude of the maximum slope was reported as 0.83. If confusion reigns, just enter 1.0, which is what has generally been done in the past. Next, MWD will ask for the order of the polynomial function to be used for the fluidity extrapolation. If you feel that the high-shear-rate end of the data is low in slope due to the presence of a significant amount of low-molecular-weight material, a higher order polynomial can describe this effect. However, it is quite likely that a very high order with a final slope significantly greater than the steepest part of the data will give a negative final coefficient. This will be picked up by MWD, and a lower order must be tried. As can be seen in Table 4.1, the higher orders tend to broaden the distribution if the selected final slope is higher than the reported maximum slope. The minor problems at high orders with the values of the distribution variables occur because the extrapolation can do things that real data cannot; avoid these problems by picking first order. Note that the first-order version is the Cross equation (see Chapter 9 for a full description).

Table 4.1 Influence of the Final Slope -ν and Polynomial Order on M_w/M_n Using the Program MWD on Typical Data Showing an Estimated Maximum Slope Magnitude of 0.83

	M_w/M_n with Order			
ν	1	2	3	4
0.80	1.36[a]	1.33[b]	1.32[b]	1.50[c]
0.85	1.43[a]	1.37[b]	1.59[c]	-[d]
0.90	1.49[a]	1.49[b]	2.04[c]	-[d]
0.95	1.56[a]	1.68[a]	2.50[c]	-[d]
1.00	1.62[a]	1.87[a]	2.97[c]	-[d]

[a] No problems; good normalization.
[b] Negative $F(m)$ at low m.
[c] Positive $F(m)$ at low m.
[d] Final slope is zero or positive.

Along with the coefficients of the polynomial, MWD displays a series of plus and minus signs corresponding to the fluidity residuals at the high-shear-rate end of the data set. A glance at these signs may help in deciding if a different order should be attempted.

The output file consists of most of the information that appears on the display terminal plus a tabulation of the interpolated and extrapolated shear rates and viscosities, m, $F(m)$, $dF(m)/dm$, log m, and $dF(\log m)/d\log m$. Finally, an estimate of M_w/M_n is printed along with the results of the normalization check. The listing of a sample output file is appended in Section 4.6.3.

4.4 Application of MWD

The first example for demonstrating the capabilities of MWD concerns a commercial polystyrene on which dynamic viscosity data had been measured [17]. Only 20 points were available, but the data, listed in Table 2 of Reference 17, extended from the Newtonian to the power-law region (Figure 4.6). After the Cox-Merz rule (Equation 4.11) was applied, a data file was set up to run with MWD; see Section 4.6.1 for the listing. A final slope magnitude of 0.91 was required to obtain the M_w/M_n of 2.1 found by SEC. The terminal dialog and the output file are reproduced in Sections 4.6.2 and 4.6.3, respectively.

Figure 4.7 shows the calculated cumulative weight distribution, F, along with its derivative, f. A replot of $F(m)$ on a logarithmic scale is shown in Figure 4.8 along with its derivative. The shape of $dF/d\log m$ is very symmetrical; therefore, the

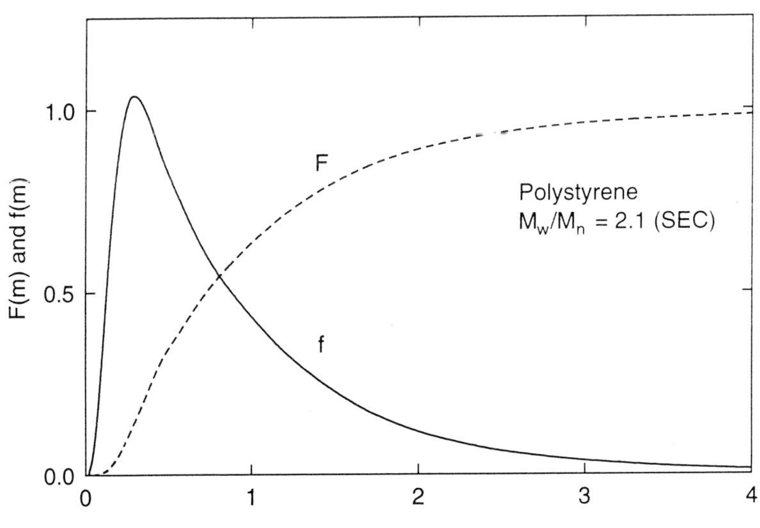

Figure 4.7 Cumulative and differential weight distributions from the data in Figure 4.6 as found by program MWD. The reduced molecular weight m is M/M_w.

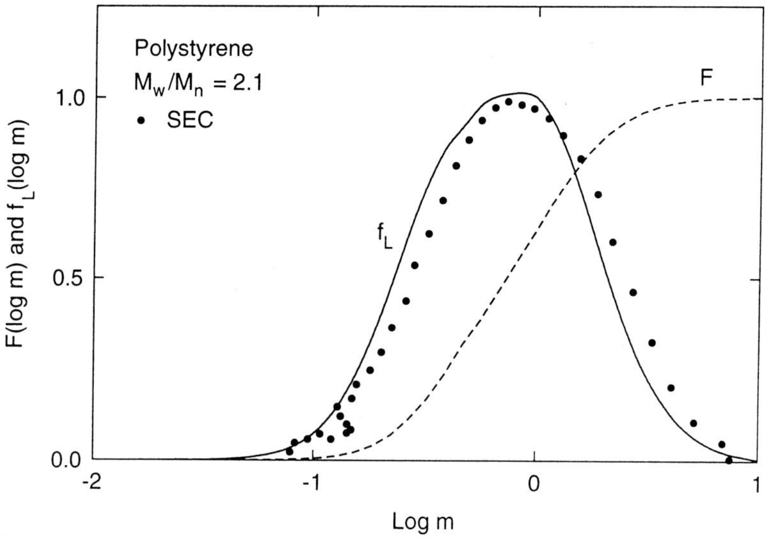

Figure 4.8 Cumulative distribution on a logarithmic scale and its derivative from the viscosity data in Figure 4.6 using a final slope magnitude of 0.91.

distribution is close to logarithmic normal. Included in Figure 4.8 is the distribution curve derived from SEC; the agreement is satisfactory.

Figure 4.4, discussed in Section 4.2, showed the original dynamic data for a blend of two narrow fractions of polystyrene [8]. Plotted in Figure 4.9 is the function $dF(\log M)/d\log M$, produced by MWD, aided by the known M_w of 426,000. This plot illustrates the influence of the final slope value on the calculated distributions. While the qualitative aspects are preserved, the results show that there is some degree of arbitrariness to the inversion of $\eta(\dot{\gamma})$ to a molecular weight distribution. One of the primary reasons for including the slope control in the program was so that the user can explore the degree of variation that is possible while not violating any fundamental constraints, such as normalization of the distribution.

The final example examines the ability of the algorithm to handle some noise. A random noise of 2% was added into the original viscosity data of the polystyrene blend [8]. The inversion still went well, producing the results in Figure 4.10. The SEC results were shown in the original publication [8] as points but are reproduced in Figure 4.10 as a spline-interpolated curve. The match is reasonable, although the emphasis by MWD on the high molecular weight end (or a de-emphasis on the low end) is evident.

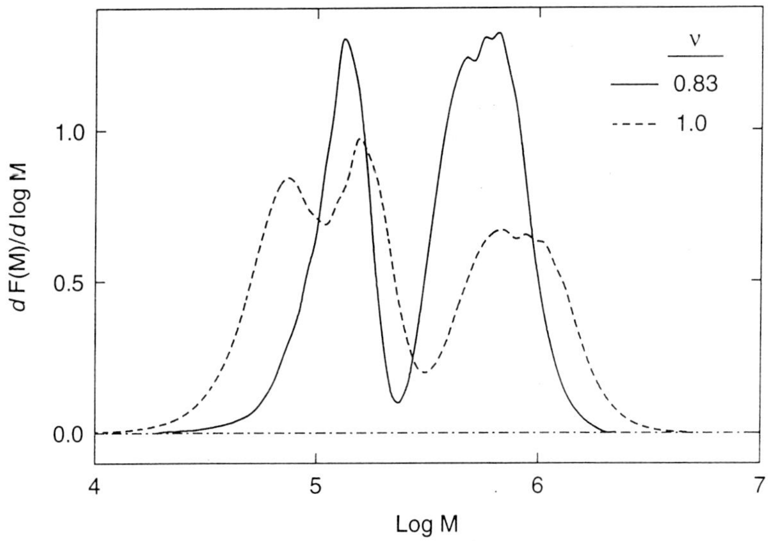

Figure 4.9 Analysis of viscosity data in Figure 4.4 by MWD. The abscissa values have been shifted using the known M_w of 426,000. The influence of the final slope value of the viscosity curve is shown.

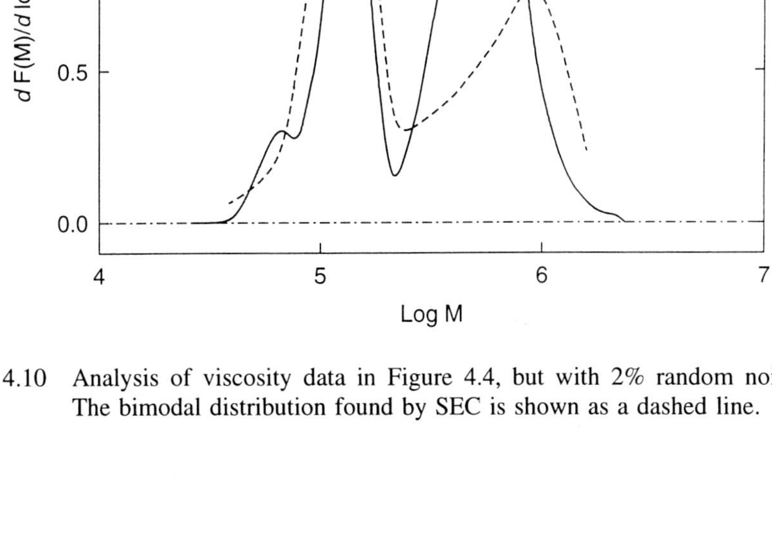

Figure 4.10 Analysis of viscosity data in Figure 4.4, but with 2% random noise added. The bimodal distribution found by SEC is shown as a dashed line.

4.5 Program Listing for MWD

Imbedded in the code are subprograms LNTER, VALUE, and EVEN inherent to the execution of MWD.

```
      PROGRAM MWD
C Attempts to find the molecular weight distribution from viscosity data
C  using a modification of the Malkin-Teishev method.
C Requires the subroutines ALINR, AMAXI, AMINI, DCURVE, POLYN, SLOPE, and
C  SORT2.
C Data file format:
C        1st record - Title, up to 79 characters.
C        2nd record - V(iscosity) or S(tress).
C        3rd record - L(ogarithmic) or N(ormal) values.
C subsequent records - shear rate (1/s), viscosity
C                 or,   shear rate (1/s), shear stress.
      IMPLICIT DOUBLE PRECISION (A-H,O-Z)
      PARAMETER(NMAX=301,NVAR=11,NLOW=60,ALF=3.4D0,X23=2.3025851)
      CHARACTER ANS*1,CRES(NLOW)*1,FORM*80,TITLE*80,TYPE*80
      DIMENSION A(NVAR),AERR(NVAR),ARES(NMAX)
      DIMENSION D(NMAX),DD(NMAX),RES(NMAX),XP(NMAX),YP(NMAX)
      DIMENSION U(-NLOW:NMAX),V(-NLOW:NMAX)
C t-statistic for 95% confidence interval with N degrees of freedom.
      T95(N)=1.96+2.3539/N+3.1902/N**2+0.7004/N**3+4.5015/N**4
10    CALL MSHEAD
      CALL UNITS(NTER,NINT,NOUT)
      WRITE(NOUT,'(/19X,A)')'MWD Analysis of Viscosity Function'
      WRITE(NOUT,'(19X,34(''-'')/)')
      READ(NINT,'(A80)')TITLE
      WRITE(NOUT,'(/A80)')TITLE
      READ(NINT,'(A80)')TYPE
      READ(NINT,'(A80)')FORM
      ITYPE=1
      IFORM=1
      DO 20 I=1,80
        IF(TYPE(I:I).EQ.'S'.OR.TYPE(I:I).EQ.'s')ITYPE=2
        IF(FORM(I:I).EQ.'N'.OR.FORM(I:I).EQ.'n')IFORM=2
20    CONTINUE
C Read in viscosity data.
      DO 30 I=1,NMAX
        READ(NINT,*,END=40)U(I),V(I)
        IF(I.EQ.NMAX)CALL STOPP(NTER,NMAX-1)
C Perform necessary transformations.
        IF(IFORM.EQ.2)THEN
          U(I)=LOG10(U(I))
          IF(ITYPE.EQ.2)THEN
            V(I)=LOG10(V(I))-U(I)
          ELSE
            V(I)=LOG10(V(I))
          ENDIF
        ELSE IF(ITYPE.EQ.2)THEN
          V(I)=V(I)-U(I)
        ENDIF
30    CONTINUE
40    NP=I-1
C Sort data into ascending shear rates.
      CALL SORT2(NP,U(1),V(1),U(1),V(1))
C At this point U and V are logs shear of rate and viscosity;
C EVEN spreads point along U axis.  RES is the working array.
      CALL EVEN(NTER,NP,U(1),V(1),RES,IER)
C U and V are logs of shear rate and viscosity; ordered and even.
```

```
C Find number of points per decade (NPPD).
      NPPD=NP/INT(U(NP)-U(1))
      IF(NPPD.LT.3)THEN
        WRITE(NTER,*)' The number of points per decade is too low!'
        STOP
      ENDIF
      WRITE(NTER,50)NP,NPPD
      WRITE(NOUT,50)NP,NPPD
50    FORMAT(/17X,'Total points =',I4/12X,'Points per decade =',I4)
C  Find zero-shear-rate viscosity using linearized Bueche-Harding equation.
      DO 60 I=1,NP
        XP(I)=1.D1**(U(I)*.75D0)
60    CONTINUE
      XLETA0=V(1)
      ERR0=1.D1
      CALL AMAXI(NP,V(1),VMAX,IVMAX)
      WRITE(NTER,70)1.D1**VMAX,IVMAX
      WRITE(NOUT,70)1.D1**VMAX,IVMAX
70    FORMAT(/3X,'Maximum measured viscosity =',1P,E11.2,2X,'at original
     & point number',I3)
      NMIN=4
      MMAX=MAX(NPPD,NMIN+4)
      DO 80 NI=NMIN,MMAX
        CALL ALINR(NI,XP(IVMAX),V(IVMAX),A)
        IF(A(2).GT.0.D0)GOTO 80
        IF(A(9)*T95(NI-2).GT.ERR0)THEN
          GOTO 90
        ELSE
          XLETA0=A(1)
          ERR0=A(9)*T95(NI-2)
          SLOP=A(2)
        ENDIF
80    CONTINUE
90    CONTINUE
      WRITE(NTER,100)1.D1**XLETA0
100   FORMAT(4X,'Zero-shear-rate viscosity =',1P,E11.2,2X,'(estimated)')
      WRITE(NTER,'(/2X,A)')'Is this estimate for the zero-shear-rate vis
     &cosity acceptable? (Y/N)'
      READ(NTER,'(A1)')ANS
      IF(ANS.EQ.'N'.OR.ANS.EQ.'n')THEN
        WRITE(NTER,*)' Enter a new estimate.'
        READ(NTER,*)XETA0
        XLETA0=LOG10(XETA0)
      ENDIF
      WRITE(NOUT,100)1.D1**XLETA0
C Add points at low shear rates.
      K=0
      J=0
      DU=U(NP)-U(NP-1)
      DO 110 I=IVMAX,-3*NPPD,-1
        U(I)=U(I+1)-DU
        V(I)=XLETA0+SLOP*1.D1**(U(I)*.75D0)
        IF(-(V(I)-V(I+1))/DU.LT.1.D-4)K=K+1
        J=J+1
        IF(K.GT.6)GOTO 120
110   CONTINUE
120   NADD=J-IVMAX
      NP=NP+NADD
C Shift data to 1 thru NP after increasing NP by number of points added.
      DO 130 I=NP,1,-1
        U(I)=U(I-NADD)
        V(I)=V(I-NADD)
130   CONTINUE
C Find maximum rough slope and point number; ask for power-law slope.
      N=3
      NST=5
      CALL SLOPE(NP,U(1),V(1),NST,N,D(1),ARES)
      CALL AMINI(NP,D(1),DMIN,IDMIN)
```

```
      WRITE(NTER,140)-DMIN,IDMIN-NADD
      WRITE(NOUT,140)-DMIN,IDMIN-NADD
140   FORMAT(/6X,'Maximum slope magnitude =',F6.2,7X,'at original',
     &        1X,'point number',I4)
      WRITE(NTER,'(/2X,A)')'Enter choice of final slope magnitude (<1).'
      READ(NTER,*)XNU
      WRITE(NOUT,'(8X,A,F6.2/)')'Final slope magnitude =',XNU
C Search for inflection at high end of data.
      NOFF=5
      N=3
      DO 160 I=NP-NOFF,NP/2,-1
        CALL POLYN(N,NP-I+1,U(I),V(I),A,AERR,ARES,SEE)
        IF(A(3).GT.0.D0.AND.AERR(3)/A(3).LT..5D0)THEN
C Good indication of an inflection; try to pin it down.
          N=4
          NPX=MAX(10,INT(1.5*REAL(NPPD)))
          DO 150 J=NP-NPX,NP/2,-1
            CALL POLYN(N,NPX,U(J),V(J),A,AERR,ARES,SEE)
            UI=-A(3)/3.D0/ABS(A(4))
            IF(UI.GE.U(J+NPX/2-2).AND.UI.LE.U(J+NPX/2+2))THEN
C If inflection is found, exclude points beyond inflection.
              NP=J+NPX/2-2
              GOTO 170
            ENDIF
150       CONTINUE
        ENDIF
160   CONTINUE
C Fit fluidity to get critical shear rate and add points at high rates.
170   WRITE(NTER,'(2X,A)')
     &     'Enter polynomial order for high shear-rate fit: (1 to 4).'
      READ(NTER,*)XNORD
      NORD=INT(XNORD+.001)
      IF(NORD.GT.4.OR.NORD.LT.1)GOTO 170
      XN=REAL(NORD)
      N=NORD+1
      NTOP=MIN(MAX(NPPD,4),NP/2)
      K=0
      DO 180 I=NP-NTOP,NP
        K=K+1
        XP(K)=10.D0**(U(I)*XNU/XN)
        YP(K)=10.D0**(-V(I))
180   CONTINUE
      CALL POLYN(N,NTOP+1,XP(1),YP(1),A,AERR,ARES,SEE)
      WRITE(NTER,'(/2X,A/)')'Results of polynomial fit:'
      WRITE(NTER,190)(I-1,A(I),T95(NTOP+1-N)*AERR(I),I=1,N)
190   FORMAT(4X,'A(',I1,') =',1P,E11.3,' +/-',1P,E10.3)
      IF(A(N).LE.0.D0)THEN
        WRITE(NTER,'(2X,A)')'Please select a lower order.'
        GOTO 170
      ENDIF
      K=0
      DO 200 I=NP,NP-NTOP,-1
        K=K+1
        IF(K.GE.30)GOTO 210
        CRES(K)='+'
        RES(K)=V(I)+LOG10(VALUE(10.D0**(U(I)*XNU/XN),N,A))
        IF(RES(K).LT.0)CRES(K)='-'
200   CONTINUE
210   NRES=K
      WRITE(NTER,'(/4X,A,30A2)')'Residuals: ',(CRES(J),J=NRES,1,-1)
      WRITE(NTER,'(16X,A)')'Shear rate ->'
      WRITE(NTER,'(/2X,A)')'Change order of polynomial? (Y/N)'
      READ(NTER,'(A1)')ANS
      IF(ANS.EQ.'Y'.OR.ANS.EQ.'y')GOTO 170
      WRITE(NOUT,'(2X,A/)')'Results of polynomial fit:'
      WRITE(NOUT,190)(I-1,A(I),T95(NTOP+1-N)*AERR(I),I=1,N)
```

```
C Use fluidity fit to add points to get high-rate end completed.
      DU=U(NP)-U(NP-1)
      K=0
      J=0
      NBACK=3
      DO 220 I=NP-NBACK,NP+4*NPPD
        U(I)=U(I-1)+DU
        V(I)=-LOG10(VALUE(1.D1**(U(I)*XNU/XN),N,A))
        IF(-(V(I)-V(I-1))/DU.GT..999*XNU)K=K+1
        J=J+1
        IF(K.GT.7)GOTO 230
220   CONTINUE
230   NP=NP+J-NBACK-1
      IF(NP.GT.NMAX)THEN
        WRITE(NTER,'(/2X,A)')'Out of storage; too many points added.'
        STOP
      ENDIF
C Calculate log of critical shear rate.
      XLSRC=(-LOG10(A(N))-XLETA0)/XNU
      WRITE(NOUT,'(/4X,A,F6.2)')'Log (critical shear rate) =',XLSRC
C Find slopes.
      N=3
      NST=7
      CALL SLOPE(NP,U(1),V(1),NST,N,D,ARES)
C Apply smoothing to slopes.
      EMP=.2D0
      CALL DCURVE(NP,D,RES,EMP)
C Ensure normalizaton and switch slopes back D from RES.
      DO 240 I=1,NP
        IF(RES(I).GT.0.D0)RES(I)=0.D0
        IF(RES(I).LT.-XNU)RES(I)=-XNU
        D(I)=RES(I)
240   CONTINUE
C Get second derivatives.
      CALL SLOPE(NP,U(1),D,NST,N,DD,ARES)
      CALL DCURVE(NP,DD,RES,EMP)
      DO 250 I=1,NP
        DD(I)=RES(I)
250   CONTINUE
      DD(1)=0.D0
      DD(NP)=0.D0
      WRITE(NOUT,260)
260   FORMAT(//T4,'Log rate',T14,'Log visc',T26,'M/Mw',T36,'F(m)',
     &        T44,'dF(m)/dm',T55,'Log m',T64,'dF/dlog m',
     &        /1X,6(2X,8('-')),2X,9('-'))
C Calculate distribution functions.
      SUM1=0.D0
      SUM2=0.D0
      SUM3=0.D0
      XMSAVE=0.D0
      FCSAVE=0.D0
      FDSAVE=0.D0
C Record results from low to high molecular weight.
      DO 270 I=NP,1,-1
        FAC=10.D0**((V(I)-XLETA0)/ALF)*10.D0**((U(I)-XLSRC)*XNU/ALF)
C Cumulative weight distribution function, FCUM.
        FCUM=1.D0+FAC/XNU*D(I)
C Log differential weight distribution function.
        FDIFL=-FAC/XNU**2*(ALF*DD(I)/X23+XNU*D(I)+D(I)**2)
C Reduced molecular weight, M/Mw.
        XM=10.D0**(-XNU*(U(I)-XLSRC)/ALF)
C Differential weight distribution function.
        FDIF=FDIFL/XM
        WRITE(NOUT,'(7F10.3)')U(I),V(I),XM,FCUM,FDIF,LOG10(XM),X23*FDIFL
C Sum for Mw/Mn.
        SUM1=SUM1+(FDIF+FDSAVE)/(XM+XMSAVE)*(XM-XMSAVE)
```

```
C Sum for differential weight distribution normalization check.
          SUM2=SUM2+(FDIF+FDSAVE)/2.D0*(XM-XMSAVE)
          XMSAVE=XM
          FDSAVE=FDIF
          FCSAVE=FCUM
270     CONTINUE
        WRITE(NOUT,'(1X,71(''-''))')
        WRITE(NTER,280)SUM1,SUM2
        WRITE(NOUT,280)SUM1,SUM2
280     FORMAT(/4X,'Mw/Mn =',F6.2,/
     &         4X,'Weight distribution normalization:',F6.2)
        CALL MSTAIL(*10)
        STOP
        END
C
        SUBROUTINE LNTER(NP,X,Y,XI,YI)
C Performs linear interpolation on NP pairs of X, Y data given the
C  abscissa for interpolation XI and returns the interpolated value YI.
        IMPLICIT DOUBLE PRECISION (A-H,O-Z)
        PARAMETER(TINY=1.D-20)
        DIMENSION X(NP),Y(NP)
C If only two points then quick return.
        IF(NP.EQ.2)THEN
          YI=(XI-X(1))*(Y(2)-Y(1))/(X(2)-X(1))+Y(1)
          RETURN
        ENDIF
C For extrapolation.
        IF(XI.LT.X(1))THEN
          I1=1
          GOTO 20
        END IF
        IF(XI.GE.X(NP))THEN
          I1=NP
          GOTO 20
        ENDIF
C  Find the two data points which bracket XI.
        I1=1
        I2=NP
        NIT=IFIX(SQRT(FLOAT(NP)))+1
        DO 10 I=1,NIT
          IMID=(I1+I2)/2
          IF(XI.GE.X(IMID))THEN
            I1=IMID
            IF(XI.LE.X(I1+1))GOTO 20
          ELSE
            I2=IMID
          END IF
10      CONTINUE
20      YI=(XI-X(I1))*(Y(I1+1)-Y(I1))/(X(I1+1)-X(I1))+Y(I1)
        RETURN
        END
C
        DOUBLE PRECISION FUNCTION VALUE(X,N,A)
C Given the coefficients A of a (N-1)th order polynomial, returns the
C  value of the polynomial evaluated at X.
        IMPLICIT DOUBLE PRECISION (A-H,O-Z)
        DIMENSION A(N)
        SUM=A(N)
        DO 10 J=N-1,1,-1
          SUM=SUM*X+A(J)
10      CONTINUE
        VALUE=SUM
        RETURN
        END
```

```
      SUBROUTINE EVEN(NTER,NP,U,V,W,IER)
C Produces interpolated points V evenly spaced on axis U. W is a working
C  vector. NTER is the terminal device unit number.
C Requires the subroutines AKIMA and LNTER and the function subprograms
C  AMAX and AMIN.
      IMPLICIT DOUBLE PRECISION (A-H,O-Z)
      PARAMETER (SMALL=1.0D-2)
      DIMENSION U(NP),V(NP),W(NP)
      DX=(AMAX(NP,U)-AMIN(NP,U))/FLOAT(NP-1)
      SHIFT=SMALL*DX
      IFLAG=0
      ISHIFT=0
C Find any nondistinct or out-of-order abscissa values.
      DO 40 I=2,NP
        IF(U(I).LT.U(I-1))THEN
          WRITE(NTER,'(/2X,A)')'ERROR:  X values are not ordered.'
          WRITE(NTER,'(2X,A)')'Use SORT2 to order data.'
          IER=2
          RETURN
        ELSE IF(ISHIFT.GE.1.AND.U(I)-U(I-1).GT.(ISHIFT+1)*SHIFT)THEN
          DO 10 J=I-ISHIFT-1,I-1
            U(J)=U(J)-SHIFT*REAL(ISHIFT)/2.D0
10        CONTINUE
          DO 20 J=I,NP
            U(J)=U(J)-SHIFT*ISHIFT
20        CONTINUE
          ISHIFT=0
        ELSE IF(U(I).EQ.U(I-1))THEN
          IF(IFLAG.EQ.0)THEN
            WRITE(NTER,'(/2X,A)')'NOTE:  X values are not distinct.'
            WRITE(NTER,'(2X,A)')'A slight abscissa shift made;  a linear
     & interpolation will be used.'
            IFLAG=1
          ENDIF
          DO 30 J=I,NP
            U(J)=U(J)+SHIFT
30        CONTINUE
          ISHIFT=ISHIFT+1
        ENDIF
40    CONTINUE
      W(1)=V(1)
      W(NP)=V(NP)
      IF(IFLAG.EQ.0)THEN
C Use minimum-curvature interpolation.
        DO 50 I=2,NP-1
          CALL AKIMA(NP,U,V,U(1)+DX*FLOAT(I-1),W(I))
50      CONTINUE
      ELSE
C Use linear interpolation.
        DO 60 I=2,NP-1
          CALL LNTER(NP,U,V,U(1)+DX*FLOAT(I-1),W(I))
60      CONTINUE
      ENDIF
      DO 70 I=1,NP
        U(I)=U(1)+DX*FLOAT(I-1)
        V(I)=W(I)
70     CONTINUE
      IER=1
      RETURN
      END
```

4.6 Sample Listings of INPUT/OUTPUT Files for MWD

4.6.1 Data file for MWD

Shown here is a file of viscosity data for a polystyrene, approximated from dynamic experimental results using the Cox-Merz rule, in the format required for running MWD. Adapted from Reference 17. Copyright © 1985 by Springer-Verlag. Reprinted by permission.

```
PS data from Pfandl & Schwarzl, Colloid Polym. Sci., 263, 328 (1985).
V
N
   .1000E-02  .2320E+05
   .2000E-02  .2320E+05
   .4000E-02  .2321E+05
   .8000E-02  .2317E+05
   .1600E-01  .2332E+05
   .3200E-01  .2275E+05
   .6400E-01  .2092E+05
   .1280E+00  .1917E+05
   .2560E+00  .1712E+05
   .5120E+00  .1474E+05
   .1000E+01  .1220E+05
   .2000E+01  .9214E+04
   .4000E+01  .6720E+04
   .8000E+01  .4536E+04
   .1600E+02  .3055E+04
   .3200E+02  .2007E+04
   .6400E+02  .1206E+04
   .1280E+03  .7302E+03
   .2560E+03  .4341E+03
   .5120E+03  .2538E+03
```

4.6.2 Terminal Dialog with MWD

Reproduced here is the terminal dialog for running MWD with the data file MWD.DAT listed in Section 4.6.1.

```
>MWD
 Enter name of data file (N, if not applicable).
MWD.DAT
 Results to be written to NEW file: MWD.RES
 OK? (Y/N)
Y

                 Total points =  20
            Points per decade =   4

    Maximum measured viscosity =   2.33E+04  at original point number  5
     Zero-shear-rate viscosity =   2.42E+04  (estimated)

 Is this estimate for the zero-shear-rate viscosity acceptable? (Y/N)
Y
```

```
     Maximum slope magnitude =    .77       at original point number  20

 Enter choice of final slope magnitude (<1).
0.91
 Enter polynomial order for high shear-rate fit:  (1 to 4).
1

 Results of polynomial fit:

   A(0) =  2.698E-04 +/- 1.440E-04
   A(1) =  1.271E-05 +/- 9.338E-07

   Residuals:  + - - - +
               Shear rate ->

 Change order of polynomial (Y/N)?
N

   Mw/Mn =  2.09
   Weight distribution normalization:  1.02

 The results are in MWD.RES
 Run another set of data? (Y/N)
N
Stop - Program terminated.
```

4.6.3 Output File from MWD

Reproduced here is the output file from MWD using the data and terminal dialog listed in Sections 4.6.1 and 4.6.2, respectively. A final slope magnitude of 0.91 was required to obtain the M_w/M_n of 2.1 found by SEC. The results are plotted in Figures 4.7 and 4.8.

```
                    MWD Analysis of Viscosity Function
                    ----------------------------------

PS data from Pfandl & Schwarzl, Colloid Polym. Sci., 263, 328 (1985).

                 Total points =  20
            Points per decade =   4

   Maximum measured viscosity =   2.33E+04  at original point number  5
    Zero-shear-rate viscosity =   2.42E+04  (estimated)

      Maximum slope magnitude =   .77       at original point number  20
        Final slope magnitude =   .91

  Results of polynomial fit:

    A(0) =  2.698E-04 +/- 1.440E-04
    A(1) =  1.271E-05 +/- 9.338E-07

    Log (critical shear rate) =   .56
```

Log rate	Log visc	M/Mw	F(m)	dF(m)/dm	Log m	dF/dlog m
7.217	-1.671	.017	.000	.000	-1.781	.000
6.916	-1.398	.020	.000	-.001	-1.700	.000
6.616	-1.124	.024	.000	.010	-1.620	.001
6.315	-.851	.029	.000	.019	-1.539	.001
6.015	-.577	.035	.000	.029	-1.459	.002
5.714	-.304	.042	.000	.045	-1.379	.004
5.414	-.031	.050	.001	.070	-1.298	.008
5.113	.243	.061	.001	.108	-1.218	.015
4.813	.516	.073	.002	.163	-1.137	.027
4.512	.789	.088	.004	.244	-1.057	.049
4.212	1.062	.106	.007	.354	-.977	.086
3.911	1.334	.127	.012	.496	-.896	.145
3.611	1.606	.153	.023	.659	-.816	.232
3.310	1.875	.184	.041	.817	-.735	.346
3.010	2.141	.221	.070	.948	-.655	.483
2.709	2.400	.266	.114	1.031	-.574	.632
2.409	2.648	.321	.168	1.031	-.494	.762
2.108	2.878	.386	.237	.970	-.414	.862
1.808	3.080	.464	.318	.864	-.333	.924
1.507	3.301	.559	.385	.763	-.253	.982
1.207	3.483	.673	.464	.650	-.172	1.007
.906	3.655	.809	.544	.544	-.092	1.014
.606	3.825	.974	.624	.447	-.011	1.003
.305	3.963	1.172	.705	.347	.069	.937
.005	4.085	1.411	.780	.255	.149	.828
-.296	4.170	1.698	.846	.173	.230	.678
-.596	4.234	2.043	.896	.111	.310	.522
-.897	4.283	2.459	.933	.067	.391	.379
-1.197	4.321	2.959	.959	.039	.471	.262
-1.498	4.357	3.561	.976	.021	.552	.174
-1.798	4.364	4.285	.987	.011	.632	.108
-2.099	4.372	5.157	.993	.005	.712	.064
-2.399	4.377	6.206	.997	.003	.793	.036
-2.700	4.379	7.469	.998	.001	.873	.019
-3.000	4.381	8.988	.999	.000	.954	.009
-3.300	4.382	10.817	1.000	.000	1.034	.001
-3.601	4.383	13.017	1.000	.000	1.115	.000

Mw/Mn = 2.09
Weight distribution normalization: 1.02

References

1. Zeichner, G. R.; Macosko, C. W.: "On-line Viscoelastic Measurements for Polymer Melt Processes," *SPE Tech. Papers* **28**, 79-81 (1982).
2. Graessley, W. W.; Glasscock, S. D.; Crawley, R. L.: "Die Swell in Molten Polymers," *Trans. Soc. Rheol.* **14**, 519-544 (1970).
3. Middleman, S.: "Effect of Molecular Weight Distribution on Viscosity of Polymeric Fluids," *J. Appl. Polym. Sci.* **11**, 417-424 (1967).
4. Cross, M. M.: "Polymer Rheology: Influence of Molecular Weight and Polydispersity," *J. Appl. Polym. Sci.* **13**, 765-774 (1969).
5. Bersted, B. H.: "An Empirical Model Relating the Molecular Weight Distribution of High-Density Polyethylene to the Shear Dependence of the Steady Shear Melt Viscosity," *J. Appl. Polym. Sci.* **19**, 2167-2177 (1975).

6. Bersted, B. H.; Slee, J. D.: "A Relationship between Steady-State Shear Melt Viscosity and Molecular Weight Distribution in Polystyrene," *J. Appl. Polym. Sci.* **21**, 2631-2644 (1977).
7. Pearson, G. H.; Garfield, L. J.: "The Effect of Molecular Weight and Weight Distribution upon Polymer Melt Rheology," *Polym. Eng. Sci.* **18**, 583-589 (1978).
8. Tuminello, W. H.; Cudré-Mauroux, N.: "Determining Molecular Weight Distributions from Viscosity versus Shear Rate Flow Curves," *Polym. Eng. Sci.* **31**, 1496-1507 (1991).
9. Malkin, A. Y.; Teishev, A. E.: "Flow Curve-Molecular Weight Distribution: Is the Solution of the Inverse Problem Possible?" *Polym. Eng. Sci.* **31**, 1590-1596 (1991).
10. Tuminello, W. H.: "Relating Rheology to Molecular Weight Properties of Polymers," *Encyclopedia of Fluid Mechanics, Polymer Processing and Flow Dynamics.* (N. P. Cheremisinoff, ed.), Gulf Publishing, Houston, 1990, Vol. 9, pp. 209-242.
11. Friedman, E. M.: "Modality of Molecular Weight Distributions," *Polym. Eng. Sci.* **30**, 569-570 (1990).
12. Akima, H.: "A New Method of Interpolation and Smooth Curve Fitting Based on Local Procedures," *J. ACM* **17**, 589-602 (1970).
13. Vinogradov, G. V.; Malkin, A. Y.; Prozovovskaya, N. V.; Kargin, V. A.: "The Rheology of Polymers. The Universal Nature of the Temperature Invariant Characteristic of the Viscosity of Polymers in the Condensed State," *Dokl. Akad. Nauk SSSR.* **154**, 890-893 (1964).
14. Ferry, J. D.: *Viscoelastic Properties of Polymers.* Wiley, New York, 1980.
15. Bueche, F.: "Influence of Rate of Shear on the Apparent Viscosity of A-Dilute Polymer Solutions, and B-Bulk Polymers," *J. Chem. Phys.* **22**, 1570-1576 (1954).
16. Bueche, F.; Harding, S. W.: "A New Absolute Molecular Weight Method for Linear Polymers," *J. Polym. Sci.* **32**, 177-186 (1958).
17. Pfandl, W.; Schwarzl, F. R.: "Relation Between Rheological Properties and the Molecular Weight Distribution of a Technical Polystyrene," *Colloid Polym. Sci.* **263**, 328-334 (1985).
18. Press, W. H.; Flannery, B. P.; Teukolsky, S. A.; Vetterling, W. T.: *Numerical Recipes in* FORTRAN: *The Art of Scientific Computing* Cambridge University Press, Cambridge, 1986.
19. Cox, W. P.; Merz, E. H.: "Correlation of Dynamic and Steady Flow Viscosity," *J. Polym. Sci.* **28**, 619-622 (1958).

Chapter 5

Superposition of Linear Properties

5.1 Introduction

Time-temperature superposition is a procedure whereby modulus (or compliance) data taken as a function of frequency or time at various temperatures are superposed to incorporate the temperature dependence into the time scale. The resulting response covers an expanded time window, which can be useful for approximating the behavior of a material at times or frequencies inaccessible with the instrument used to obtain the original data. In addition to superposition, the program `SHIFTT` addresses the errors involved in carrying out the procedure. While the following discussion deals with rheological variables, `SHIFTT` can be adapted to any data that obey the time-temperature principle, including reduced viscosity, $\eta(\dot{\gamma})/\eta_o$.

Although empirical in its origins, time-temperature superposition has had rigorous theoretical justification, mainly by Ferry and his co-workers, as described in Ferry [1]. According to their point of view, temperature can be regarded as a volume-increasing variable similar to a negative isotropic pressure. The increase in free volume facilitates molecular movement and results in a multiplicative shift in all relaxation processes. It is this shift that the rheologist needs for the superposition of data obtained at two different temperatures.

The ability to superpose viscoelastic data, also referred to as thermorheological simplicity, has been demonstrated for a wide variety of single-phase polymers. Lack of the ability to superpose is often a strong indication of two-phase behavior or a phase change brought about by the changing temperature. Chemical changes can also interfere with superposition.

The temperature dependence of the shift factor has been most often compared with the prediction of the Williams-Landel-Ferry (WLF) empirical equation [2]:

$$\log a_T = -\frac{C_1^g(T - T_g)}{C_2^g + T - T_g} \tag{5.1}$$

where a_T is the shift factor required to superpose data taken at temperature T onto data taken at the glass transition T_g, while C_1^g and C_2^g are parameters related to the free volume of the polymer. The WLF equation has been most successful when applied to shift factors for data in the range $T_g < T < T_g + 100$ K. Kaelble [3] suggested a modification to extend this range below T_g:

$$\log a_T = -\frac{C_1^g (T - T_g)}{C_2^g + |T - T_g|} \tag{5.2}$$

For higher temperatures, better results are obtained with the Arrhenius equation

$$\log a_T = b_1 \left(\frac{1}{T} - \frac{1}{T_R} \right) \tag{5.3}$$

or the Fox equation [4]

$$\log a_T = b_1 \left(\frac{1}{T^{b_2}} - \frac{1}{T_R^{b_2}} \right) \tag{5.4}$$

where T_R is a reference temperature to which the shifts of data at temperature T are made.

Traditionally, the shifting process has been accomplished by plotting the modulus-time data sets on logarithmic coordinates and moving the sets taken at temperature T along the horizontal axis to form a smooth continuous curve with the set taken at the reference temperature. Prior to this, the modulus data are often normalized for density ρ and temperature as suggested by the Rouse theory [5] and the kinetic theory of rubber elasticity [6]. The result is a modulus that would be expected if the temperature were, in fact, at the reference temperature:

$$G(T_R, t) = \frac{T_R}{T} \frac{\rho(T_R)}{\rho(T)} G\left(T, \frac{t}{a_T}\right) \tag{5.5}$$

If both components of the complex modulus are shifted with identical values of the shift factor, tan δ should reduce to a master curve. Concentrating on tan δ instead of a dimensional modulus means no density or temperature adjustment should be required. Any lack of fit is subsequently resolved by a shift along the log(tan δ) axis [1]. While the program allows a superposition of tan δ data to be made, there is no provision for vertical shifts. Vertical shifts are usually small and require extraordinarily extensive and precise data to be significant. Also, the meaning of these shifts has had little theoretical justification [7].

It can be appreciated that the shifting process has some characteristics of extrapolation, with the positive aspect that the shape of a section of the extrapolated curve is known. Two possible superposition schemes are illustrated in Figure 5.1. First, an extrapolation of the data at the higher temperature, T_2, is established by using some function. Data from the lower temperature, T_1, is shifted along the frequency scale until a least-squares optimum fit is obtained. However, this algorithm ignores important information; the shape of the T_1 data. It does not exploit the shape of the lower-temperature data to influence the form of the function used

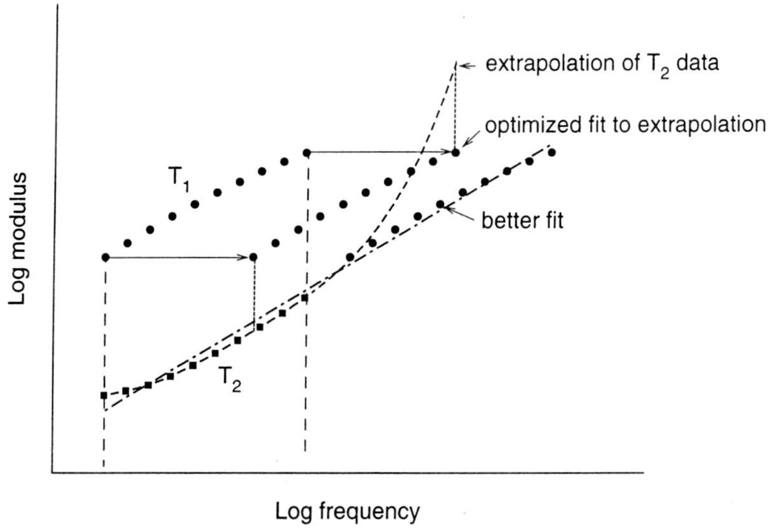

Figure 5.1 Illustration of the shifting operation along the frequency axis. The shift with the "better fit" corresponds to that calculated by Equation 5.8 in the text.

to generate the fit. A better algorithm would shift the T_1 data until an optimum overall fit to some smooth curve was obtained.

Clearly an important ingredient in any extrapolation recipe is the function to be used. What functions should be chosen for the superposition of data sets onto a smooth curve? An incorrect choice could produce a bias in the results. An autocorrelation function may be considered: poorly merged sets will show a high degree of data alternation. While possessing the virtue of bias elimination introduced by the choice of a curve to fit the data, this function works well only with evenly spaced data with significant overlap and requires a sorting operation after every iteration, which adds significantly to the computational effort.

Looking at the original Rouse molecular theory [5] for guidance suggests that the following forms might be useful for dynamic viscoelastic data:

$$G'(\omega) = G_0 + \sum_{i=1}^{\infty} \frac{G_i (\omega\tau_i)^2}{1 + (\omega\tau_i)^2} \tag{5.6}$$

$$G''(\omega) = \sum_{i=1}^{\infty} \frac{G_i \, \omega\tau_i}{1 + (\omega\tau_i)^2} \tag{5.7}$$

where G_0 is the limiting low-frequency modulus and G_i and τ_i are the partial moduli and associated relaxation times, respectively. Unfortunately, values for these are not known a priori, and one is quickly faced with a multiple-variable nonlinear problem which could become badly conditioned in certain parts of the viscoelastic response. (Note that the G_i from the original Rouse theory were assumed to be all equal to $\rho k_B T$ as suggested by Equation 5.5; in Equations 5.6 and 5.7 they are used as the empirical fitting parameters.)

An easy way out is to resort to an empirical function (e.g., a power series), and this may not be as terrible as one might expect. Power series produce oscillatory residuals, which are very badly behaved toward the extremes of the data set. But, in the case of superposition, the center of the data composed of the two sets at T_1 and T_2 is where the performance is critical. As suggested by Figure 5.1, even a straight line can do reasonably well. With some care concerning the order of the polynomial and the weighting factors (heavier weighting near the center produces smoother fits), a reasonable superposition reflecting both the magnitude and the curvature of the data should be obtained.

Another approach is to find the shifts which would give a best fit simultaneously to shifted data from all temperatures by using a power series of high order. While this sounds attractive, it does not take too much algebra to show that this exercise results in sets of equations which are nonlinear in the parameters. These could, in principle, be solved with a nonlinear equation solver, but this is not likely to be any more accurate than the one-dimensional search used in `SHIFTT` and is likely to exhibit severe convergence problems.

5.2 Description of the Algorithms

5.2.1 SHIFTT

The program `SHIFTT` works using a reliable, straightforward algorithm. Starting at the high end of the temperature scale, data sets are examined as temperature pairs, and the shift required for superposition is calculated. The master curve is subsequently formed by the addition of these pairwise shifts. The suspicion that there may be a systematic addition of bias does not prove to be warranted. On the positive side, the method estimates the superposition error through the addition of errors from each pairwise operation, the shape of the master curve produced is independent of the reference temperature, and widely spaced data sets are handled capably.

The superposition procedure is performed in the subroutine `SHIFT`. The first step is to obtain an initial estimate of the shift by using the solution of the following problem: what shift will minimize the standard error of estimate, s_ε, if both sets are shifted to lie on a straight line? The answer is

$$\log a_T = \frac{\bar{y}_{T_L} - \bar{y}_{T_H}}{\beta_1} \tag{5.8}$$

where $\bar{y}_{T_L}$ and $\bar{y}_{T_H}$ are the average log moduli at the low and high temperatures, respectively (Σ log $G(T_i)/n_i$, where n_i is the number of data points at T_i), and β_1 is the slope of the straight line fitted to the two sets of data when there is no shift. With this initial guess, a polynomial is fitted to the two sets of shifted data with the subroutine `POLYN`. The order of the polynomial is the minimum value between the arguments n+1 or Δf, where n is the total number of points and Δf is the number of decades covered by the data. Iteration of the shift factor is continued until a minimum in the weighted least squares of the residuals about the polynomial is found. The weighting used here is Lorentzian, centered about the midpoint of the two sets. The weighting helps to reduce the impact of the high residuals produced by polynomials near the extremes of a data set. The objective function, supplied in the subroutine `FAT`, is optimized with the IMSL subroutine `DUVMIF`, which uses a quadratic interpolation method to find the minimum point of a smooth, univariate function [8].

The error in each pairwise shifting operation is calculated from

$$\pm\ t_{95,\,\nu}\ s_{\varepsilon} \sqrt{C_{ii}} \tag{2.9}$$

where $t_{95,\,\nu}$ is the t-statistic as a function of the degrees of freedom, ν, s_{ε} is the standard error of estimate for the data set, and C_{ii} is the diagonal element for the estimated covariance matrix of standard errors in the fitted parameters [9]. The covariance matrix elements are determined in the subroutine `ERROR` from the inverse of the curvature matrix. The curvature matrix is one-half the Hessian matrix. The Hessian, comprising all second partial derivatives of the objective function with respect to the parameters, is approximated by using forward differences by the IMSL subroutine `DFDHES` [8].

At the conclusion of the pairwise shifting operation and in the event that the reference temperature specified does not correspond to any of the isotherms in the data file, the final shift is interpolated to the reference temperature by using Akima's method [10].

Incorporated into `SHIFTT` is the subprogram `RHOR` which provides the density and temperature correction suggested by Equation 5.5. Two options are available to handle density changes. In the more precise first option, expansion coefficient data below and above the glass transition are necessary. To approximate the correction, the second option requires an estimate of the glass transition of the material, and, subsequently, uses the estimates for the expansion coefficient α,

$$\alpha = 0.09T_g^{-1}, \qquad T < T_g \tag{5.9}$$

and

$$\alpha = 0.20T_g^{-1}, \qquad T > T_g \tag{5.10}$$

suggested by Van Krevelen [11].

As a final step, the algorithm computes the two empirical WLF parameters C_1 and C_2 corresponding to the reference temperature T_R. The calculation is often done by the transformation of Equation 5.1 into the form of a straight line. One such transform is

$$-\frac{1}{\log a_T} = \frac{1}{C_1} + \frac{C_2}{C_1(T - T_R)} \tag{5.11}$$

Another transform [1] results in the expression

$$-\frac{T - T_R}{\log a_T} = \frac{C_2}{C_1} + \frac{T - T_R}{C_1} \tag{5.12}$$

However, disadvantages exist with the transformations. In Equation 5.11, undue emphasis is placed on data close to T_R. In Equation 5.12, the errors in the left-hand side build near $T = T_R$.

In SHIFTT, calculation of the WLF parameters is done by nonlinear techniques. Aside from providing better precision, the nonlinear procedure enjoys uniformity of the errors in the dependent variable, $\log a_T$, and an additional degree of freedom at $T = T_R$ for estimating the parameters. The two-dimensional function, included in the subroutine FWLF, is minimized in the IMSL routine DUMINF by using a quasi-Newton method and a finite-difference gradient [8]. The initial estimates for the parameters are determined from the relations

$$C_1 = \frac{C_1^g C_2^g}{C_2} \tag{5.13}$$

$$C_2 = C_2^g + T_R - T_g \tag{5.14}$$

The universal values $C_1^g = 17.44$ and $C_2^g = 51.6$ K are used as the parameters corresponding to T_g. Clearly, the optimization will run into convergence problems if insufficient data are available. Shift factors for at least six temperatures in the range of T_g and $T_g + 100$ K should be available, and this criterion is checked before any fit is attempted. (Those familiar with programming can easily alter the subroutine FWLF to incorporate other relationships, such as those proposed in Equations 5.2 to 5.4, or specify the temperature range or weighting parameters.) The errors for the WLF parameters are calculated from Equation 2.9.

To provide an alternative statistical method for characterizing the error involved in the superposition procedure, the Monte Carlo simulation method [9] is featured as an option. The need for this type of operation arises when only a few sets of time data at each temperature – the extreme of which is two – are available, thereby resulting in an inaccurate error of estimate. According to this method, the shifting procedure must be repeated several times with data that have been altered randomly within a fixed distribution. The determination of the variance of the distribution is a problem. For large data sets it can be estimated by looking at the noise at the center of the set; otherwise it must be entered as a parameter. The subroutine `NOISE` generates data sets from the original file containing pseudorandom noise through the IMSL statistical routine `DRNNOA` [12]. The superposition is repeated via `SHIFT`, and the error is calculated through `STAT`.

5.2.2 CONVRT

For dynamic experiments, the best results are obtained if the frequency scan is initiated only after the test specimen is thermally soaked to eliminate temperature gradients. In contrast, data obtained during a temperature scan result in different temperatures for a given set of frequencies even at low scan rates. This type of data would give `SHIFTT` acute indigestion because the isochrones are not superposable along the temperature axis. An auxiliary program, `CONVRT`, is available to alleviate the difficulties with this type of experiment. The moduli data are interpolated, with the `AKIMA` cubic spline subroutine [10], to a common temperature, and a file is created which is usable by `SHIFTT`. `AKIMA` is a minimum-curvature method; the version listed (Appendix B) has been modified from the original code [13] to allow slight extrapolation necessary at both ends of the temperature spectrum. `CONVRT` is set up to handle up to 10 frequencies and minimize data entry via the keyboard terminal. Otherwise, customizing is recommended to suit the original data file.

5.3 Execution of the Programs

5.3.1 SHIFTT

The execution of `SHIFTT` requires the subroutines `ERROR` and `STAT` from Appendix B and the IMSL routine `DUMINF` [8]. The subprograms `FWLF`, `NOISE`, `RHOR`, and `SHIFT`, inherent to running `SHIFTT`, are included in the code listed in Section 5.5.1. Refer to Appendix A for the program dependencies. For example, `ERROR` calls `MATINV` (Appendix B.11) and the IMSL routine `DFDHES` [8].

Before `SHIFTT` can be used, a data file must be set up by using a text processor or editor that outputs ASCII code or by the auxiliary program `CONVRT`. The first record is the title for the file; insert anything up to 79 characters containing identification information. Data may be entered as the dimensional numbers (any units) or in the corresponding common logarithmic values, but be consistent.

Because there is no definitive way of telling the difference from the data alone, indicate the form of the data in the second record of the data file. Enter `N` for dimensional numbers or `L` for logarithmic values.

The next record must contain two entries separated by a delimiter (a blank space or a comma). The first entry is the temperature (°C) at which data was collected, and the next is an integer designating the number of points gathered at this temperature. For instance, if this number is 10, the next 10 records will carry the data, three entries in each record. For dynamic data, the first entry will be the frequency, the second will be the real (storage) part of the complex modulus, and the third will be the imaginary (loss) part of the complex modulus. Data from a transient (creep or stress relaxation) experiment are entered as the time and the modulus (J^{-1} for creep) in place of the frequency and the real part of the complex modulus, respectively. The integer 1 is placed in the third column, in place of the imaginary part of the complex modulus, as a flag for the program. For each isotherm, the time or frequency data should be entered in increasing order. The process is repeated for all the isotherms in order of increasing temperature. An example of the data file format is listed in Section 5.6.1.

To increase the versatility in running `SHIFTT`, one can enter some options via the keyboard. The first question deals with the problem of shifting with a polynomial fit, especially for widely spaced data. A default value for tolerance for wild shifts is set at four decades of time. If the shift is greater than this, the program assumes that a shift to an inappropriate portion of the polynomial curve occurred, and the shift is reset to the straight-line estimate. When this occurs, the estimate of the error is assumed to scale with the shift factor. A message will appear indicating the temperature at which the default value for error was used.

The second question asks for the reference temperature and the glass transition temperature (°C). A strict application of the universality concept to amorphous polymers dictates that these two temperatures should be the same, although empirical extensions of the WLF equation have used reference temperatures considerably above T_g, resulting in different values for the WLF parameters [7]. Refer to Equations 5.13 and 5.14 for the appropriate transformations of C_1 and C_2. The supplied reference temperature is checked against the minimum and maximum temperatures in the data file. If the entry is outside this range, a choice for a new reference temperature or an exit from the program is required. For data resulting from dynamic experiments, T_g is most often assumed to be that corresponding to the peak in the loss modulus at 1 Hz [7]. If this information is unknown beforehand, an educated guess would be sufficient; the superposition itself does not depend very strongly on the choice of the glass-transition temperature.

The third question deals with the option for density and temperature corrections described in Equation 5.5. If normalization of the modulus is desired, an inquiry as to the availability of expansion coefficient data is made. If data are accessible, the corresponding values of α (K^{-1}) below and above T_g are entered. Otherwise, corrections using Equations 5.9 and 5.10 are made by using the previously entered value of the glass-transition temperature, which predicates a more accurate estimate of T_g.

The last question addresses the estimation of errors incurred upon shifting the moduli by performing a Monte Carlo simulation. As described in the previous section, this option requires knowledge of the variance in the modulus (in decades). To skip this procedure, enter 0.

At this point, the shifts will be carried out. In the case of dynamic experiments, the components of the complex modulus are shifted separately and subsequently averaged. Tables of the shift factors with the corresponding errors and the reduced variables are written in the output file. The fitting of the WLF equation is attempted, and the resulting parameters and respective 95% confidence intervals are printed out. If the Monte Carlo simulation is chosen, a summary of the errors in the shift factors for each isotherm will appear at the end of the file. Sample listings of a terminal dialog and output file are given in Sections 5.6.2 and 5.6.3, respectively.

5.3.2 CONVRT

The program `CONVRT` for handling multiplexed time-temperature scans requires the external subroutine `AKIMA` [10,13]. The first record in the data file is for the standard title of up to 79 characters. The second record is to indicate the number of frequencies available, with a maximum of 10. The third record should provide the frequencies in sequential ascending order to accommodate `SHIFTT`. From the fourth record down, the data in dimensional units are to be listed starting with the first frequency indicated in the third record. Enter the temperature (^{o}C), the storage modulus, and the loss modulus. The resulting file is in the form needed for executing `SHIFTT`.

5.4 Application of SHIFTT

The application and versatility of `SHIFTT` can be seen most easily by examining its performance with illustrative data sets; these are summarized in Table 5.1. The first example treats the viscoelastic response of a polycarbonate (M_w = 40,000). Torsional-creep ($G_c(t) > 10^8$ Pa) and stress-relaxation experiments [14] are presented in Figure 5.2 based on tabulated data taken from a textbook problem [15]. In comparison to the original publication, two isotherms (147.2 ^{o}C and 148.9 ^{o}C) were completely omitted in Reference 15, while the remaining isotherms contained fewer data points.

Table 5.1 Published Data Sets Used to Illustrate the Application of SHIFTT

Material	Experiment	Instrument	Time Range, decades	Ref.
Polycarbonate (M_w = 40,000)	Torsional creep/ stress relaxation	Custom	3	14,15
Polyisobutylene (M_v = 1.35 × 10^6)	Dynamic shear	Fitzgerald transducer	3	17,18
NBS Nonlinear Test Fluid #1	Shear creep	(unspecified)	5	21
Plasticized PVC	Dynamic torsion	Rheometrics mechanical spectrometer	2	22
Neoprene	Dynamic torsion	Rheometrics System 4	4	22

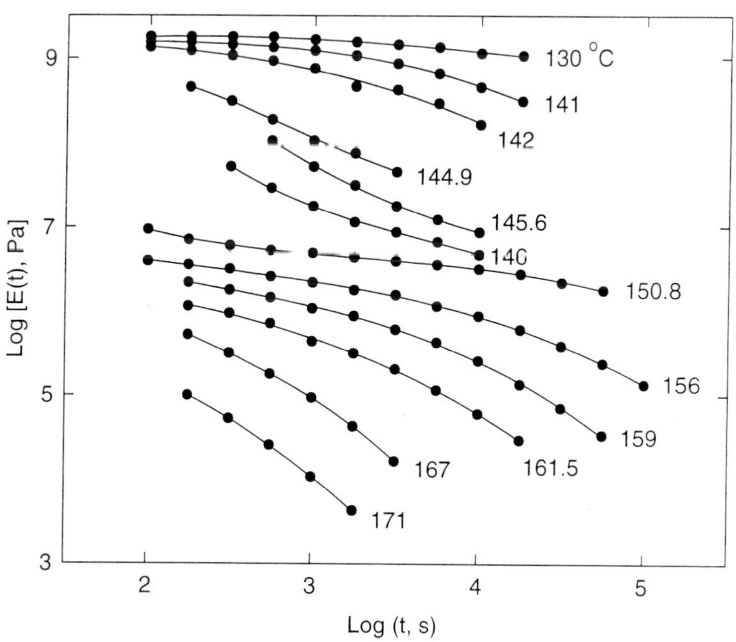

Figure 5.2 Torsional-creep and stress-relaxation data for polycarbonate, M_w = 40,000 [15]. Adapted from Ref. 15. Copyright © 1983 by John Wiley & Sons, Inc. Reprinted by permission.

The application of SHIFTT to the polycarbonate data, reduced to 150.8 °C, is shown in Figure 5.3. The listings of the input/output files are provided in Section 5.6. The density correction was applied by using reported coefficient-of-expansion values [14, Table II]. Shown in the Figure 5.3 inset is a comparison of 95% confidence intervals computed by the algorithm with the result worked out by hand [16]. The traditional manual method for superposition is to select a reference temperature and shift horizontally the next closest isotherm until a smooth overlap is obtained. Successive data sets are added to the growing curve. Thus, each shift will depend on the nature of all the sets that preceded it, unlike the pairwise shifting used by SHIFTT. From the Figure 5.3 inset, it is apparent that the result found manually hugs the lower bound of the 95% confidence interval. This result emphasizes the effect of error propagation when superposition is attempted manually. The calculated WLF parameters are $C_1 = 16.39 \pm 5.97$ and $C_2 = 54.3 \pm 24.1$ K, and are in good agreement with the published results $C_1 = 16.14$ and $C_2 = 56$ K [14]. Particularly noteworthy are the large errors in the parameters.

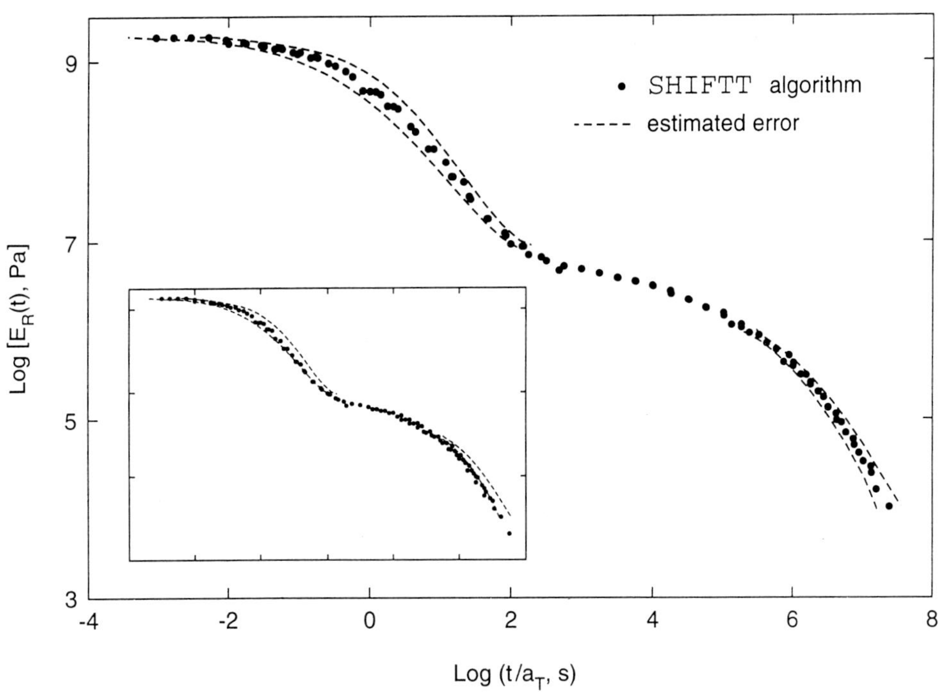

Figure 5.3 Master curve and corresponding 95% confidence intervals in the shift factors for polycarbonate (M_w = 40,000) [14,15] reduced to 150.8 °C. Inset is a comparison of the problem worked out by hand (adapted from Ref. 16; reprinted by permission) with the confidence intervals estimated by SHIFTT.

The second illustration, featuring the dynamic properties of polyisobutylene (M_v = 1.35 × 10^6) [17], emphasizes the computation of the WLF parameters. The numerical data were found in the form of components of the complex compliance, while the reduced variables and shift factors were presented in a subsequent paper [18]. The result for the storage modulus, G', is plotted in Figure 5.4 at a reference temperature of 25 °C. A comparison of the shift factors, based on the average between both moduli components, is shown in Table 5.2. As in the preceding example, not all the isotherms are available from the original paper [17]; however, the agreements between the values of log a_T are satisfactory.

For the polyisobutylene data between $T_g < T < T_g + 100$, the calculated WLF parameters from SHIFTT, where T_g was taken as 205 K [1, Table 11-II], are C_1 = 7.05 ± 0.52 and C_2 = 176 ± 9 K. Reported values [1] vary from 8.61 and 200.4 (Table 11-II) to 9.08 and 209 (Section 3, Appendix D), the latter pair being cited as the best values based on results from several investigators [17,19,20]. To account for the discrepancy, we took the published values of the shift factor and applied regression analysis to Equation 5.12 without the restriction of the temperature window $T_g < T < T_g + 100$. The result, C_1 = 9.05 ± 0.89 and C_2 = 207 ± 25, corresponds well with the cited best values. To illustrate the differences in precision between the linear and nonlinear techniques, we removed the temperature window restriction. The result from SHIFTT was C_1 = 8.28 ± 0.53 and C_2 = 194 ± 9; statistically, these values are not different from those cited in Table 11-II of Reference 1.

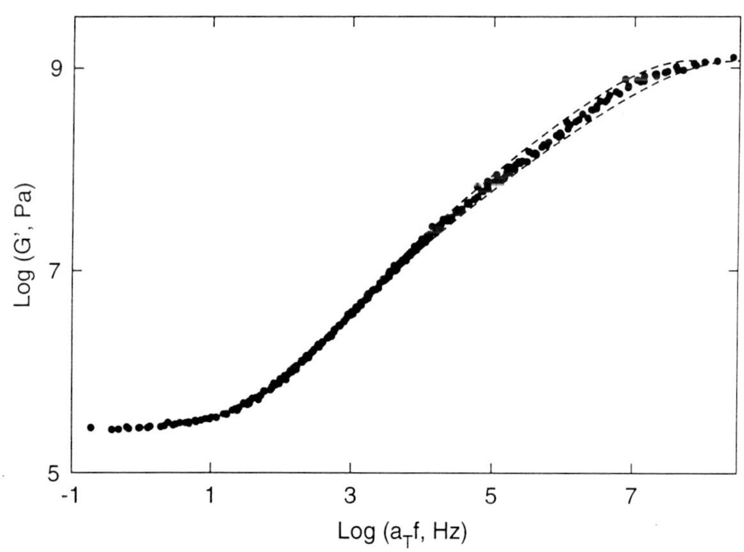

Figure 5.4 Dynamic shear modulus for a polyisobutylene (M_v = 1.35 × 10^6) reduced to 25 °C [17,18]. The 95% confidence intervals are the widest where there is a drastic change in the modulus plus noise in the data. Adapted from Ref. 17. Copyright © 1953 American Institute of Physics. Reprinted by permission.

Table 5.2 Comparison of Shift Factors for Polyisobutylene Data Reduced to 25 °C [17,18]

T, °C	log a_T SHIFTT	log a_T Reported[a]
-44.6	4.69 ± 0.34[b]	4.75
-40.4	4.14 ± 0.27	4.23
-34.7	3.61 ± 0.23	3.69
-33.4	–[c]	3.44
-29.8	3.16 ± 0.19	3.22
-28.5	–	3.06
-25.0	2.78 ± 0.16	2.84
-23.0	–	2.68
-19.9	2.41 ± 0.14	2.45
-14.8	2.06 ± 0.13	2.08
-9.9	1.75 ± 0.11	1.78
-5.0	1.48 ± 0.09	1.51
-0.1	1.20 ± 0.08	1.22
4.9	0.94 ± 0.06	0.98
9.8	0.69 ± 0.05	0.72
14.8	0.46 ± 0.03	0.51
19.9	0.23 ± 0.02	0.27
25.0	0.00	0.00
30.2	-0.23 ± 0.02	-0.21
34.9	-0.42 ± 0.04	-0.42
40.0	-0.59 ± 0.08	-0.62
50.0	-0.91 ± 0.11	-0.99
59.8	-1.22 ± 0.16	-1.31
80.0	-1.81 ± 0.25	-1.90
99.9	-2.32 ± 0.37	-2.29

[a] Adapted from Table II in Ref. 18. Copyright © 1953 American Institute of Physics. Reprinted by permission.
[b] 95% confidence interval.
[c] Numerical data unavailable from Ref. 17.

The third example, shown in Figure 5.5, is for the compliance of 7% anionic polystyrene ($M_w = 4.4 \times 10^5$) in a mixed solvent (65% *m*-tricresyl phosphate and 35% Aroclor), usually referred to as NBS Nonlinear Test Fluid #1, between -45 and -10 °C [21]. The tabulated recoverable shear-creep compliance data were first converted to the required format for `SHIFTT` before the reduction to -25 °C was performed. The shift factors generated by `SHIFTT` are compared in Table 5.3 with

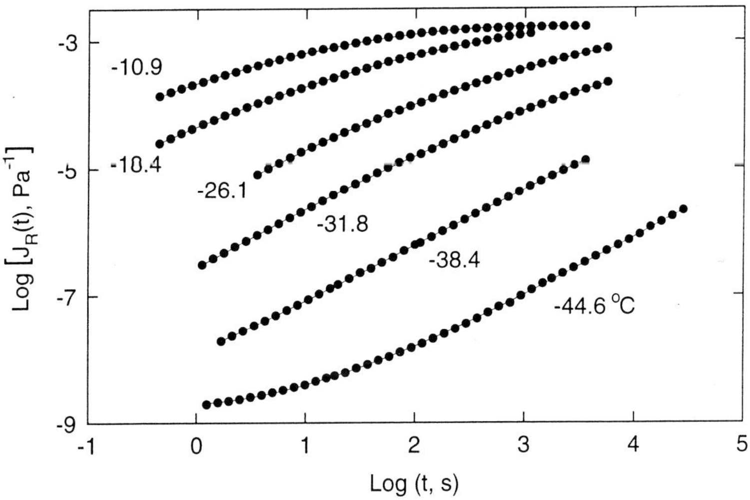

Figure 5.5 Recoverable shear-creep compliance isotherms for NBS Nonlinear Test Fluid #1. Adapted from Ref. 21. Copyright © 1980 by Academic Press, Inc. Reprinted by permission.

the reported values. While at first glance the shift factors appear to be distinctly different, upon careful examination it becomes clear that the difference arises near the reference temperature. In all probability, the source of discrepancy is the method used to interpolate to a reference temperature at which there is no actual data. The magnitude of the shifts between isotherms are in excellent agreement.

Most notable in the first three cases cited are the narrow 95% confidence intervals for the shifted time, which is almost guaranteed by the extreme data overlap. The widest interval occurs where there is a drastic change in the modulus.

The last two data sets [22] illustrate the algorithm's capability of handling widely spaced data. The results for a plasticized poly(vinyl chloride), from a dynamic shear experiment conducted between -30 and 20 °C at 10° intervals and a frequency range of 0.16 to 10 s^{-1}, are presented in Figure 5.6. An even more extreme case, shown in Figure 5.7, was derived from data for a neoprene elastomer. The complete data set, collected from -37 to 9 °C at 5° intervals, was compared to the results obtained if all the data between the initial and final isotherms were omitted. For the 9 °C isotherm, the shift factor for the full set was -5.78 ± 0.20 as compared to -5.43 ± 0.21 for the incomplete set. This procedure was also carried on the NBS Nonlinear Test Fluid #1 data, wherein three out of the six isotherms were omitted. The results are included in Table 5.3 and show remarkable agreement with the complete set. Extreme care should be taken if only sparse data are available; the optimization in the algorithm must then be constrained to avoid the negative shifts that are possible with polynomials.

Table 5.3 Comparison of Shift Factors for Recoverable Compliance Data of NBS Nonlinear Test Fluid # 1 Reduced to -25 °C [21]

	log a_T		
T, °C	SHIFTT	Reported[a]	Edited[b]
-44.6	4.90 ± 0.02[c]	5.00	4.84 ± 0.01
-38.4	3.02 ± 0.01	3.11	-
-31.8	1.37 ± 0.01	1.47	-
-26.1	0.22 ± 0.00	0.33	0.23 ± 0.00
-25.0	0.00 ± 0.00	0.00	0.00 ± 0.00
-18.4	-1.28 ± 0.00	-1.15	-
-10.9	-2.39 ± 0.01	-2.25	-2.38 ± 0.00

[a] Adapted from Table II in Ref. 21. Copyright © 1980 by Academic Press, Inc. Reprinted by permission.
[b] Three isotherms were omitted to test the algorithm's capability on widely spaced data.
[c] 95% confidence interval.

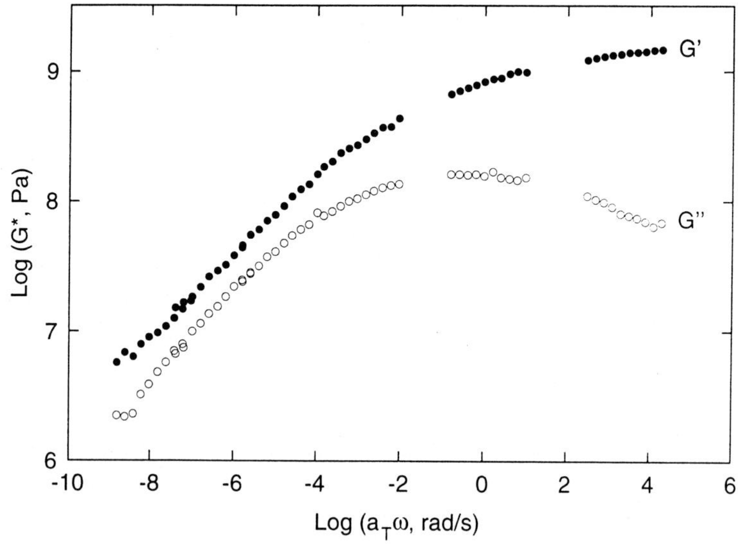

Figure 5.6 Dynamic shear modulus master curve for a plasticized poly(vinyl chloride) reduced to -20 °C [22].

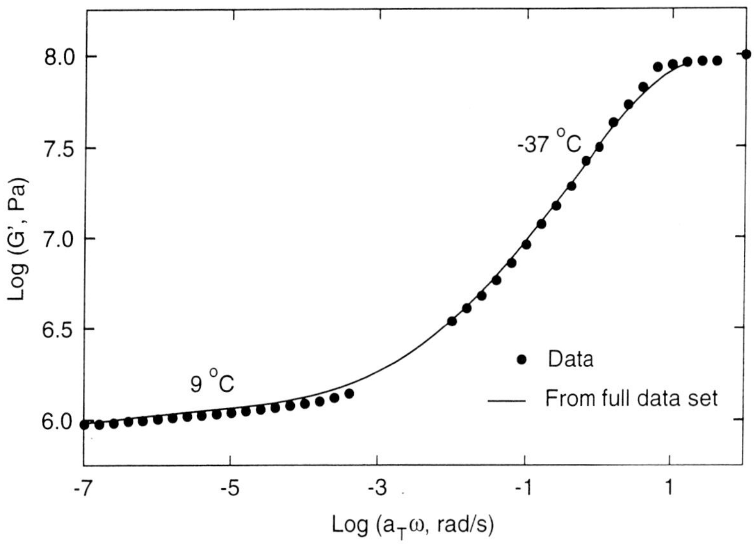

Figure 5.7 Storage modulus master curve for neoprene, T_R = -37 °C, based on data taken at -37 and 9 °C and compared to results for the full data set taken at 5° intervals [22].

5.5 Program Listings

5.5.1 SHIFTT

When using Microsoft® FORTRAN, you need the metacommand $LARGE to specify that all the arrays can span more than one segment (64K). To avoid using $LARGE, decrease the value of the integer NMAX, found in the PARAMETER statements, to 20. Note that NMAX defines the allowable number of isotherms and the number of data points for each isotherm. Imbedded into the code are the subprograms FAT, FAT1, FWLF, NOISE, RHOR, SHIFT, and SUMS inherent to the execution of SHIFTT.

```
$LARGE
      PROGRAM SHIFTT
C Superposes dynamic or transient viscoelastic data.
C For dynamic data, superposes real and imaginary components separately.
C Requires the subroutines ERROR, STAT, and STOPP and IMSL routine DUMINF.
C Data file format:
C   1st record - Title, up to 79 characters.
C   2nd record - enter L if data in form of log(base 10); otherwise, N.
C   3rd record - lowest temperature (C), number of frequencies (or times).
C   4th record - lowest frequency (Hz), modulus (2 numbers, for dynamic; for
C                transient experiments, use 1.0 for all third entries).
C   5th record - next frequency, etc. for lowest temperature.
```

```
      IMPLICIT DOUBLE PRECISION (A-H,O-Z)
      PARAMETER (NMAX=51,NPAR=2,NPD=10,EPSFCN=0.2221D-15)
      CHARACTER ALPHA*1,ANS*1,FORM*80,TITLE*80,TRED*1,TRANS*1
      DIMENSION A(NPAR),AGUESS(NPAR),ASCALE(NPAR),AERR(NPAR),COV(NPAR)
      DIMENSION F(NMAX,NMAX),GP(NMAX,NMAX),GPP(NMAX,NMAX)
      DIMENSION H(NPAR,NPAR),HI(NPAR,NPAR),IPARAM(7),RPARAM(7)
      DIMENSION SAT(NMAX,5),SGP(NMAX,NMAX),SGPP(NMAX,NMAX)
      COMMON /BLOC1/ AORIG,X(NMAX*2),Y(NMAX*2),I,ND,NF(NMAX),NP
      COMMON /BLOC2/ AT(NMAX),ATTOL,T(NMAX),TG,TR,NT
      EXTERNAL FAT,FDHES,FWLF
C t-statistic for 95% confidence interval with N degrees of freedom.
      T95(N)=1.96+2.3539/N+3.1902/N**2+0.7004/N**3+4.5015/N**4
10    CALL MSHEAD
      CALL UNITS(NTER,NINT,NOUT)
      WRITE(NTER,*)
     &      ' Default (4 decades) for wild shifts acceptable? (Y/N)'
      READ(NTER,'(A1)')ANS
      IF(ANS.EQ.'Y'.OR.ANS.EQ.'y')THEN
        ATTOL=4.D0
      ELSE
        WRITE(NTER,*)' Enter tolerance for wild shifts (decades).'
        READ(NTER,*)ATTOL
      ENDIF
      WRITE(NOUT,20)
20    FORMAT(/T17,'SHIFTT Time-Temperature Superposition'/T17,37('-')/)
      READ(NINT,'(A80)')TITLE
      WRITE(NOUT,'(A80/)')TITLE
      READ(NINT,'(A80)')FORM
      IFORM=1
      DO 30 I=1,80
        IF(FORM(I:I).EQ.'N'.OR.FORM(I:I).EQ.'n')IFORM=2
30    CONTINUE
      DO 40 I=1,NMAX
        READ(NINT,*,END=50)T(I),NF(I)
        IF(I.EQ.NMAX)CALL STOPP(NTER,NMAX-1)
        IF(NF(I).GT.NMAX-1)CALL STOPP(NTER,NMAX-1)
        T(I)=T(I)+273.2
        READ(NINT,*)(F(I,J),GP(I,J),GPP(I,J),J=1,NF(I))
40    CONTINUE
50    NT=I-1
C IFLAG=1, no loss modulus GPP data available.
C IFLAG=2, GPP available, storage modulus GP data used for shifting.
C IFLAG=3, GPP available, GPP data used for shifting.
C ITYPE denotes frequency (1) or time (2) scale.
      IFLAG=1
      ITYPE=1
      IF(GPP(1,1).NE.GPP(2,1).OR.GPP(2,1).NE.GPP(NT,1))IFLAG=2
      IF(GP(NT/2,1).GT.GP(NT/2,NF(NT/2)))ITYPE=2
C Ask for reference temperature (TR) and glass transition (TG).
60    WRITE(NTER,*)
     &      ' Input reference and glass transition temperatures (C).'
      READ(NTER,*)TR,TG
      TR=TR+273.2
      IF(TR.GT.T(NT).OR.TR.LT.T(1))THEN
        WRITE(NTER,*)' Reference temperature out of range of data.'
        WRITE(NTER,*)' Reenter? (Y/N)'
        READ(NTER,'(A1)')TRANS
        IF(TRANS.EQ.'Y'.OR.TRANS.EQ.'y')GOTO 60
      ENDIF
      TG=TG+273.2
      WRITE(NOUT,70)TG,TG-273.2,TR,TR-273.2
70    FORMAT(5X,    'Glass Temperature =',F7.1,' K',F8.1,' C'/
     &       1X,'Reference Temperature =',F7.1,' K',F8.1,' C'/)
C Provide option for the density, temperature correction.
      WRITE(NTER,*)' Apply density, temperature correction? (Y/N)'
      READ(NTER,'(A1)')TRED
C Take logs of data and apply density and temperature corrections.
      IF(TRED.EQ.'Y'.OR.TRED.EQ.'y')THEN
```

```
         WRITE(NTER,*)' Are expansion coefficients available? (Y/N)'
         READ(NTER,'(A1)')ALPHA
         IF(ALPHA.EQ.'Y'.OR.ALPHA.EQ.'y')THEN
           WRITE(NTER,*)' Enter expansion coefficient below Tg (1/K).'
           READ(NTER,*)AG
           WRITE(NTER,*)' Enter expansion coefficient above Tg (1/K).'
           READ(NTER,*)AR
         ELSE
           AG=9.0D-2/TG
           AR=2.0D-1/TG
         ENDIF
       ENDIF
       DO 90 I=1,NT
         TFAC=0.D0
         IF(TRED.EQ.'Y'.OR.TRED.EQ.'y')
      &    TFAC=LOG10(TR/T(I)/RHOR(T(I),TG,TR,AG,AR))
         DO 80 J=1,NF(I)
           IF(IFORM.EQ.2)THEN
             F(I,J)=LOG10(F(I,J))
             GP(I,J)=LOG10(GP(I,J))
             GPP(I,J)=LOG10(GPP(I,J))
           ENDIF
           GP(I,J)=GP(I,J)+TFAC
           IF(ITYPE.EQ.1.AND.IFLAG.EQ.2)GPP(I,J)=GPP(I,J)+TFAC
80       CONTINUE
90     CONTINUE
       IF(ITYPE.EQ.1)WRITE(NOUT,100)
100    FORMAT(/,21X,'Results using Storage Modulus')
       IF(ITYPE.EQ.2)WRITE(NOUT,110)
110    FORMAT(/,15X,'Results using Transient Modulus')
C Get shift pairs, AT(I), where I applies to higher temperature.  SHIFT
C  carries all common blocks.  IMC=2, is flag for Monte Carlo simulation.
       IMC=1
       WRITE(NTER,'(/2X,A/)')'Computing!'
       CALL SHIFT(ITYPE,IFLAG,IMC,NTER,NOUT,F,GP,GPP)
       WRITE(NOUT,'(/2X,A/)')'Variable units identical to those of data.'
C Estimate WLF constants if number of temperatures NTWLF in the range of
C  TG < T < TG+100 is greater than six.
       NTWLF=0
       DO 120 I=1,NT
         IF(T(I).GE.TG.AND.T(I)-TG.LE.1.D2)NTWLF=NTWLF+1
120    CONTINUE
       IF(NTWLF.GT.6)THEN
C Set up call to IMSL routine DUMINF. Initial estimates based on the
C  transformation of universal constants from TG to TR.
         AGUESS(2)=51.6-TG+TR
         AGUESS(1)=17.44*51.6/AGUESS(2)
         ASCALE(1)=1.D0
         ASCALE(2)=1.D0
         FSCALE=1.D0
         IPARAM(1)=0
         CALL DUMINF(FWLF,NPAR,AGUESS,ASCALE,FSCALE,IPARAM,RPARAM,A,FX)
C Calculate errors in WLF constants.
         CALL ERROR(FWLF,NPAR,A,ASCALE,FX,H,HI,COV)
         DO 130 I=1,NPAR
           AERR(I)=T95(NTWLF-NPAR)*SQRT(FX/FLOAT(NTWLF-NPAR)*COV(I))
130      CONTINUE
         WRITE(NTER,'(/2X,A/)')'Estimate of WLF Constants'
         WRITE(NOUT,'(/2X,A/)')'Estimate of WLF Constants'
         WRITE(NTER,140)A(1),AERR(1),A(2),AERR(2)
         WRITE(NOUT,140)A(1),AERR(1),A(2),AERR(2)
140      FORMAT(5X,'C1 =',F7.2,' +/- ',F7.2/5X,'C2 =',F7.2,' +/- ',F7.2/)
       ELSE
         WRITE(NOUT,*)' Insufficient data above Tg for WLF calculation.'
       ENDIF
C Run Monte Carlo estimate of error on shifts; flag IMC = 2.
       WRITE(NTER,*)' Enter estimate of error in moduli data (in decades)
      & or 0. to skip.'
       READ(NTER,*)S
```

```
      IF(S.GT.1.D-8)THEN
C Save original data.
        DO 150 I=1,NT
        DO 150 J=1,NF(I)
          SGP(I,J)=GP(I,J)
          SGPP(I,J)=GPP(I,J)
150     CONTINUE
        DO 170 J=1,5
          CALL NOISE(S,NT,NF,SGP,SGPP,GP,GPP)
          IF(IFLAG.EQ.3)IFLAG=2
          CALL SHIFT(ITYPE,IFLAG,2,NTER,NOUT,F,GP,GPP)
          DO 160 I=1,NT
            SAT(I,J)=AT(I)
160       CONTINUE
170     CONTINUE
        DO 190 I=1,NT
          DO 180 J=1,5
            X(J)=SAT(I,J)
180       CONTINUE
C Estimate errors in the shift factors.
          CALL STAT(5,X,AVE,SD)
          Y(I)=SD
190     CONTINUE
        WRITE(NOUT,200)
200     FORMAT(//T5,'Monte Carlo Error Analysis'//T21,'Estimated'/T5,
     &         'Temp., K (C)',T20,'Error in AT'/T4,14('-'),T20,11('-'))
        WRITE(NOUT,210)(T(I),T(I)-273.2,Y(I),I=1,NT)
210     FORMAT(T3,F6.1,' (',F6.1,')',T20,F7.2)
      ENDIF
      CALL MSTAIL(*10)
      STOP
      END
C
      SUBROUTINE FAT(N,AT,F)
C Evaluates the function (shift factor) to be minimized.
C Requires the subroutines POLYN and STAT.
      IMPLICIT DOUBLE PRECISION(A-H,O-Z)
      PARAMETER (NMAX=51,NPAR=10)
      DIMENSION B(NPAR),BERR(NPAR),R(NMAX*2),XT(NMAX*2),W(NMAX*2)
      COMMON /BLOC1/ AORIG,X(NMAX*2),Y(NMAX*2),I,ND,NF(NMAX),NP
      SAVE W,ISAVE
C Check for correct sign; if not return with full SSE.
      IF(AORIG*AT.LT.0.)THEN
        CALL STAT(NP,Y,AVE,SD)
        F=SD**2*(NP-N)
        RETURN
      ENDIF
      DO 10 J=1,NF(I)
        XT(J)=X(J)
10    CONTINUE
      DO 20 J=NF(I)+1,NP
        XT(J)=X(J)+AT
20    CONTINUE
C Get residuals R(NP) from polynomial fit and weighting factors, W(NP).
      CALL POLYN(ND,NP,XT,Y,B,BERR,R,SEE)
      IF(I.NE.ISAVE)THEN
        ISAVE=I
        XAVE=(XT(NF(I))+XT(NF(I)+1))/2.D0
        XWIDE=(XT(NP)-XT(1))/2.D0
        DO 30 J=1,NP
          W(J)=1.D0/(1.D0+((XT(J)-XAVE)/XWIDE)**2)
30      CONTINUE
      ENDIF
      F=0.D0
      DO 40 J=1,NP
        F=F+R(J)**2*W(J)
40    CONTINUE
      RETURN
      END
```

```
      DOUBLE PRECISION FUNCTION FAT1(AT)
C Computes function value (shift factor) to be minimized via subroutine FAT.
      IMPLICIT DOUBLE PRECISION (A-H,O-Z)
      CALL FAT(1,AT,F)
      FAT1=F
      RETURN
      END
C
      SUBROUTINE FWLF(N,C,F)
C Evaluates the function (WLF equation, Eq. 5.1) to be minimized.
      IMPLICIT DOUBLE PRECISION(A-H,O-Z)
      PARAMETER (NMAX=51)
      DIMENSION C(N)
      COMMON /BLOC2/ AT(NMAX),ATTOL,T(NMAX),TG,TR,NT
      F=0.D0
      DO 10 I=1,NT
        IF(T(I).GE.TG.AND.T(I)-TG.LE.1.D2)THEN
            F=F+(AT(I)+C(1)*(T(I)-TR)/(C(2)+T(I)-TR))**2
        ENDIF
10    CONTINUE
      RETURN
      END
C
      SUBROUTINE NOISE(S,NT,NF,SGP,SGPP,GP,GPP)
C Generates data sets with random noise for Monte Carlo estimate of error
C  in shifts. Input are standard deviation of data in decades (S); number
C  of temperatures (NT) and frequencies at each temperature (NF); original
C  values of modulus components (SGP and SGPP).  GP and GPP are modulus
C  values containing random noise.  Requires the IMSL routine DRNNOA.
      IMPLICIT DOUBLE PRECISION (A-H,O-Z)
      PARAMETER (NMAX=50)
      DIMENSION GP(NMAX,NMAX),GPP(NMAX,NMAX),NF(NMAX)
      DIMENSION R(NMAX*NMAX+1),SGP(NMAX,NMAX),SGPP(NMAX,NMAX)
      NP=0
      DO 10 I=1,NT
      DO 10 J=1,NF(I)
        NP=NP+1
10    CONTINUE
C Check for type of data.
      IFLAG=1
      ITYPE=1
      IF(SGP(NT/2,1).GT.SGP(NT/2,NF(NT/2)))ITYPE=2
      IF(SGPP(1,1).NE.SGPP(2,1).OR.SGPP(2,1).NE.SGPP(NT,1))IFLAG=2
      CALL DRNNOA(2*NP,R)
      K=1
      DO 20 I=1,NT
      DO 20 J=1,NF(I)
        GP(I,J)=SGP(I,J)+R(K)*S
        IF(ITYPE.NE.2.AND.IFLAG.NE.1)GPP(I,J)=SGPP(I,J)+R(K+1)*S
        K=K+2
20    CONTINUE
      RETURN
      END
C
      DOUBLE PRECISION FUNCTION RHOR(T,TG,TR,AG,AR)
C Returns ratio of density at temperature T to that at reference TR.
C Unless provided, expansivities are approximations from Van Krevelen.
      IMPLICIT DOUBLE PRECISION (A-H,O-Z)
      IF(T.GE.TG.AND.TR.GE.TG)THEN
        RHOR=EXP(-AR*(T-TR))
      ELSE IF(T.LE.TG.AND.TR.LE.TG)THEN
        RHOR=EXP(-AG*(T-TR))
      ELSE IF(T.GE.TG.AND.TR.LE.TG)THEN
        RHOR=EXP(-AR*(T-TG))*EXP(-AG*(TG-TR))
      ELSE IF(T.LE.TG.AND.TR.GE.TG)THEN
        RHOR=EXP(-AR*(TG-TR))*EXP(-AG*(T-TG))
      ENDIF
      RETURN
      END
```

```
      SUBROUTINE SHIFT(ITYPE,IFLAG,IMC,NTER,NOUT,F,GP,GPP)
C Performs superposition procedure.
C Requires the subroutines ALINR, AKIMA, and ERROR and the
C  IMSL routine DUVMIF.
      IMPLICIT DOUBLE PRECISION (A-H,O-Z)
      PARAMETER (NMAX=51,NPAR=1,NPD=10,EPSFCN=0.2221D-15)
      DIMENSION A(11),ASCALE(NPAR),BT(NMAX),COV(NPAR),ERRAT(NMAX)
      DIMENSION ERRBT(NMAX),F(NMAX,NMAX),GP(NMAX,NMAX),GPP(NMAX,NMAX)
      DIMENSION H(NPD,NPD),HI(NPD,NPD)
      EXTERNAL FAT,FAT1,DFDHES
      COMMON /BLOC1/ AORIG,X(NMAX*2),Y(NMAX*2),I,ND,NF(NMAX),NP
      COMMON /BLOC2/ AT(NMAX),ATTOL,T(NMAX),TG,TR,NT
C t-statistic for 95% confidence interval with N degrees of freedom.
      T95(N)=1.96+2.3539/N+3.1902/N**2+0.7004/N**3+4.5015/N**4
C If IMC = 2, Monte Carlo simulation in effect.
10    DO 50 I=NT,2,-1
        DO 20 J=1,NF(I)
          X(J)=F(I,J)
          Y(J)=GP(I,J)
20      CONTINUE
        NP=NF(I)+NF(I-1)
        DO 30 J=NF(I)+1,NP
          K=J-NF(I)
          X(J)=F(I-1,K)
          Y(J)=GP(I-1,K)
30      CONTINUE
C Get estimate for shift factor AT.
        CALL ALINR(NP,X,Y,A)
        SYI=SUMS(NP,1,1,NF(I),Y)
        SYJ=SUMS(NP,1,NF(I)+1,NP,Y)
        AT(I)=-(SYI/NF(I)-SYJ/NF(I-1))/A(2)
C Estimate shift using the Kaelble modification of the WLF equation.
        ATWLF=(-1.)**(ITYPE+1)*(17.4*(T(I)-TG)/(51.6+ABS(T(I)-TG))-
     &         17.4*(T(I-1)-TG)/(51.6+ABS(T(I-1)-TG)))
C Check for incorrect sign of shift due to widely spaced data.
        IF(AT(I)*ATWLF.LT.0.)AT(I)=ATWLF
        ND=MIN(NP**.333+2,SNGL(MAX(ABS(F(I-1,NF(I-1))-F(I-1,1))+2,
     &         ABS(F(I,NF(I))-F(I,1))+2,ABS(ATWLF))))
C Set up call to IMSL DUVMIF.
        AGUESS=AT(I)
        STEP=1.D-3
        MAXFN=500
        BOUND=ATTOL
        XACC=1.D-4
        CALL DUVMIF(FAT1,AGUESS,STEP,BOUND,XACC,MAXFN,ATI)
        FX=FAT1(ATI)
C Estimate error in shift factor.
        ASCALE(1)=1.D0
        CALL ERROR(FAT,NPAR,ATI,ASCALE,FX,H,HI,COV)
C Check for wild shifts.
        IF(ABS(ATI-AT(I)).LT.ATTOL)THEN
          AT(I)=ATI
C Estimate error in the shift factor ERRAT.
          IF(IMC.EQ.1)
     &    ERRAT(I)=T95(NP-NPAR)*SQRT(FX/FLOAT(NP-NPAR)*COV(NPAR))
        ELSE
C Assume error scales with shift.
          IF(IMC.EQ.1)THEN
            ERRAT(I)=ERRAT(I+1)*EXP(ABS(AT(I))-ABS(AT(I+1)))
            WRITE(NTER,40)T(I)
40          FORMAT(' A default error estimate occurred at ',F6.1,' K'/)
          ENDIF
        ENDIF
50    CONTINUE
C Switch shift to lower temperature of each pair.
      DO 60 I=1,NT-1
        IF(IMC.EQ.1)ERRAT(I)=ERRAT(I+1)
        AT(I)=AT(I+1)
60    CONTINUE
```

```
      AT(NT)=0.D0
      IF(IMC.EQ.1)ERRAT(NT)=ERRAT(NT+1)
      DO 70 I=NT-1,1,-1
        IF(IMC.EQ.1)ERRAT(I)=ERRAT(I+1)+ERRAT(I)
        AT(I)=AT(I+1)+AT(I)
70    CONTINUE
C Interpolate to get shift at reference temperature, TR.
      CALL AKIMA(NT,T,AT,TR,ATR)
      IF(IMC.EQ.1)CALL AKIMA(NT,T,ERRAT,TR,ERRATR)
C Subtract shift at TR from shifts at all temperatures.
      DO 80 I=1,NT
        IF(IMC.EQ.1)ERRAT(I)=ABS(ERRAT(I)-ERRATR)
        AT(I)=AT(I)-ATR
80    CONTINUE
      IF(IMC.EQ.1)THEN
C Print out headings and results.
        IF(ITYPE.EQ.1)THEN
          IF(IFLAG.EQ.2.OR.IFLAG.EQ.3)THEN
            WRITE(NOUT,90)
90          FORMAT(//T50,'Log Storage',T64,'Log  Loss'/T5,'Temp., K (C)'
     &      ,T23,'Log Shift',T38,'Log Freq.',T52,'Modulus',T65,'Modulus'
     &      /2(3X,(14('-'))),T38,9('-'),T50,11('-'),T64,9('-'))
          ELSE
            WRITE(NOUT,100)
100         FORMAT(//T50,'Log Storage'/T5,'Temp., K (C)',T23,'Log Shift'
     &      ,T38,'Log Freq.',T52,'Modulus'/
     &      2(3X,(14('-'))),T38,9('-'),T50,11('-'))
          ENDIF
        ELSE
          WRITE(NOUT,110)
110       FORMAT(//T50,'Log Transient'/T5,'Temp., K (C)',T23,'Log Shift'
     &    ,T38,'Log Time',T53,'Modulus'/
     &    2(3X,(14('-'))),T38,8('-'),T50,13('-'))
        ENDIF
        DO 150 I=1,NT
          IF(ITYPE.EQ.1)WRITE(NOUT,120)T(I),T(I)-273.2,AT(I),ERRAT(I)
          IF(ITYPE.EQ.2)WRITE(NOUT,120)T(I),T(I)-273.2,-AT(I),ERRAT(I)
120       FORMAT(/T3,F6.1,' (',F6.1,')',T20,F6.2,' +/-',F5.2)
        DO 150 J=1,NF(I)
          IF(ITYPE.EQ.1)THEN
            IF(IFLAG.EQ.1)WRITE(NOUT,130)F(I,J)+AT(I),GP(I,J)
            IF(IFLAG.EQ.2)WRITE(NOUT,130)F(I,J)+AT(I),GP(I,J),GPP(I,J)
            IF(IFLAG.EQ.3)WRITE(NOUT,130)F(I,J)+AT(I),GPP(I,J),GP(I,J)
130         FORMAT(T38,F7.3,T52,F6.3,T65,F6.3)
          ENDIF
          IF(ITYPE.EQ.2)WRITE(NOUT,140)F(I,J)+AT(I),GP(I,J)
140       FORMAT(T38,F7.3,T53,F6.3)
150     CONTINUE
      ENDIF
C Redo with the loss modulus, GPP. If GPP not available, skip over.
      IF(IFLAG.NE.1.OR.ITYPE.EQ.1)THEN
C If GPP calculations have been done, print results.
        IF(IFLAG.EQ.2)THEN
C Exchange GP for GPP.
          DO 160 I=1,NT
            BT(I)=AT(I)
            ERRBT(I)=ERRAT(I)
          DO 160 J=1,NF(I)
            GTEMP=GP(I,J)
            GP(I,J)=GPP(I,J)
            GPP(I,J)=GTEMP
160       CONTINUE
C Switch IFLAG to prevent another iteration.
          IFLAG=3
          IF(IMC.EQ.1)THEN
            WRITE(NOUT,'(//22X,A)')'Results using Loss Modulus'
          ENDIF
          GOTO 10
        ENDIF
```

```
C Average shifts obtained for the components of the complex modulus.
        IF(IMC.EQ.1)THEN
          WRITE(NOUT,'(//24X,A)')'Combined Shift Results'
          WRITE(NOUT,90)
        ENDIF
        DO 170 I=1,NT
          AT(I)=(AT(I)+BT(I))/2.D0
          IF(IMC.EQ.1)THEN
            ERRAT(I)=(ERRAT(I)+ERRBT(I))/2.D0
            WRITE(NOUT,120)T(I),T(I)-273.2,AT(I),ERRAT(I)
C Note that this prints in order GP, GPP.
            WRITE(NOUT,130)(F(I,J)+AT(I),GPP(I,J),GP(I,J),J=1,NF(I))
          ENDIF
170     CONTINUE
      ENDIF
C If data is from stress relaxation, switch signs of shift factors.
      IF(ITYPE.EQ.2)THEN
        DO 180 I=1,NT
          AT(I)=-AT(I)
180     CONTINUE
      ENDIF
      END
C
      DOUBLE PRECISION FUNCTION SUMS(NP,ND,NI,NF,X)
C Returns sums of degree ND for an input vector X(NI) to X(NF).
      IMPLICIT DOUBLE PRECISION (A-H,O-Z)
      DIMENSION X(NP)
      SUMS=0.D0
      DO 10 I=NI,NF
        SUMS=SUMS+X(I)**ND
10    CONTINUE
      RETURN
      END
```

5.5.2 CONVRT

```
      PROGRAM CONVRT
C Converts multiplexed time-temperature sweeps from dynamic experiments
C  to a form required with the time-temperature superposition program
C  SHIFTT.
C Requires the subroutines AKIMA and STOPP.
C Data file format:
C   1st record - title, up to 79 characters.
C   2nd record - number of frequencies (maximum=10).
C   3rd record - each frequency in sequential order.
C   4th record - temperature (C), E', E" for first frequency
C   5th record - next data point for first frequency in order of
C                increasing temperature;
C                repeat in sequential order of frequencies.
      IMPLICIT DOUBLE PRECISION (A-H,O-Z)
      PARAMETER (NMAX1=10,NMAX2=201)
      CHARACTER TITLE*80
      DIMENSION EP(NMAX1,NMAX2),EP1(NMAX2),EPP(NMAX1,NMAX2),EPP1(NMAX2)
      DIMENSION F(NMAX1),NT(NMAX1),T(NMAX1,NMAX2),TAV(NMAX2),T1(NMAX2)
10    CALL MSHEAD
      CALL UNITS(NTER,NINT,NOUT)
      READ(NINT,'(A80)')TITLE
      WRITE(NOUT,'(A80)')TITLE
      WRITE(NOUT,'(A1)')'N'
      READ(NINT,*)NF
      IF(NF.GT.NMAX1)CALL STOPP(NTER,NMAX1)
      WRITE(NTER,'(/2X,A)')'Computing!'
      READ(NINT,*)(F(I),I=1,NF)
      READ(NINT,*)T(1,1),EP(1,1),EPP(1,1)
```

```
      T(1,1)=T(1,1)+273.2
      DO 30 I=1,NF
        DO 20 J=2,NMAX2
          READ(NINT,*,END=40)T(I,J),EP(I,J),EPP(I,J)
          T(I,J)=T(I,J)+273.2
          IF(T(I,J).LT.T(I,J-1))THEN
            NT(I)=J-1
            T(I+1,1)=T(I,J)
            EP(I+1,1)=EP(I,J)
            EPP(I+1,1)=EPP(I,J)
            GOTO 30
          ENDIF
          IF(J.EQ.NMAX2)CALL STOPP(NTER,NMAX2-1)
20      CONTINUE
30    CONTINUE
40    NT(I)=J-1
      NTM=NT(1)
      DO 50 I=2,NF
         IF(NTM.GT.NT(I))NTM=NT(I)
50    CONTINUE
      DO 100 J=1,NTM
        TA=0.D0
        DO 60 I=1,NF
          TA=TA+T(I,J)
60      CONTINUE
        TAV(J)=TA/FLOAT(NF)
        WRITE(NOUT,'(1X,F6.1,2X,I2)')TAV(J)-273.2,NF
        DO 90 I=1,NF
          DO 70 K=1,NTM
            T1(K)=T(I,K)
C Search on log values of modulus.
            EP1(K)=LOG10(EP(I,K))
            EPP1(K)=LOG10(EPP(I,K))
70        CONTINUE
C Check if interpolation is necessary.
          IF(ABS(T1(J)-TAV(J)).LT..1)THEN
            E1=EP1(J)
            E2=EPP1(J)
          ELSE
            CALL AKIMA(NTM,T1,EP1,TAV(J),E1)
            CALL AKIMA(NTM,T1,EPP1,TAV(J),E2)
          ENDIF
          WRITE(NOUT,80)F(I),1.D1**E1,1.D1**E2
80        FORMAT(1X,F7.2,1P,2E12.3)
90      CONTINUE
100   CONTINUE
      CALL MSTAIL(*10)
      STOP
      END
```

5.6 Sample Listings of INPUT/OUTPUT Files for **SHIFTT**

5.6.1 Data File for **SHIFTT**

Shown here is an abridged data file in the format required for running SHIFTT. The rest of the numerical data may be obtained from the tabulation available in pp. 54-55 of Reference 15. The entire data set is plotted in Figure 5.2. Adapted from Reference 15.

```
Polycarbonate Stress Relaxation; Mercier et al., JAPS, 9, 447 (1965).
L
130.  10
  2.00 10.25  1
  2.25 10.25  1
  2.50 10.25  1
  2.75 10.25  1
  3.00 10.22  1
  3.25 10.19  1
  3.50 10.16  1
  3.75 10.13  1
  4.00 10.06  1
  4.25 10.03  1
141.  10
  2.00 10.19  1
  2.25 10.19  1
  2.50 10.16  1
  2.75 10.13  1
  3.00 10.09  1
  3.25 10.03  1
  3.50  9.94  1
  3.75  9.82  1
  4.00  9.66  1
  4.25  9.50  1
142.   9
  2.00 10.13  1
  2.25 10.09  1
  2.50 10.03  1
  2.75  9.97  1
  3.00  9.88  1
  3.25  9.67  1
  3.50  9.63  1
  3.75  9.47  1
  4.00  9.22  1
144.9  6
  2.25  9.66  1
  2.50  9.50  1
  2.75  9.28  1
  3.00  9.03  1
  3.25  8.88  1
  3.50  8.66  1
145.6  6
  2.75  9.03  1
  3.00  8.72  1
  3.25  8.50  1
  3.50  8.25  1
  3.75  8.09  1
  4.00  7.94  1
146.   7
  2.50  8.72  1
  2.75  8.47  1
  3.00  8.25  1
  3.25  8.06  1
  3.50  7.94  1
  3.75  7.82  1
  4.00  7.67  1
150.8  12
  2.00  7.97  1
  2.25  7.85  1
  2.50  7.78  1
  2.75  7.72  1
  3.00  7.69  1
  3.25  7.64  1
  3.50  7.59  1
  3.75  7.55  1
  4.00  7.50  1
  4.25  7.44  1
  4.50  7.34  1
  4.75  7.25  1
```

5.6.2 Terminal Dialog with **SHIFTT**

Reproduced here is the terminal dialog for running SHIFTT with the data file SHIFTT.DAT, the partial listing of which is shown in Section 5.6.1.

```
>SHIFTT
 Enter name of data file (N, if not applicable).
SHIFTT.DAT
 Results to be written to NEW file: SHIFTT.RES
 OK (Y/N)?
Y
 Default (4. decades) for wild shifts acceptable? (Y/N)
Y
 Input reference and glass transition temperatures (C).'
150.8
145.0
 Apply density, temperature correction? (Y/N)
Y
 Are expansion coefficients available? (Y/N)
Y
 Enter expansion coefficient below Tg (1/K)
0.0003
 Enter expansion coefficient above Tg (1/K)
0.000576

 Computing!

 Estimate of WLF Constants

    C1 -   16.39 +/-    5.97
    C2 =   54.29 +/-   24.11

 Enter estimate of error in moduli data (in decades) or 0. to skip
0.

 The results are in SHIFTT.RES
 Run another set of data? (Y/N)
N
Stop - Program terminated.
```

5.6.3 Output File from **SHIFTT**

Reproduced here is the output file from SHIFTT using the data and terminal dialog listed in Sections 5.6.1 and 5.6.2, respectively. The results are plotted in Figure 5.3.

```
                    SHIFTT Time-Temperature Superposition
                    -------------------------------------

Polycarbonate Stress Relaxation; Mercier et al., JAPS 9, 447 (1965).

     Glass Temperature =  418.2 K   145.0 C
 Reference Temperature =  424.0 K   150.8 C
```

Results using Transient Modulus

Temp., K (C)	Log Shift	Log Time	Log Transient Modulus
403.2 (130.0)	5.03 +/- .40		
		-3.034	10.268
		-2.784	10.268
		-2.534	10.268
		-2.284	10.268
		-2.034	10.238
		-1.784	10.208
		-1.534	10.178
		-1.284	10.148
		-1.034	10.078
		-.784	10.048
414.2 (141.0)	4.00 +/- .33		
		-2.000	10.198
		-1.750	10.198
		-1.500	10.168
		-1.250	10.138
		-1.000	10.098
		-.750	10.038
		-.500	9.948
		-.250	9.828
		.000	9.668
		.250	9.508
415.2 (142.0)	3.35 +/- .27		
		-1.350	10.137
		-1.100	10.097
		-.850	10.037
		-.600	9.977
		-.350	9.887
		-.100	9.677
		.150	9.637
		.400	9.477
		.650	9.227
418.1 (144.9)	2.17 +/- .18		
		.084	9.665
		.334	9.505
		.584	9.285
		.834	9.035
		1.084	8.885
		1.334	8.665
418.8 (145.6)	1.84 +/- .12		
		.912	9.034
		1.162	8.724
		1.412	8.504
		1.662	8.254
		1.912	8.094
		2.162	7.944
419.2 (146.0)	1.32 +/- .10		
		1.181	8.724
		1.431	8.474
		1.681	8.254
		1.931	8.064
		2.181	7.944
		2.431	7.824
		2.681	7.674
424.0 (150.8)	.00 +/- .00		
		2.000	7.970
		2.250	7.850
		2.500	7.780
		2.750	7.720
		3.000	7.690
		3.250	7.640
		3.500	7.590

```
                                          3.750          7.550
                                          4.000          7.500
                                          4.250          7.440
                                          4.500          7.340
                                          4.750          7.250
429.2 ( 156.0)     -1.51 +/-   .05
                                          3.508          7.586
                                          3.758          7.546
                                          4.008          7.496
                                          4.258          7.406
                                          4.508          7.336
                                          4.758          7.246
                                          5.008          7.186
                                          5.258          7.056
                                          5.508          6.936
                                          5.758          6.776
                                          6.008          6.586
                                          6.258          6.376
                                          6.508          6.126
432.2 ( 159.0)     -2.26 +/-   .07
                                          4.509          7.334
                                          4.759          7.244
                                          5.009          7.154
                                          5.259          7.024
                                          5.509          6.934
                                          5.759          6.774
                                          6.009          6.624
                                          6.259          6.404
                                          6.509          6.124
                                          6.759          5.844
                                          7.009          5.524
434.7 ( 161.5)     -2.87 +/-   .10
                                          5.119          7.052
                                          5.369          6.962
                                          5.619          6.842
                                          5.869          6.632
                                          6.119          6.492
                                          6.369          6.302
                                          6.619          6.052
                                          6.869          5.772
                                          7.119          5.462
440.2 ( 167.0)     -3.70 +/-   .15
                                          5.946          6.708
                                          6.196          6.488
                                          6.446          6.238
                                          6.696          5.950
                                          6.946          5.618
                                          7.196          5.208
444.2 ( 171.0)     -4.38 +/-   .17
                                          6.629          5.985
                                          6.879          5.705
                                          7.129          5.395
                                          7.379          5.015
                                          7.629          4.615

Variable units identical to those of data.


Estimate of WLF Constants

   C1 =  16.39 +/-    5.97
   C2 =  54.29 +/-   24.11
```

References

1. Ferry, J. D.: *Viscoelastic Properties of Polymers.* Wiley, New York, 1980, Chapter 11.
2. Williams, M. L.; Landel, R. F.; Ferry, J. D.: "The Temperature Dependence of Relaxation Mechanisms in Amorphous Polymers and Other Glass-Forming Liquids," *J. Amer. Chem. Soc.* **77**, 3701-3707 (1955).
3. Kaelble, D. H.: *Computer-Aided Design of Polymers and Composites.* Marcel Dekker, New York, 1985, pp. 145-147.
4. Fox, T. G.; Loshaek, S.: "Influence of Molecular Weight and Degree of Crosslinking on the Specific Volume and Glass Temperature of Polymers," *J. Polym. Sci.* **15**, 371-390 (1955).
5. Rouse, P. E., Jr.: "A Theory of the Linear Viscoelastic Properties of Dilute Solutions of Coiling Polymers," *J. Chem. Phys.* **21**, 1272-1280 (1953).
6. Tobolsky, A. V.; Andrews, R. D.: "Systems Manifesting Superposed Elastic and Viscous Behavior," *J. Chem. Phys.* **13**, 3-27 (1943).
7. Nielsen, L. E.: *Mechanical Properties of Polymers and Composites.* Marcel Dekker, New York, 1974, Vol. 1.
8. IMSL MATH/LIBRARY *User's Manual.* IMSL, Houston, 1991, Vol. 3.
9. Press, W. H.; Flannery, B. P.; Teukolsky, S. A.; Vetterling, W.T.: *Numerical Recipes in* FORTRAN: *The Art of Scientific Computing.* Cambridge University Press, Cambridge, 1986.
10. Akima, H.: "A New Method of Interpolation and Smooth Curve Fitting Based on Local Procedures," *J. ACM* **17**, 589-602 (1970).
11. Van Krevelen, D. W.: *Properties of Polymers.* Elsevier Scientific, New York, 1976, Chapter 4.
12. IMSL STAT/LIBRARY *User's Manual.* IMSL, Houston, 1991, Vol. 3.
13. Ruckdeschel, F. R.: BASIC *Scientific Subroutines.* McGraw-Hill, New York, 1981, Vol. II.
14. Mercier, J. P.; Aklonis, J. J.; Litt, M.; Tobolsky, A. V.: "Viscoelastic Behavior of the Polycarbonate of Bisphenol A," *J. Appl. Polym. Sci.* **9**, 447-459 (1965).
15. Aklonis, J. J.; MacKnight, W. J.: *Introduction to Polymer Viscoelasticity.* Wiley, New York, 1983, Chapter 3.
16. Weiss, R. A.: private communication.
17. Fitzgerald, E. R.; Grandine, L. D., Jr.; Ferry, J. D.: "Dynamic Mechanical Properties of Polyisobutylene," *J. Appl. Phys.* **24**, 650-655 (1953).
18. Ferry, J. D.; Grandine, L. D., Jr.; Fitzgerald, E. R.: "The Relaxation Distribution Function of Polyisobutylene in the Transition from Rubber-Like to Glass-Like Behavior," *J. Appl. Phys.* **24**, 911-916 (1953).
19. Plazek, D. J.; Vrancken, M. N.; Berge, J. W.: "A Torsion Pendulum for Dynamic and Creep Measurements of Soft Viscoelastic Materials," *Trans. Soc. Rheol.* **2**, 39-51 (1958).

20. Philippoff, W.: "Mechanical Investigations of Elastomers in a Wide Range of Frequencies," *J. Appl. Phys.* **24**, 685-689 (1953).
21. Plazek, D. J.: "Viscoelastic and Steady-State Rheological Response," *Methods of Experimental Physics.* Academic Press, New York, 1980, Vol. 16c.
22. Shaw, M. T.: "Time-Temperature Superposition Algorithms and Expected Error," *SPE Tech. Papers* **34**, 1182-1184 (1988).

Chapter 6

Squeeze-Flow Analysis

6.1 Introduction

The program SQUEEZ offered in this chapter is designed to convert raw data from almost any squeezing device into useful forms for the many available analyses. Because the various non-Newtonian approaches to the analysis of squeezing flow data have not enjoyed universal success, SQUEEZ takes a conservative stance and provides analyses only in the form of preparation of data files for viscoelastic analysis by FITSTR (Chapter 10) and for inelastic analyses using the power-law solution and by an integral approximation which takes into account the variation of viscosity along the flow direction. The integral approximation is very similar to the Cross-Kaye approximation described in Chapter 7. In addition, the analysis of Tanner [1] for a third-order fluid (shear stresses only) is attempted.

A squeeze-flow experiment is easy to do; one simply inserts a sample between flat, parallel plates and moves the plates together. There are two fairly standard modes for the squeezing operation: constant force applied to the plates and constant approach velocity. The former can be effected by a constant-force actuator (air cylinder, dead weight) pushing on the plates, while the latter is easily accomplished in a mechanical testing machine or a similar device.

The geometry can feature either completely filled circular plates, where the excess sample spills over the edge, or plates of large extent compared to the radius of the sample (Figure 6.1). The former has the advantage that the forces do not build up so fast, but the disadvantage of uncertainty in the nature of flow at the edge [2].

The advantages of squeeze flow are important. It is one of the few flows where the loading of the specimen and the cleaning of the equipment are relatively simple. If the polymer sticks to the plates, they can be protected by a release layer, for example, foil or paper. Curing and cross-linking of the sample do not endanger the equipment, and highly filled samples can be easily accommodated. Complete adhesion or slip at the plates can be achieved quite easily by controlling the lubricity of the surface; this has proven to be important in the handling of delicate food samples [3], for example. With copious lubrication, the shear in the sample is eliminated and uniform biaxial extension results [4,5].

For the constant-radius setup, degraded polymer is extruded out of the gap where it no longer contributes to the response of the plates. The opposite is the case for the constant-volume setup; here, the outer layers are emphasized, and the method becomes a sensitive, although qualitative, method for detecting compositional changes.

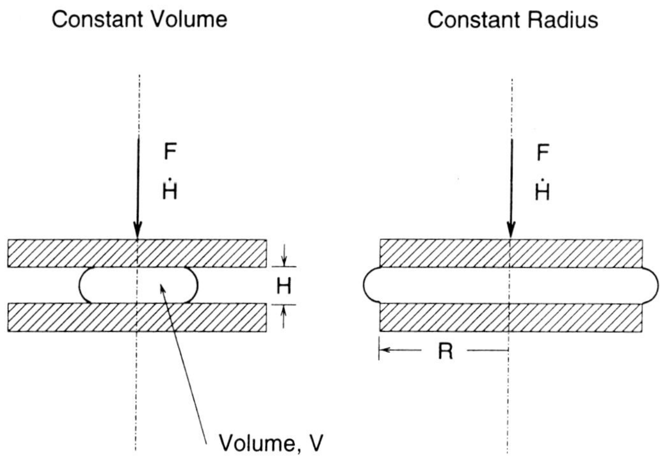

Figure 6.1 Squeeze-flow geometry showing constant-volume and constant-radius cases.

Squeeze flow mimics the process of compression molding, which is of importance in many fabrication operations, including the processing of sheet molding compounds (SMC) [6]. By stacking coated films, squeeze-flow measurements can be carried out on the coating despite its thinness [7].

In spite of the simplicity of the geometry and operation, the analysis of squeezing flow is not simple; for starters, it is nonuniform and unsteady. The forces that are measured at the plates will involve all the complexities of transient biaxial and shear flows which are varying in intensity throughout the sample. Unlike torsional flow or capillary flow, the unsteady aspect of squeeze flow does not allow one to correct for the nonuniformities in the flow. While analyses of viscoelastic flows have been attempted, most are complex and nonanalytic in nature. For these reasons, the practical analyses of squeezing flows have been limited to generalized Newtonian fluids. Regardless, the transient aspects of the squeezing-flow experiment can yield erratic results when the squeezing rates are high; some of these results [7-9] are illustrated in Figure 6.2.

The generalized Newtonian fluid is one in which each component of the stress tensor is strictly proportional to the corresponding component of the rate-of-deformation tensor. The proportionality constant, the viscosity, is allowed to depend on the invariants of the rate-of-deformation tensor, usually only the second invariant. Several expressions have been developed for this dependence; see Chapter 9 for examples.

Appropriate for a Newtonian fluid is the Stefan equation [10]

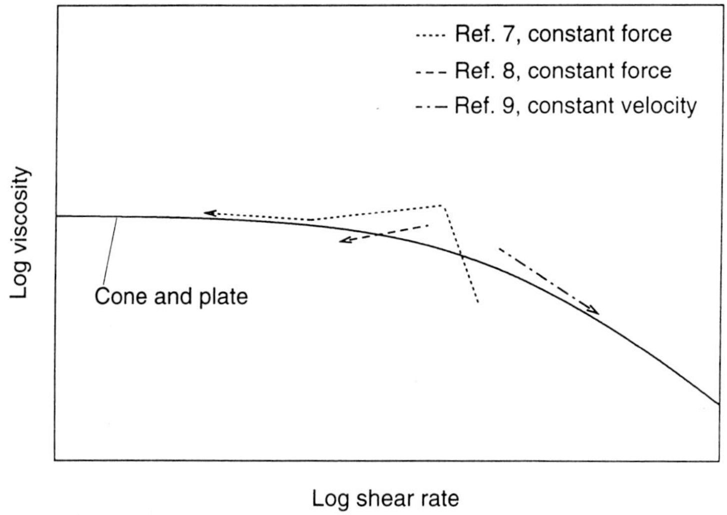

Figure 6.2 Schematic of effects reported in attempting to reproduce the viscosity function with a single squeeze-flow experiment. (The direction of increasing time is indicated by the arrows.)

$$F = \frac{3\pi R^4 \eta_A(-\dot{H})}{2H^3} \tag{6.1}$$

where F is the squeezing force, R is the plate radius, η_A is the apparent viscosity, H is the gap between plates, and $-\dot{H}$ is the approach velocity. If the gap between the plates is treated as a slit die, as illustrated in Figure 6.3, the wall shear stress σ_w determines the radial pressure gradient dP/dr which, upon integration over the plate area, gives the force F. Pursuing this analysis, we can write Equation 6.1 in terms of an apparent, Newtonian wall shear stress and shear rate at the position $r = R$. For the apparent stress the expression is

$$\sigma_A = \frac{2FH}{\pi R^3} \tag{6.2}$$

and for the apparent shear rate the expression is

$$\dot{\gamma}_A = \frac{3R(-\dot{H})}{H^2} \tag{6.3}$$

One intriguing possibility for squeezing-flow analysis is evident from the similarities between squeezing flow and torsional flow between parallel plates. In Chapter 7,

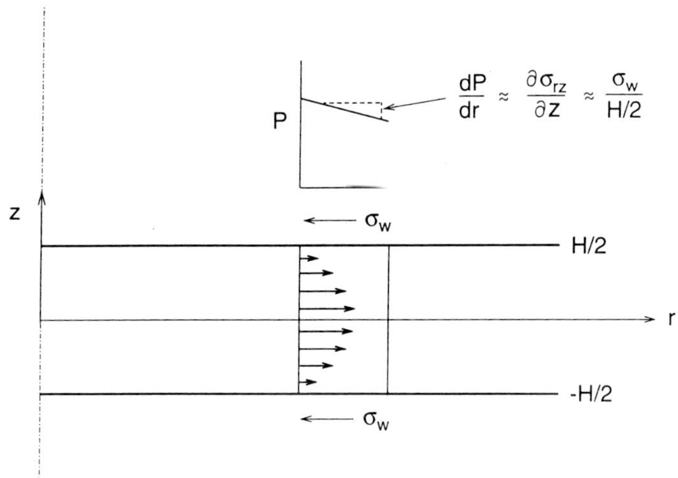

Figure 6.3 Schematic of radial flow during squeezing with the steady-state and lubrication approximations in effect. With some fluids, the flow may have a pronounced velocity component in the axial direction (other than that due to the very small squeezing velocity) because of the dependence of viscosity on shear rate.

we note that the expression for torque M developed during torsional flow start-up can be expressed as

$$M = 2\pi R^3 \int_0^1 \sigma_{21}(x)\, x^2\, dx \tag{6.4}$$

whereas the force for squeezing flow can be expressed as [11]

$$F = \frac{4\pi R^3}{H} \int_0^1 \sigma_{21}(x)\, x^2\, dx \tag{6.5}$$

where $x = \dot{\gamma}/\dot{\gamma}_R = r/R$. Using a first-order Gaussian moment integration [12,13], we obtain

$$\int_0^1 \sigma_{21}(x)\, x^2\, dx \cong \frac{1}{3}\sigma_{21}\left(\frac{3}{4}R\right) \tag{6.6}$$

For the squeezing flow this suggests that the apparent viscosity

$$\eta\left(\dot{\gamma} = \frac{3}{4}\frac{3R(-\dot{H})}{H^2}\right) \cong \frac{2FH^3}{3\pi R^4(-\dot{H})} \tag{6.7}$$

can be assigned to a shear rate of three-fourths of the outer corner shear rate with some confidence. As with torsional flow, this assumption should apply during start-up as well. The apparent stress $2FH/\pi R^3$ can then be compared with the stress that would result if the material were subjected to a homogeneous but unsteady start-up experiment with a shear rate varying along the way but always equal to the $9R\dot{H}/4H^2$ of the squeeze-flow experiment. In contrast to the foregoing argument, Tichy and Winer [14] found that using half of the maximum Newtonian shear rate provided reasonable results.

For *named* generalized non-Newtonian fluids, many studies are available, especially for power-law fluids [15-17]. A technique for handling Ellis fluids has been reported [8]. For arbitrary generalized Newtonian fluids, a computational framework has been provided [18].

For a linear viscoelastic material a solution has been presented by Tanner [1], while Shirodkar and Middleman [19], Isayev and Azari [20], and Kramer [21] have worked with single-integral constitutive equations. Solutions have also been presented for ordered fluids [1,22,23]. Most of these analyses have had only limited testing.

The result found by Tanner [1] for a linear viscoelastic model is

$$F = \frac{3\pi R^4}{2} \int_0^t G(t-s)\left(H^{-3}\frac{dH}{ds}\right) ds \tag{6.8}$$

where $G(t-s)$ is the relaxation modulus of the melt. Equation 6.8 can be written in an equivalent form by normalizing the gap by the initial gap and noting that $(dH/dt)/H^3$ is $0.5d(H^{-2})/dt$. The result is

$$\frac{4F(t)}{3\pi R^3}\left(\frac{H_0}{R}\right)^2 = \int_0^t G(t-s)\,\frac{d}{ds}\left(1 - \frac{1}{\lambda^2}\right) ds \tag{6.9}$$

where $\lambda = H/H_0$. The form of Equation 6.9 is identical to the transverse stress generated via planar extensional flow. Thus, the capabilities of `FITSTR` in Chapter 10 can be applied to find the relaxation function parameters. Clearly, an appropriate question is how far this approach might extend into the nonlinear regime.

6.2 Description of the Algorithm SQUEEZ

In view of the state of affairs for analyzing squeezing flows, we decided to emphasize the data-handling aspects rather than the analysis, although the analyses suggested earlier are attempted. `SQUEEZ` should be regarded as a program for rheologists who wish to ease the task of correcting raw gap-force-time data from squeezing experiments of any type: constant force, constant velocity, constant radius, constant volume, or variations of these parameters. The corrections include instrument compliance and sample inertia. The corrected data are then available for further processing by using the selected analyses featured in SQUEEZ or one of the many analyses in the literature.

The program needs files of gap reading or force taken at uniform intervals of time. If required, the gap reading is converted to an absolute gap by using supplied calibration points. Integrated with this step is the correction for compliance of the instrument. If the run is nominally constant force, a second file must be available containing a calibration run: gap versus time, but where the gap is closed. This information is needed because many force actuators have significant rise times. Because the instrument has some compliance, the gap indicator will deflect slightly in proportion to the force exerted by the actuator during the closed-gap run. The data in this file are used as a look-up function to predict the actual force applied to the moving plate during a real run.

At this point gap (or force) values are read in from the input file. Gap readings are converted to actual gap after corrections for instrument compliance are applied. Derivatives of the gaps are needed, but before these are taken the gap data from constant-force experiments are transformed to the variable $H^{-2} - H_0^{-2}$, which should be more linear with time than the gap itself. H_0 is the initial gap. Additionally, random noise is reduced by using the smoothing routine DCURVE described in Chapter 4. The first derivatives are then taken using the subroutine SLOPEV, which is a five-point quadratic, moving-strip method for evenly spaced data. In the constant-force experiment, the derivatives of the transformed variable are again smoothed before the inverse transform is done.

With the first derivatives in hand, the force readings are corrected for instrument compliance and sample inertia. Finite-difference second derivatives have been found adequate for this step. The correction for sample inertia uses half the estimated mass of the sample; a density of 1 g/cm^3 is assumed. After the forces are corrected, the gaps are readjusted for instrument compliance.

Next, SQUEEZ provides options for working up the data. One option invokes the three-fourths rule and the power-law function to estimate the non-Newtonian viscosity. The power-law approximation is accomplished in terms of a shift in the apparent shear rate by the amount

$$\frac{\log[3n(n+3)/4(2n+1)] + (1-n)\log[(2n+1)/3n]}{1-n} \tag{6.10}$$

where n is the power-law index. The resulting shift allows the apparent viscosity from the Stefan equation to be used and is somewhat analogous to the application of the three-fourths rule (Equation 6.7). Also listed with this option is the value of the mean shear rate $3R/2H^2$, which has also been found useful [14]. If the data are appropriate (positive η_o, τ), SQUEEZ will provide the estimates and the 95%

confidence intervals for the zero-shear-rate viscosity, η_o, and time constant, τ, assuming a third-order fluid.

The final option is simply the writing of the corrected data to the output file. If the mode is one of constant velocity, constant radius (filled plates), `SQUEEZ` can prepare a separate file, via the routine `RESFIL`, that is ready to run with the program `FITSTR` in Chapter 10. `FITSTR` attempts to fit time-strain-stress data with a nonlinear constitutive model. In this instance, use Model 1 (viscoelastic solid with the Kohlrausch relaxation function) and lock out the nonlinear parameter x_{NL} by choosing identical bracketing values; that is, 2 and 2. With these steps, the linear viscoelastic relaxation modulus will be calculated from Equation 6.9.

6.3 Execution of **SQUEEZ**

The execution of `SQUEEZ` requires the subroutines `ALINR` and `DCURVE` found in Appendix B. The subprograms `FOPEN`, `FORCE`, `RESFIL`, and `SLOPEV`, inherent to running `SQUEEZ`, are included in the code listed in Section 6.5. Note that `FOPEN` and `RESFIL` are set up for Microsoft® FORTRAN; modifications may be required for other compilers.

The squeeze-flow data file is structured according to the nature of the experiment. However, there are common features. The first record is set aside for the title of the file; insert anything up to 79 characters containing identification information. The second record carries one alphanumeric character describing the nature of the experiment: `F` or `V` for constant force or velocity, respectively. All data must be entered in consistent units. For example, when using SI units, use meter, Newton, and second to denote dimensions of length, force, and time, respectively.

The next three records are gap calibration data. This information must be acquired by using two calibration shims, one of which can be of zero thickness. One of the shims must be run at two known forces to provide the instrument compliance. Three entries appear on each record: the actual gap, the reading of the gap-measuring device, and the force applied to the gap to obtain the readings. With one shim, two different forces must be applied, preferably at the extremes of forces appearing in the run. As noted in Section 6.2, it is good practice to have one of the actual gaps close to the smallest gap seen in the experiment; zero gap is a good choice. These three records provide the path whereby the gap and force readings can lead to the actual gap seen by the fluid during the experiment. Preferably, nothing on the instrument should be changed (including the temperature) between the acquisition of these data and the running of the material. As shown in Section 6.4, these numbers are very important.

The sixth record changes slightly with the type of experiment; it contains all the geometric information. For constant-force experiments, the first entry is the force applied; for constant-velocity experiments, this spot is occupied by the velocity.

Next comes the plate radius, the volume of the sample (enter a value much larger than the actual volume if the experiment is constant radius from start to finish), the mass of the moving parts of the instrument, the time interval between data points, and the zero-time gap. It is assumed by SQUEEZ that the zero-time gap is taken at zero force. With deadweight actuators, the load on the top plate must be borne by a latch until the sample is fully relaxed.

Do not worry if the experiment starts out as constant volume, but the material later spills over the edge; SQUEEZ will keep the calculation going. In this case, enter the true initial volume.

The next and following records are the gap readings, in constant-force experiments, or the force readings, in constant-velocity experiments. This must be the actual force; the force transducer is assumed to be calibrated. A sample data file of a constant-force experiment is provided in Section 6.6.1.

If the experiment is nominally constant force as designated by an F entry on the second record, SQUEEZ will be expecting (and ask for via FOPEN) a second file containing information about the force transient. The first record of the auxiliary file contains the time interval. The second and subsequent records contain the gap readings for the closed-gap (no sample) calibration run. The reason the gap reading changes is that the frame of the instrument is compliant; the gap reading serves as a convenient force transducer. Because only the time dependence of these gap readings is of concern, the gap entries may be in any units. See Section 6.6.2 for an example. Enter 0 for the time interval in the auxiliary file if you wish to assume that the actuator reaches its full force instantly. In this case, no further entries are needed.

With the force transient read in, SQUEEZ can compute a fully corrected force; that is, the force which is actually applied to the fluid. If the assumptions of quasi-steady-state flow and generalized Newtonian behavior are in effect, it does not really matter whether the force is constant or not. However, it is important that it be known.

6.4 Application of SQUEEZ

The first example using SQUEEZ is taken from the literature [7] and concerns a low-density polyethylene (LDPE) melt at 150 °C. The data were taken on a constant-force instrument with 50-mm-diameter plates. The squeezing was done at forces ranging from 0.25 to 8 kN. The data, from Figure 3 of Reference 7, were digitized at constant time intervals for use with SQUEEZ (Figure 6.4). The input/output files of the digitized data for 0.25 kN are listed in Section 6.6. It was assumed that the actuator reached its full force instantly.

In concert with the procedures mentioned in original article, an estimate of the non-Newtonian viscosity via the power-law approximation was chosen. The results are plotted in Figure 6.5 and compared with the capillary data taken on the same

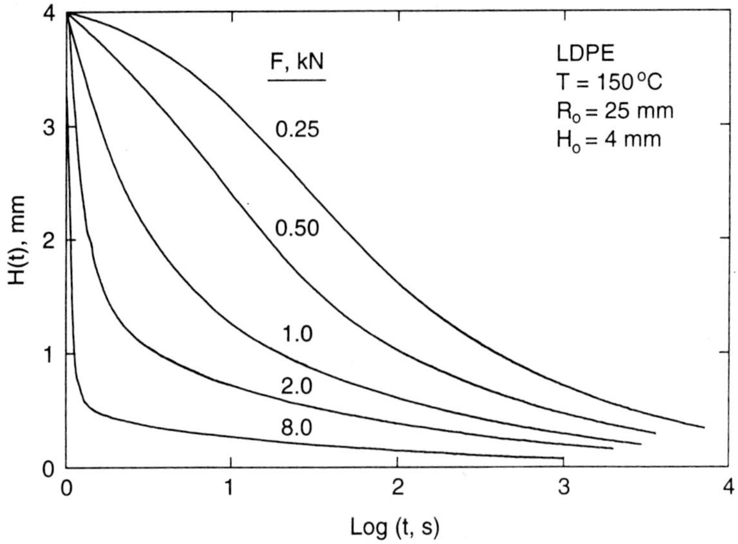

Figure 6.4 Constant-force squeezing-flow data for a low-density polyethylene melt (M_w = 5.3×10^5) at 150 °C. Adapted from Ref. 7. Copyright © 1992 by Hüthig & Wepf, Basel. Reprinted by permission.

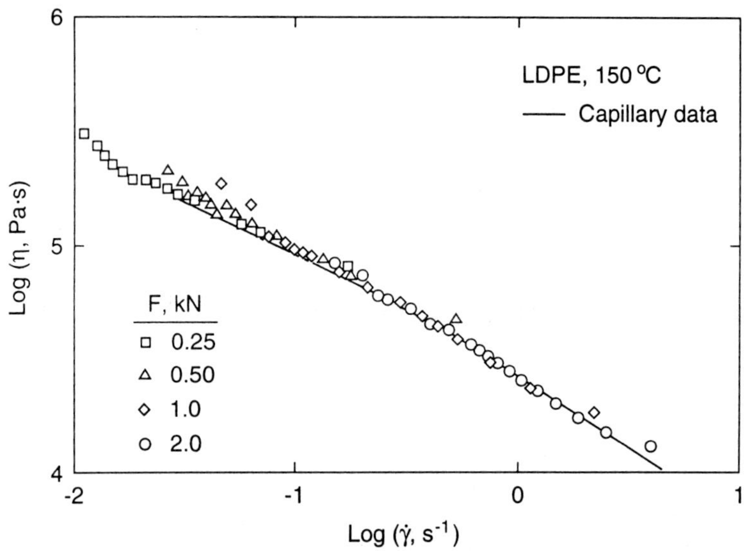

Figure 6.5 The squeeze-flow results for the LDPE data in Figure 6.4. The data points, edited for clarity, are from SQUEEZ using the power-law approximation and are compared with capillary data (–) adapted from Ref. 7. Copyright © 1992 by Hüthig & Wepf, Basel. Reprinted by permission.

material. The agreement of the squeezing-flow data and the conventional data is reasonable, as was found in the original article. While the overprediction of viscosity at the beginning of the run (high shear rates) is due to stress overshoot [7,15,16], the overprediction near the end is less easily explained. If the instrument compliance were undercorrected, the viscosity would be underpredicted. Pressure effects would be felt throughout the run in a constant-force experiment (but not in a constant-velocity experiment). The edge effects may be important [2].

In Figure 6.6, we have plotted constant-velocity data for a silicone gum (General Electric SE-30, $M_w = 5 \times 10^5$) run at room temperature. The squeezing was done with velocities ranging from 4.2 to 85 μm/s using 57-mm-diameter plates. As can be seen, the forces climb rapidly as the gap narrows. With SQUEEZ, the analyses of the data corresponding to velocities of 4.2, 21, and 85 μm/s result in the viscosity functions shown in Figure 6.7. The three-fourths rule (Equation 6.7) gives a better agreement than the apparent viscosity derived from the Stefan equation (Equation 6.1) when compared with published data taken from steady simple shear and dynamic experiments [24]. For the data corresponding to a velocity of 21 μm/s, the program was asked to produce the data file shown in Section 6.6.5. The file was used by FITSTR, from Chapter 10, to derive the relaxation modulus $G(t)$ from Equation 6.9 with the constraint that the modulus follows the Kohlrausch relaxation function (Equation 10.5).

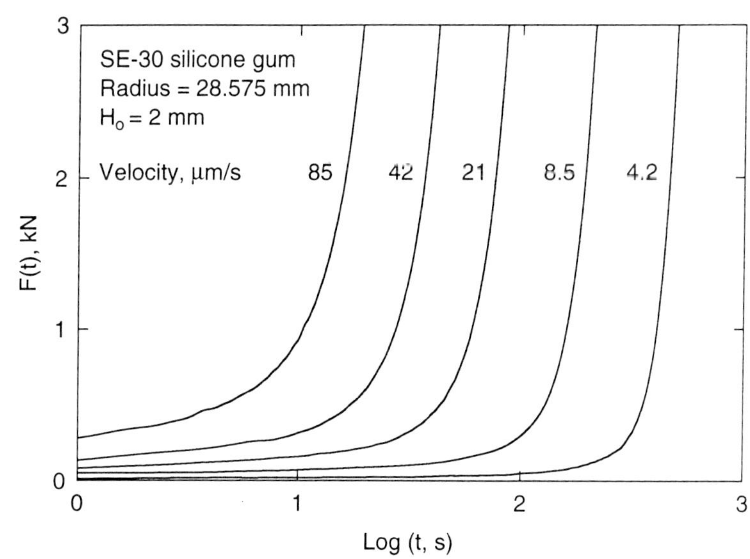

Figure 6.6 Constant-velocity squeezing flow data for a silicone gum (General Electric SE-30, $M_w = 5 \times 10^5$) at room temperature.

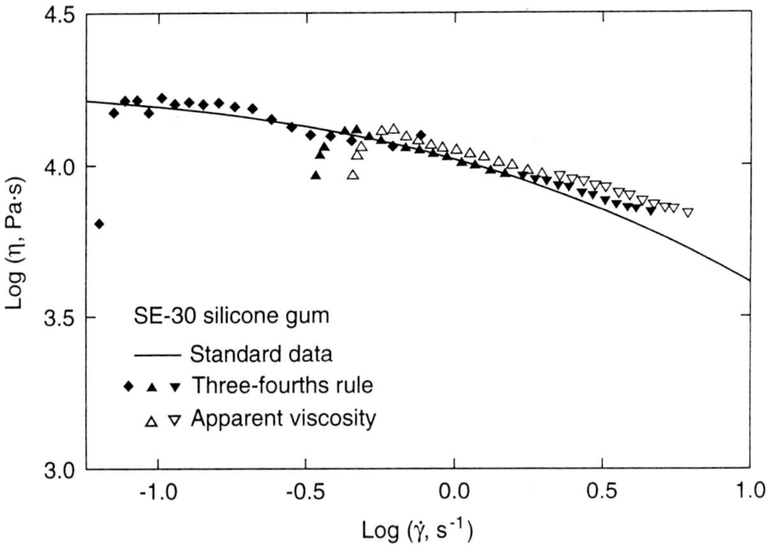

Figure 6.7 The viscosity functions calculated by SQUEEZ using the data in Figure 6.6 for velocities of 4.2 μm/s (♦), 21 μm/s (▲), and 85 μm/s (▼). The unfilled symbols represent the apparent viscosities derived from the Stefan equation while the filled symbols are derived from the three-fourths rule.

On running the data file shown in Section 6.6.5 with FITSTR, the Kohlrausch relaxation function parameters for the silicone gum were found to be $G = 44.1 \pm 5.7$ kPa, $B = 1.12 \pm 0.05$, and $\tau = 0.12 \pm 0.003$ s. The relaxation modulus is plotted in Figure 6.8 and compared with those obtained from published transient and dynamic data [24]. The transient data were taken from a 50% step-strain experiment whereas the relaxation function was derived from the dynamic data by using Schwarzl's conversion formula [25]. The results are in reasonable agreement as far as magnitude is concerned, but the relaxation function derived from the squeezing flow appears narrower.

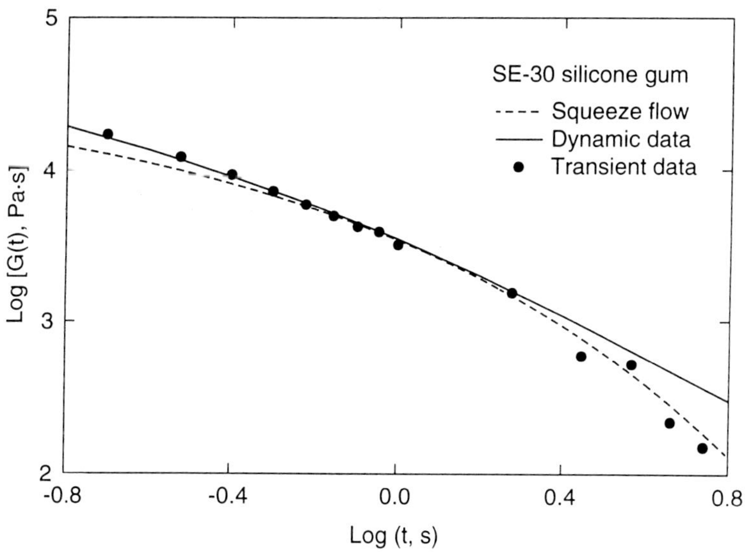

Figure 6.8 Relaxation modulus derived from SQUEEZ and FITSTR (Chapter 10) using the constant-velocity (21 μm/s) data and Equation 6.9. The dashed line was calculated from dynamic data [24] using the approximation of Schwarzl [25]. The transient data (●) were obtained from a 50% step-strain experiment [24].

6.5 Program Listing for **SQUEEZ**

Imbedded into the code are the subprograms FOPEN, FORCE, RESFIL, and SLOPEV inherent to the execution of SQUEEZ.

```
      PROGRAM SQUEEZ
C Handles squeeze-flow data from constant-force or constant-velocity
C  experiments.
C Requires the subroutines ALINR, DCURVE, and STOPP.
C Data file format:
C         1st record - Title, insert anything up to 79 characters.
C         2nd record - V(constant velocity) or F(constant force)
C     next 3 records - gap calibration data; at least 2 different forces
C                      and two different gaps: each record should contain
C                      actual gap, gap reading, applied force.
C         6th record - geometric information: first entry is applied force
C                      or approach velocity; subsequent entries are plate
C                      radius, sample volume (use large number for constant
C                      radius), mass of moving parts of instrument, time
C                      interval between data points, and zero-time gap.
C subsequent records - gap or force data at uniform time intervals.
      IMPLICIT DOUBLE PRECISION (A-H,O-Z)
      PARAMETER (NGCD=4,NMAX1=501,NMAX2=11,PI=3.141592654,TINY=1.D-9)
      CHARACTER ANS*1,REC*80,TITLE*80
      DIMENSION F(NMAX1),GCAL(NMAX1),H(NMAX1),HDOT(NMAX1),T(NMAX1)
      DIMENSION SRN(NMAX1),SRPL(NMAX1),VIS(NMAX1),X(NMAX1),Y(NMAX1)
      DIMENSION A(NMAX2),U(NMAX2),V(NMAX2)
      DIMENSION FCAL(NGCD),HACT(NGCD),HOBS(NGCD)
```

```
C t-statistic for 95% confidence interval with N degrees of freedom.
      T95(N)=1.96+2.3539/N+3.1902/N**2+0.7004/N**3+4.5015/N**4
C Shift factor for power-law; S is the power-law index.
      FAC(S)=10.**((LOG10((S+3.)*3.*S/4./(2.*S+1.))+
     &       (1.-S)*LOG10((2.*S+1.)/3./S))/(1.-S))
10    CALL MSHEAD
      CALL UNITS(NTER,NINT,NOUT)
      WRITE(NOUT,20)
20    FORMAT(25X,'Squeeze-Flow Analysis'/25X,21('-'))
      READ(NINT,'(A80)')TITLE
      WRITE(NOUT,'(/A80/)')TITLE
      IGEOM=1
      IMODE=1
      READ(NINT,'(A80)')REC
      DO 30 I=1,80
        IF(REC(I:I).EQ.'F'.OR.REC(I:I).EQ.'f')IMODE=2
30    CONTINUE
      READ(NINT,*)(HACT(I),HOBS(I),FCAL(I),I=1,NGCD-1)
      HACT(NGCD)=HACT(1)
      HOBS(NGCD)=HOBS(1)
      FCAL(NGCD)=FCAL(1)
      DO 40 I=1,NGCD-1
        IF(ABS(HACT(I)-HACT(I+1)).LT.TINY)
     &    CFRAME=ABS((HOBS(I)-HOBS(I+1))/(FCAL(I)-FCAL(I+1)))
40    CONTINUE
      WRITE(NOUT,'(/2X,A,1P,E10.3)')'Frame compliance:',CFRAME
      DO 50 I=1,NGCD-1
        IF(ABS(HACT(I)-HACT(I+1)).GT.TINY)THEN
          X1=HOBS(I)+CFRAME*FCAL(I)
          X2=HOBS(I+1)+CFRAME*FCAL(I+1)
          Y1=HACT(I)
          Y2=HACT(I+1)
          GS=ABS((Y2-Y1)/(X2-X1))
          GI=Y1-GS*X1
        ENDIF
50    CONTINUE
      WRITE(NOUT,60)GI,GS
60    FORMAT(3X,'Gap calibration:'/8X,'Intercept =',1P,E10.3/
     &       12X,'Slope =',E10.3)
C CFRAME corrects gap reading, not actual gap.
      IF(IMODE.EQ.2)THEN
        IFLAG=1
        WRITE(NTER,'(/2X,A)')'Actuator rise time calibration.'
        WRITE(NTER,'(2X,A)')'Requires no-sample gap data file.'
        NOF=NOUT+6
        CALL FOPEN(NTER,NOF)
        READ(NOF,*)TICAL
        IF(TICAL.LT.TINY)THEN
          IFLAG=0
          GOTO 90
        ENDIF
        DO 70 I=1,NMAX1
          READ(NOF,*,END=80)GCAL(I)
          IF(I.EQ.NMAX1)CALL STOPP(NTER,NMAX1-1)
70      CONTINUE
80      NCAL=I-1
      ENDIF
90    CONTINUE
C Assemble corrected time-force-gap arrays.
      IF(IMODE.EQ.1)THEN
C Read crosshead speed XHS, radius R, volume VOL, moving mass XMASS, time
C  interval TI, and initial gap H0.  H0 is a zero-force value, and is the
C  gap reading (not actual gap).
        READ(NINT,*)XHS,R,VOL,XMASS,TI,H0
C Correct H0.
        H0=H0*GS+GI
        IF(PI*R**2*H0.GT.VOL)IGEOM=2
```

```
C Read force data.
          DO 100 I=1,NMAX1
            READ(NINT,*,END=110)F(I)
            IF(I.EQ.NMAX1)CALL STOPP(NTER,NMAX1-1)
            T(I)=REAL(I)*TI
            H(I)=H0
C Correct gaps for instrument frame compliance.
            IF(F(I).GT.TINY)H(I)=H0-XHS*T(I)+CFRAME*F(I)*GS
100       CONTINUE
110       NP=I-1
        ELSE
C Read constant force FSTD, R, VOL, XMASS, TI, and H0.
          READ(NINT,*)FSTD,R,VOL,XMASS,TI,H0
          IF(PI*R**2*H0.GT.VOL)IGEOM=2
          H0=H0*GS+GI
C Read gap data.
          DO 120 I=1,NMAX1
            READ(NINT,*,END=130)H(I)
            IF(I.EQ.NMAX1)CALL STOPP(NTER,NMAX1-1)
            T(I)=REAL(I)*TI
            IF(IFLAG.EQ.0)THEN
              F(I)=FSTD
            ELSE
C Correct force at early times using rise-time data GCAL.
              F(I)=FORCE(T(I),FSTD,NCAL,GCAL,TICAL)
            ENDIF
C Correct gaps for frame compliance.
            H(I)=GI+GS*(H(I)+CFRAME*F(I))
120       CONTINUE
130       NP=I-1
        ENDIF
        IF(NP.LT.12)THEN
          WRITE(NTER,*)
     &         ' Insufficient number of data for SQUEEZ to proceed.'
          STOP
        ENDIF
C Find second derivatives and do inertial corrections.
        IF(IMODE.EQ.1)THEN
          DO 140 I=1,NP
            Y(I)=H(I)
140       CONTINUE
        ELSE
C Transform gap to linearize (if fluid is Newtonian).
          DO 150 I=1,NP
            Y(I)=1.D0/H(I)**2-1.D0/H0**2
150       CONTINUE
        ENDIF
C Reduce random noise.
        EMP=.3
        CALL DCURVE(NP,Y,X,EMP)
C Find dH/dt.
        CALL SLOPEV(NP,T,X,Y)
C Reduce random noise.
        CALL DCURVE(NP,Y,HDOT,EMP)
C Reverse transform.
        IF(IMODE.EQ.2)THEN
          DO 160 I=1,NP
            HDOT(I)=-0.5D0*H(I)**3*HDOT(I)
160       CONTINUE
        ENDIF
C Check for +HDOT (while physically possible, it will be excluded).
        DO 170 I=2,NP
          IF(HDOT(I).GE.0.)HDOT(I)=HDOT(I-1)
170     CONTINUE
C Correct for inertia, both force and gap (through frame compliance).
C Use 1/2 mass of sample and assume density = 1.0.
```

```
      DO 180 I=1,NP-1
        SMASS=MIN(0.5*VOL,.5*PI*R**2*H(I))
        XMASS=XMASS+SMASS
        FINERT=-XMASS*(HDOT(I+1)-HDOT(I))/TI
        F(I)=F(I)+FINERT
        H(I)=H(I)+CFRAME*FINERT*GS
180   CONTINUE
C Write results to appropriate files.
      WRITE(NTER,'(/2X,A)')
     &     'Estimate viscosity-shear rate results? (Y/N)'
      READ(NTER,'(A1)')ANS
      IF(ANS.EQ.'Y'.OR.ANS.EQ.'y')THEN
        WRITE(NTER,190)
        WRITE(NOUT,190)
190     FORMAT(/T5,'Viscosity approximations:'//T28,'Shear rate'/
     &         T16,33('-'),/T18,'3/4ths',T28,'Power-law',T53,'Apparent'
     &         /T7,'Time',T19,'Rule',T29,'Function',T41,'Average',T52,
     &          'Viscosity'/5(3X,9('-')))
        DO 210 I=1,NP
          RMIN=MIN(R,SQRT(VOL/PI/H(I)))
          SRN(I)=3.D0*RMIN*(-HDOT(I))/H(I)**2
C Estimate and verify power-law slope.
          NX=11
          IP=I
          IF(I.LT.6)IP=6
          IF(IP.GT.NP-5)IP=NP-5
          DO 200 J=1,NMAX2
            U(J)=LOG(H(IP-6+J))
            V(J)=LOG(-HDOT(IP-6+J))
200       CONTINUE
          CALL ALINR(NX,U,V,A)
          XN=1./(A(2)-2.)
          IF(XN.LE..17)XN=.17
          IF(XN.GE..999)XN=.999
          SRPL(I)=FAC(XN)*SRN(I)
          VIS(I)=2.D0*H(I)**3*F(I)/3.D0/PI/RMIN**4/(-HDOT(I))
          WRITE(NOUT,260)T(I),.75*SRN(I),SRPL(I),.5*SRN(I),VIS(I)
          IF(I/(NP/6)*(NP/6).EQ.I)
     &      WRITE(NTER,260)T(I),.75*SRN(I),SRPL(I),.5*SRN(I),VIS(I)
210     CONTINUE
        WRITE(NOUT,220)
220     FORMAT(/3X,'The corresponding log values of the preceding',
     &          1X,'data:'/)
        WRITE(NOUT,'(5F12.3)')(LOG10(T(I)),LOG10(SRN(I))-.1249,
     &         LOG10(SRPL(I)),LOG10(SRN(I))-.301,LOG10(VIS(I)),I=1,NP)
      ENDIF
C If data are appropriate, estimate third-order fluid parameters.
      WRITE(NTER,'(/2X,A)')'Require third-order fluid analysis? (Y/N)'
      READ(NTER,'(A1)')ANS
      IF(ANS.EQ.'Y'.OR.ANS.EQ.'y')THEN
        DO 230 I=1,NP
          RMIN=MIN(R,SQRT(VOL/PI/H(I)))
          X(I)=SRN(I)**2
          Y(I)=2.D0*H(I)**3*F(I)/3.D0/PI/RMIN**4/(-HDOT(I))
230     CONTINUE
        CALL ALINR(NP,X,Y,A)
        IF(A(2).GE.0..OR.A(1).LE.0.)THEN
          WRITE(NTER,'(/2X,A)')
     &        'There is a problem with the third-order fluid analysis.'
        ELSE
          ALFA=SQRT(-2.5D0*A(2)/A(1))
          ERRA=ALFA*(A(9)/A(1)-A(10)/A(2))/2.D0
          WRITE(NTER,240)A(1),A(9)*T95(NP-2),ALFA,ERRA*T95(NP-2)
          WRITE(NOUT,240)A(1),A(9)*T95(NP-2),ALFA,ERRA*T95(NP-2)
240       FORMAT(/T4,'Third-order fluid results:'//5X,
     &        'Zero-shear-rate viscosity =',1P,E11.3,' +/-',E11.3/
     &        17X,'Time constant =',E11.3,' +/-',E11.3)
        ENDIF
      ENDIF
```

```
      WRITE(NTER,'(/2X,A)')
     &     'Include corrected squeeze-flow data in output file? (Y/N)'
      READ(NTER,'(A1)')ANS
      IF(ANS.EQ.'Y'.OR.ANS.EQ.'y')THEN
        WRITE(NOUT,250)
250     FORMAT(/T2,'Corrected squeeze-flow data:'//T53,'Apparent',T65,
     &         'Apparent'/T6,'Time',T17,'Gap (H)',T29,'-dH/dt',T42,
     &         'Force',T55,'Rate',T66,'Stress'/6(3X,9('-')))
        DO 270 I-1,NP
          RMIN=MIN(R,SQRT(VOL/PI/H(I)))
          Y(I)=2.D0*H(I)*F(I)/PI/RMIN**3
          WRITE(NOUT,260)T(I),H(I),-HDOT(I),F(I),SRN(I),Y(I)
260       FORMAT(1P,6E12.3)
270     CONTINUE
        WRITE(NOUT,220)
        WRITE(NOUT,'(6F12.3)')(LOG10(T(I)),LOG10(H(I)),LOG10(-HDOT(I)),
     &                LOG10(F(I)),LOG10(SRN(I)),LOG10(Y(I)),I=1,NP)
      ENDIF
      IF(IMODE.EQ.1.AND.IGEOM.EQ.1)THEN
        WRITE(NTER,'(/2X,A)')
     &       'Create file for viscoelastic analysis? (Y/N)'
        READ(NTER,'(A1)')ANS
        IF(ANS.EQ.'Y'.OR.ANS.EQ.'y')THEN
          NRF=NOF+1
          CALL RESFIL(NTER,NRF)
          WRITE(NRF,280)TITLE
280       FORMAT(1X,A/ '1.   3   0'/'0.   1.  0.')
          AREA=.75*PI*R**2*(R/H0)**2
          FT=(T(NP)/T(1))**0.1
          J=0
          DO 290 I=1,NP
            IF(T(I).GE.T(1)*FT**J)THEN
              WRITE(NRF,'(3E11.3)')T(I),H0/H(I),F(I)/AREA
              J=J+1
            ENDIF
290       CONTINUE
          CLOSE (UNIT=NRF)
        ENDIF
      ENDIF
      CALL MSTAIL(*10)
      STOP
      END
C
      SUBROUTINE FOPEN(NTER,NUNIT)
C Handles opening of auxiliary file for a program compiled using
C  Microsoft FORTRAN.
      LOGICAL FEXIST
      CHARACTER FDAT*33
      WRITE(NTER,'(2X,A)')'Enter filename below.'
10    READ(NTER,'(A33)')FDAT
      INQUIRE(FILE=FDAT,EXIST=FEXIST)
      IF(FEXIST)THEN
        OPEN(UNIT=NUNIT,FILE=FDAT,STATUS='OLD')
        RETURN
      ELSE
        WRITE(NTER,'(2X,A,A33)')'Cannot find ',FDAT
        WRITE(NTER,'(2X,A)')'Re-enter name of data file.'
        GOTO 10
      ENDIF
      END
C
      DOUBLE PRECISION FUNCTION FORCE(T,FSTD,NCAL,GCAL,TICAL)
C Interpolates force according to known actuator response in GCAL.
      IMPLICIT DOUBLE PRECISION (A-H,O-Z)
      PARAMETER(TINY=1.E-4)
      DIMENSION GCAL(NCAL)
      I=INT(T/TICAL+TINY)+1
      RANGE=GCAL(1)-GCAL(NCAL)
```

```
      IF(I.GE.NCAL-1)THEN
        FORCE=FSTD
      ELSE
C Interpolate linearly.
        FORCE=FSTD*(GCAL(1)-(GCAL(I)+(GCAL(I+1)-GCAL(I))*
     &         (T-(I-1)*TICAL)/TICAL))/RANGE
      ENDIF
      RETURN
      END
C
      SUBROUTINE RESFIL(NTER,NRF)
C Handles preparation for results file to be run with FITSTR.
      LOGICAL REXIST,ROPEN
      CHARACTER ANS*1,CROW*80,CSTAT*3,RRES*33
10    WRITE(NTER,*)' Enter name for results file.'
      READ(NTER,'(A33)')RRES
      INQUIRE(FILE=RRES,EXIST=REXIST,OPENED=ROPEN)
      IF(ROPEN)THEN
        WRITE(NTER,*)
     &      ' The file ',RRES(1:INDEX(RRES,' ')),' is already attached.'
        WRITE(NTER,*)' Retry.'
        GOTO 10
      ELSE
        IF(REXIST)THEN
          CSTAT='OLD'
        ELSE
          CSTAT='NEW'
        ENDIF
        WRITE(NTER,*)' Results to be written to ',CSTAT,' file: ',RRES
        WRITE(NTER,*)' OK (Y/N)?'
        READ(NTER,'(A1)')ANS
        IF(ANS.EQ.'Y'.OR.ANS.EQ.'y')THEN
          OPEN(NRF,FILE=RRES,STATUS=CSTAT)
          IF(CSTAT.EQ.'OLD')THEN
            DO 20 I=1,100000
              READ(NRF,'(A80)',END=30)CROW
20          CONTINUE
30          BACKSPACE NRF
          ENDIF
          RETURN
        ENDIF
      ENDIF
      END
C
      SUBROUTINE SLOPEV(NP,X,Y,D)
C Returns slope for evenly spaced, ordered points using 5-point formulaes,
C  second-order formulae.  NP = number of X,Y pairs; D = derivative set.
      IMPLICIT DOUBLE PRECISION (A-H,O-Z)
      DIMENSION X(NP),Y(NP),D(NP)
      H= X(2)-X(1)
      HD=H*20.
      D(1)=(-21.*Y(1)+13.*Y(2)+17.*Y(3)-9.*Y(4))/HD
      D(NP)=-(-21.*Y(NP)+13.*Y(NP-1)+17.*Y(NP-2)-9.*Y(NP-3))/HD
      D(2)=(-11.*Y(1)+3.*Y(2)+7.*Y(3)+Y(4))/HD
      D(NP-1)=-(-11.*Y(NP)+3.*Y(NP-1)+7.*Y(NP-2)+Y(NP-3))/HD
      HC=H*10.
      DO 10 I=3,NP-2
        D(I)=(-2.*Y(I-2)-Y(I-1)+Y(I+1)+2.*Y(I+2))/HC
10    CONTINUE
      RETURN
      END
```

6.6 Sample Listings of INPUT/OUTPUT Files for SQUEEZ

6.6.1 Data File for SQUEEZ

Shown here is a data file, from a constant-force squeeze-flow experiment, in the format required for running SQUEEZ.

```
LDPE, Mw = 530000, T = 150 C; 0.25 kN data digitized from Fig. 3, Ref. 7.
0.000  0.000  0.000
0.001  0.001  0.000
0.000 -0.0002 100.0
250. 0.025 1.0 1.0 32. 0.004
 2.406E-03
 1.909E-03
 1.649E-03
 1.491E-03
 1.375E-03
 1.288E-03
 1.220E-03
 1.160E-03
 1.112E-03
 1.071E-03
 1.037E-03
 1.003E-03
 9.772E-04
 9.526E-04
```

6.6.2 Auxiliary Data File for SQUEEZ

Shown here are typical force-buildup data in the required format when processing data taken from constant-force squeeze-flow experiments.

```
 0.01
0.0030
0.0025
0.0020
0.0018
0.0017
0.00168
0.00165
0.00165
```

6.6.3 Terminal Dialog with SQUEEZ

Reproduced here is the terminal dialog for running SQUEEZ with the data file SQUEEZ1.DAT is listed in Section 6.6.1. It was assumed that the actuator reached its full force instantly such that the auxiliary file SQUEEZ2.DAT is set up with a value of 0 (time interval) as the first, and only, record.

```
>SQUEEZ
 Enter name of data file (N, if unavailable)
SQUEEZ1.DAT
 Results to be written to NEW file: SQUEEZ1.RES
 OK? (Y/N)
Y

 Actuator rise time calibration.
 Requires no-sample gap data file.
 Enter filename below.
SQUEEZ2.DAT

 Estimate viscosity-shear rate results? (Y/N)
Y

    Viscosity approximations:

                        Shear rate
             ----------------------------------
              3/4ths      Power-law                 Apparent
    Time       Rule        Function     Average     Viscosity
  ---------  ---------   ---------   ---------   ---------
  6.400E+01  9.652E-02   7.518E-02   6.435E-02   1.907E+05
  1.280E+02  6.281E-02   4.892E-02   4.187E-02   2.422E+05
  1.920E+02  4.478E-02   3.488E-02   2.986E-02   3.050E+05
  2.560E+02  3.443E-02   2.732E-02   2.295E-02   3.684E+05
  3.200E+02  2.779E-02   2.216E-02   1.853E-02   4.318E+05
  3.840E+02  2.322E-02   1.852E-02   1.548E-02   4.944E+05
  4.480E+02  1.961E-02   1.564E-02   1.308E-02   5.658E+05

 Require third-order fluid analysis? (Y/N)
Y

   Third-order fluid results:

     Zero-shear-rate viscosity =  4.270E+05 +/-  6.410E+04
                 Time constant =  8.330E+00 +/-  2.892E+00

 Include corrected squeeze-flow data in output file? (Y/N)
Y

 The results are in SQUEEZ1.RES
 Run another set of data? (Y/N)
N
Stop - Program terminated.
```

6.6.4 Output File from **SQUEEZ**

Reproduced here is the output file from SQUEEZ using the data and terminal dialog listed in Sections 6.6.1 and 6.6.3, respectively.

```
                     Squeeze-Flow Analysis
                     ---------------------

LDPE, Mw = 530000, T = 150 C; 0.25 kN data digitized from Fig. 3, Ref. 7.


  Frame compliance: 2.000E-06
   Gap calibration:
        Intercept = 0.000E+00
            Slope = 1.000E+00
```

Viscosity approximations:

Time	Shear rate: 3/4ths Rule	Shear rate: Power-law Function	Shear rate: Average	Apparent Viscosity
3.200E+01	1.297E-01	1.010E-01	8.646E-02	1.712E+05
6.400E+01	9.652E-02	7.518E-02	6.435E-02	1.907E+05
9.600E+01	7.659E-02	5.965E-02	5.106E-02	2.144E+05
1.280E+02	6.281E-02	4.892E-02	4.187E-02	2.422E+05
1.600E+02	5.243E-02	4.084E-02	3.495E-02	2.732E+05
1.920E+02	4.478E-02	3.488E-02	2.986E-02	3.050E+05
2.240E+02	3.902E-02	3.073E-02	2.601E-02	3.367E+05
2.560E+02	3.443E-02	2.732E-02	2.295E-02	3.684E+05
2.880E+02	3.078E-02	2.454E-02	2.052E-02	4.001E+05
3.200E+02	2.779E-02	2.216E-02	1.853E-02	4.318E+05
3.520E+02	2.535E-02	2.022E-02	1.690E-02	4.632E+05
3.840E+02	2.322E-02	1.852E-02	1.548E-02	4.944E+05
4.160E+02	2.138E-02	1.705E-02	1.425E-02	5.278E+05
4.480E+02	1.961E-02	1.564E-02	1.308E-02	5.658E+05

The corresponding log values of the preceding data:

1.505	-.887	-.996	-1.063	5.233
1.806	-1.015	-1.124	-1.191	5.280
1.982	-1.116	-1.224	-1.292	5.331
2.107	-1.202	-1.311	-1.378	5.384
2.204	-1.280	-1.389	-1.456	5.436
2.283	-1.349	-1.457	-1.525	5.484
2.350	-1.409	-1.512	-1.585	5.527
2.408	-1.463	-1.564	-1.639	5.566
2.459	-1.512	-1.610	-1.688	5.602
2.505	-1.556	-1.654	-1.732	5.635
2.547	-1.596	-1.694	-1.772	5.666
2.584	-1.634	-1.732	-1.810	5.694
2.619	-1.670	-1.768	-1.846	5.723
2.651	-1.707	-1.806	-1.884	5.753

Third-order fluid results:

```
Zero-shear-rate viscosity =  4.270E+05 +/-  6.410E+04
            Time constant =  8.330E+00 +/-  2.892E+00
```

Corrected squeeze-flow data:

Time	Gap (H)	-dH/dT	Force	Apparent Rate	Apparent Stress
3.200E+01	2.906E-03	1.947E-05	2.500E+02	1.729E-01	2.960E+04
6.400E+01	2.409E-03	9.958E-06	2.500E+02	1.287E-01	2.454E+04
9.600E+01	2.149E-03	6.288E-06	2.500E+02	1.021E-01	2.189E+04
1.280E+02	1.991E-03	4.426E-06	2.500E+02	8.374E-02	2.028E+04
1.600E+02	1.875E-03	3.277E-06	2.500E+02	6.991E-02	1.910E+04
1.920E+02	1.788E-03	2.545E-06	2.500E+02	5.971E-02	1.821E+04
2.240E+02	1.720E-03	2.052E-06	2.500E+02	5.203E-02	1.752E+04
2.560E+02	1.660E-03	1.687E-06	2.500E+02	4.590E-02	1.691E+04
2.880E+02	1.612E-03	1.422E-06	2.500E+02	4.104E-02	1.642E+04
3.200E+02	1.571E-03	1.219E-06	2.500E+02	3.706E-02	1.600E+04
3.520E+02	1.537E-03	1.065E-06	2.500E+02	3.380E-02	1.566E+04
3.840E+02	1.503E-03	9.326E-07	2.500E+02	3.096E-02	1.531E+04
4.160E+02	1.477E-03	8.294E-07	2.500E+02	2.851E-02	1.505E+04
4.480E+02	1.453E-03	7.357E-07	2.500E+02	2.615E-02	1.480E+04

The corresponding log values of the preceding data:

```
1.505    -2.537   -4.711   2.398    -.762   4.471
1.806    -2.618   -5.002   2.398    -.890   4.390
1.982    -2.668   -5.201   2.398    -.991   4.340
2.107    -2.701   -5.354   2.398   -1.077   4.307
2.204    -2.727   -5.485   2.398   -1.155   4.281
2.283    -2.748   -5.594   2.398   -1.224   4.260
2.350    -2.764   -5.688   2.398   -1.284   4.244
2.408    -2.780   -5.773   2.398   -1.338   4.228
2.459    -2.793   -5.847   2.398   -1.387   4.215
2.505    -2.804   -5.914   2.398   -1.431   4.204
2.547    -2.813   -5.973   2.398   -1.471   4.195
2.584    -2.823   -6.030   2.398   -1.509   4.185
2.619    -2.831   -6.081   2.398   -1.545   4.177
2.651    -2.838   -6.133   2.398   -1.583   4.170
```

6.6.5 Data File Prepared by SQUEEZ for Use with FITSTR

For the data listed here, the relaxation modulus was derived from the Chapter 10 program FITSTR by using the Kohlrausch function, and is shown in Figure 6.8.

```
  Silicone gum (GE SE-30, Mw=500000)
 1.   3   0
 0.   1.  0.
      .950E+00    .101E+01    .188E+03
      .190E+01    .102E+01    .229E+03
      .285E+01    .103E+01    .255E+03
      .380E+01    .104E+01    .281E+03
      .665E+01    .107E+01    .326E+03
      .950E+01    .110E+01    .367E+03
      .152E+02    .118E+01    .446E+03
      .238E+02    .133E+01    .575E+03
      .380E+02    .165E+01    .959E+03
      .598E+02    .248E+01    .232E+04
```

References

1. Tanner, R. I.: *Engineering Rheology.* Clarendon Press, Oxford, 1985.
2. Bird, R. B.; Armstrong, R. C.; Hassager, O.: *Dynamics of Polymeric Liquids.* Wiley-Interscience, New York, 1987, Vol. 1, pp. 539-547 (viscoelastic fluid), pp. 20-22 (Newtonian fluid), pp. 189-192 (power-law fluid).
3. Bagley, E. B.; Christianson, D. D.; Wolf, W. J.: "Frictional Effects in Compressional Deformation of Gelatin and Starch Gels and Comparison of Material Response in Simple Shear, Torsion, and Lubricated Uniaxial Compression," *J. Rheol.* **29**, 103-108 (1985).
4. Chatraei, SH.; Macosko, C. W.; Winter, H. H.: "Lubricated Squeezing Flow: A New Biaxial Extensional Rheometer," *J. Rheol.* **25**, 433-443 (1981).
5. Winter, H. H.; Holly, E.: "Shear and Biaxial Extension Rheology of Polyethylene – Comparison of LLDPE and LDPE," *SPE Tech. Papers* **31**, 613-615 (1985).

6. Rosich, M.; Allen, P.: "Analyzing Variations in SMC Formulations," *Plastics Eng.* **47**(4), 29-30 (1991).
7. Laun, H. M.: "Rheometry towards Complex Flows: Squeeze Flow Technique," *Makromol. Chem., Macromol. Symp.* **56**, 55-66 (1992).
8. Shaw, M. T.; Giovanetti, R. C.: "Analysis of an Ellis Fluid by Squeezing Flow," *SPE Tech. Papers* **27**, 190-192 (1981).
9. Winther, G.; Almdal, K.; Kramer, O.: "Determination of Polymer Melt Viscosity by Squeezing Flow with Constant Plate Velocity," *J. Non-Newtonian Fluid Mech.* **39**, 119-136 (1991).
10. Stefan, J.: "Versuche über die scheinbare Adhäsion," *Sitzungsber. K. Akad. Wiss. Math. Natur. Wien* **69**, (Part 2) 713-735 (1874).
11. Co, A.; Bird, R. B.: "Analysis of Squeezing Flow between Parallel Disks Using the Goddard-Miller Model," *Rheology Research Center Report No. 31,* University of Wisconsin, 1974.
12. Press, W. H.; Flannery, B. P.; Teukolsky, S. A.; Vetterling, W. T.: *Numerical Recipes in* FORTRAN: *The Art of Scientific Computing* Cambridge University Press, Cambridge, 1986.
13. Abramowitz, M.; Stegun, I. A. (eds.): *Handbook of Mathematical Functions with Formulas, Graphs and Mathematical Tables* Dover, New York, 1965, p. 921.
14. Tichy, J. A.; Winer, W. O.: "An Investigation into the Influence of Fluid Viscoelasticity in a Squeeze Film Bearing," *J. Lubr. Technol., Trans. ASME* **100**, 56-64 (1978).
15. Leider, P. J.; Bird, R. B.: "Squeezing Flow Between Parallel Disks. I. Theoretical Analysis," *Ind. Eng. Chem. Fundam.* **13**, 336-341 (1974).
16. Leider, P. J.: "Squeezing Flow between Parallel Disks. II. Experimental Results," *Ind. Eng. Chem. Fundam.* **13**, 342-346 (1974).
17. Grimm, R. J.: "Squeezing Flows of Polymeric Liquids," *AIChE J.* **24**, 427-439 (1978).
18. Brindley, G.; Davies, J. M.; Walters, K.: "Elastico-Viscous Squeeze Films. Part I," *J. Non-Newtonian Fluid Mech.* **1**, 19-37 (1976).
19. Shirodkar P.: Middleman, S.: "Lubrication Flows in Viscoelastic Liquids. I. Squeezing Flow between Approaching Parallel Rigid Planes," *J. Rheol.* **26**, 1-17 (1982).
20. Isayev, A. I.; Azari, A. D.: "Viscoelastic Effect in Compression Molding of Elastomers: Shear-Free Squeezing Flow," *Rubber Chem. Technol.* **59**, 868-882 (1986).
21. Kramer, J. M.: "Large Deformations of Viscoelastic Squeeze Films," *Appl. Sci. Res.* **30**, 1-16 (1974).
22. Phan-Thien, N.; Dudek, J.; Boger, D. V.; Tirtaatmadja, V.: "Squeeze Film Flow of Ideal Elastic Liquids," *J. Non-Newtonian Fluid Mech.* **18**, 227-254 (1985).

23. Phan-Thien, N.; Tanner, R. I.: "Journal Lubrication and Normal Stress Measurement," *J. Non-Newtonian Fluid Mech.* **9**, 107-117 (1981).
24. *Rheometrics System Four Operations Manual.* Rheometrics, Piscataway, New Jersey.
25. Schwarzl, F. R.: "Numerical Calculation of Stress Relaxation Modulus from Dynamic Data for Linear Viscoelastic Materials," *Rheol. Acta* **14**, 581-590 (1975).

Chapter 7

Parallel-Plate Torsional Flow

7.1 Introduction

With most commercial rotational instruments, the choices of fixtures are often limited to the cone and plate, the Couette (cup and bob), and the parallel plates. Of these fixtures, the Couette geometry is particularly difficult to load with high-viscosity fluids; thus, the cone and plate and the parallel plates are the most popular fixtures for steady, transient and dynamic measurements on melts.

Many experimentalists avoid the parallel-plate geometry because of the inhomogeneity of the flow, although the popular capillary geometry has the same problem, plus a few more (see Chapter 8). The parallel-plate fixtures, defined in Figure 7.1, do offer important advantages over the cone and plate, some of which are the following:

- The distances between plates can be brought to very small gaps. A small gap suppresses secondary flows and reduces inertial corrections and thermal effects by providing excellent heat transfer. These factors combine to allow the geometry to be used at higher shear rates.
- While the axial thrust on the parallel plates is proportional to the difference between the first and second normal stress functions (N_1 and N_2, respectively), rather than just N_1 as in the cone and plate, $N_1 - N_2$ can be used quite effectively by continuum and molecular theorists. It is just as easy to match a constitutive equation to $N_1 - N_2$ as it is to N_1 (see Chapter 10). Although it always is advantageous to have two independent normal stress functions, this requires pressure-profile measurements in either the cone-and-plate or the parallel-plate geometries.
- The parallel plates, being flat and parallel, have advantages for mounting optics and applying electrical and magnetic fields.
- For some studies in which shear rate is an important independent variable, the rate gradient in the radial direction may be an advantage because the effect of rate can be examined in a single specimen [1,2].
- For filled systems, the gap between the plates may be adjusted relative to filler size [3-5].
- The precision of a flat surface is much easier to check than a conical surface.
- For very viscous samples, it is convenient to prepare blanks of the material to load into the gap. Disc-shaped blanks can be made simply and with high accuracy.

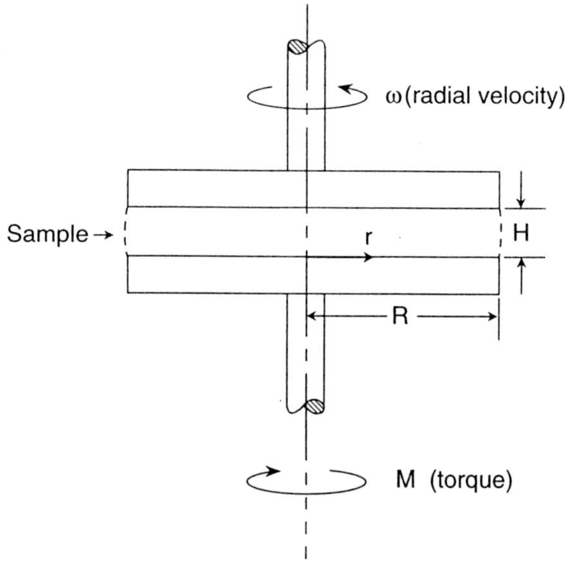

Figure 7.1 Schematic depiction of the parallel-plate geometry.

- Surface and edge effects can be probed systematically by varying the gap and radius.
- The flat surfaces of the parallel plates are much easier to clean, a small point perhaps with most melts but very important with some sticky formulations!

Having sold the parallel-plate geometry, what about the problem of inhomogeneous flow? Although the flow is viscometric, the shear rate varies linearly with radius, thus

$$\dot{\gamma} = \omega \frac{r}{H} \tag{7.1}$$

where $\dot{\gamma}$ is the shear rate, ω is the angular velocity, r is the radius, and H is the gap. This means that information about the stresses produced at each radius, and contributing to the observed torque, can be probed by merely slowing the rotation rate. The result is an equation that looks much like the Rabinowitsch correction for capillary flows (see Equation 8.3) and has no assumptions beyond the usual ones of no slip and viscometric flow. The equation is expressed as

$$\eta(\dot{\gamma}_R) = \frac{2M}{\pi R^3 \dot{\gamma}_R}\left[\frac{3}{4} + \frac{1}{4}\frac{d\ln M}{d\ln \dot{\gamma}_R}\right] \tag{7.2}$$

where η is the viscosity, M is the torque, and $\dot{\gamma}_R$ is the shear rate at the plates' outside radius R. (The plates are always assumed to be entirely filled.)

This relationship presents few problems except that the derivative tends to reduce the shear-rate range over which reliable results can be obtained. One would like the derivatives to be a minor source of error; unfortunately, this is rarely true. Worse, the errors introduced by differentiation can be systematic in nature such that the curvature of the viscosity function is in error. This type of error could have serious consequences for the determination of characteristic times, for example. The situation for transient flows is more complex, as will be discussed.

As for $N_1 - N_2$, the corrections for nonuniformity result in a similar correction, namely

$$N_1(\dot{\gamma}_R) - N_2(\dot{\gamma}_R) = \frac{4F}{\pi R^2}\left[\frac{1}{2} + \frac{1}{2}\frac{d\ln F}{d\ln \dot{\gamma}_R^2}\right] \tag{7.3}$$

where F is the axial force needed to keep the plates at the original gap. Both Equations 7.2 and 7.3 are written in a form that brings the correction term to unity as the shear rate goes to zero.

Particularly important for characterizing viscoelastic melts are transient data. The relevant quantities are the shear stress growth function, $\sigma_{21}^+(t, \dot{\gamma})$, and the difference between the first and second normal stress growth functions, $N_1^+(t, \dot{\gamma}) - N_2^+(t, \dot{\gamma})$, as the melt is sheared from rest. However, our discussion has not covered transient measurements done with parallel plates. What is the story for such flows?

7.1.1 Transient Flows

The situation for transient flows in the parallel plates has been analyzed by Soskey and Winter [6] and Ganani and Powell [7], among others. The same corrections as those governed by Equation 7.2 are performed. For example, the relaxation modulus becomes

$$G(t, \gamma_R) = \frac{2M}{\pi R^3 \gamma_R}\left[\frac{3}{4} + \frac{1}{4}\frac{d\ln M}{d\ln \gamma_R}\right] \tag{7.4}$$

where γ_R is the strain at the plate rim ($\gamma_R = \omega t R/H$ and $\gamma = \gamma_R\, r/R$). The transient shear stresses are most conveniently written as

$$\sigma_{21}^+(t, \dot{\gamma}_R) = \frac{2M}{\pi R^3}\left[\frac{3}{4} + \frac{1}{4}\frac{d\ln M(t, \dot{\gamma}_R)}{d\ln \dot{\gamma}_R}\right] \tag{7.5}$$

The derivative in Equation 7.5 must be taken at both the time, t, and rate, $\dot{\gamma}_R$, corresponding to stress being found. Unfortunately, the logarithmic plots of M versus $\dot{\gamma}_R$ are not linear at high strains (see Figure 7.5).

One gains the impression that the analysis of flows between parallel plates is unavoidably connected with finding derivatives. Or is it?

7.1.2 A Derivative-Free Approximation

A perceptive approach to torsional flows in parallel plates was presented by Cross and Kaye in 1987 [8]. Their argument results in a very simple correction scheme. To obtain the true or real viscosity, determine the apparent viscosity directly from the expression

$$\eta_A = \frac{2M}{\pi R^3 \dot{\gamma}_R} \tag{7.6}$$

The apparent viscosity is then assigned to a different shear rate, less than that at the rim ($\dot{\gamma}_R$). The authors showed that this shear rate is simply $3\dot{\gamma}_R/4$.

Does this approach work? Cross and Kaye did show that the errors were small for some fluids, but they did not show how well it works for fluids that start out Newtonian and become power-law at high shear rates. The results are still very good. To demonstrate, we assumed a realistic viscosity function: a summation of two Cross models [9,10] which gives the curve a little texture. Using this analytical function, we calculated the torque using the fundamental force balance and a high-accuracy integration of stress over radius with the help of program STEADY described later in this chapter. The apparent viscosity was then calculated from Equation 7.6.

The comparison of the real and apparent viscosities is shown in Figure 7.2, which demonstrates that it would wrong (e.g., 18%) to assume they were equivalent. Importantly, however, the texture is not lost by integrating over the range of shear rates found in the parallel-plate geometry; this is emphasized in the Figure 7.2 inset where the the slopes of the viscosity curves are compared.

We applied the Cross-Kaye approximation by shifting the shear rate associated with the apparent viscosity by 3/4 (or left by 0.1249 decade). The match was excellent. Figure 7.3 shows a comparison of the two curves; the solid curve is the assumed viscosity function, and the dashed curve is the apparent viscosity shifted to the left by 0.1249 decade. The error is shown on the same drawing.

The results of Ganani and Powell suggest that transient torques, at least in cases where the stress overshoot is large, can give correction terms for Equation 7.5 which are nonconstant and even nonmonotonic with time. How does influence the Cross-Kaye approximation? In their results, Ganani and Powell showed that use of the

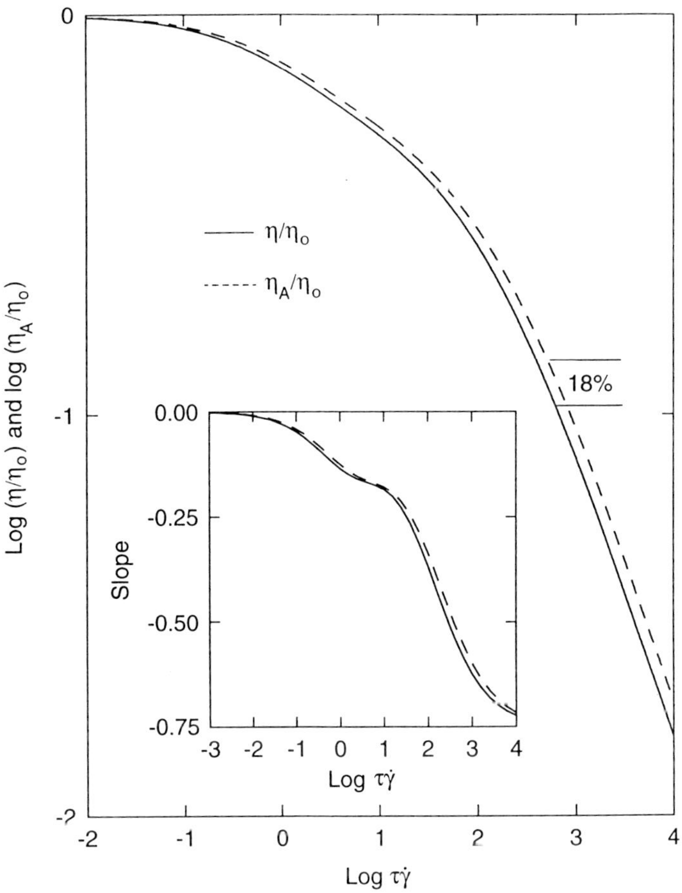

Figure 7.2 Illustration of the expected torque output (plotted as $(2M/\pi R^3)/\dot{\gamma}_R = \eta_A$) for a non-Newtonian fluid with a slight inflection. As the inset illustrates, the inflection is not lost.

apparent stress (proportional to M) leads to less overshoot than if the real stress were calculated. However, the Cross-Kaye approximation would assign this stress to a lower shear rate. Is the correction enough to account for the difference? As we show, it does a reasonably good job.

Our way of appreciating the *magic* of 3/4 in this regard is to examine the integral

$$M(t) = \frac{2\pi R^3}{\dot{\gamma}_R^3} \int_0^{\dot{\gamma}_R} \sigma_{21}^{+}(t, \dot{\gamma})\, \dot{\gamma}^2\, d\dot{\gamma} \tag{7.7}$$

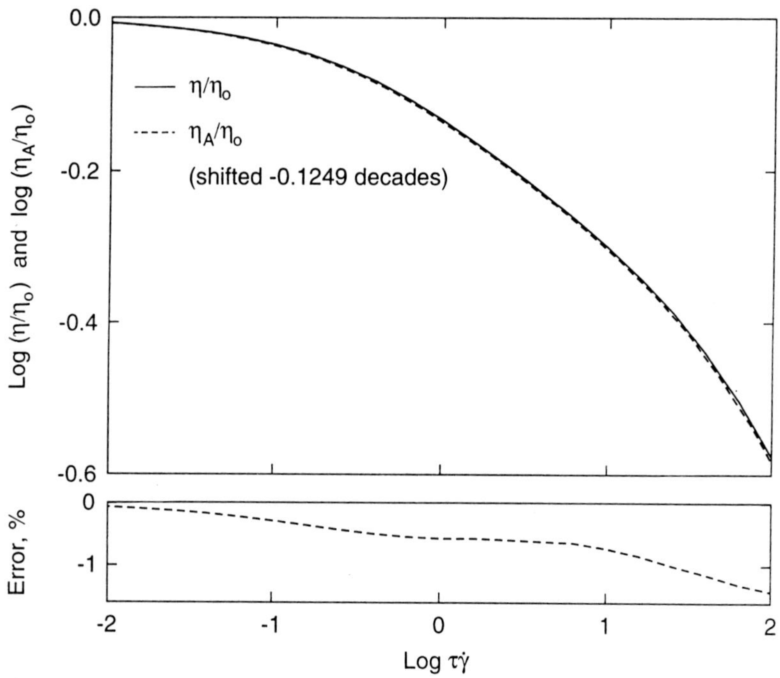

Figure 7.3 Application of the Cross-Kaye approximation to data shown in Figure 7.2. The "error" scale, the percent difference between the true and corrected viscosities, is plotted in the inset at the bottom.

in terms of numerical integration. First, the integral is rewritten by substituting $x = \dot{\gamma}/\dot{\gamma}_R$:

$$M(t) \;=\; 2\pi R^3 \int_0^1 \sigma_{21}^+(t, \dot{\gamma}_R x)\, x^2\, dx \tag{7.8}$$

If the flow is in the linear regime, the stress is proportional to rate, $\sigma_{21}^+(t, \dot{\gamma}_R x) \cong x\sigma_{21}^+(t, \dot{\gamma}_R)$, and the integral can be rewritten as

$$M(t) \;=\; 2\pi R^3 \int_0^1 \frac{\sigma_{21}^+(t, \dot{\gamma}_R x)}{x}\, x^3\, dx \;\cong\; 2\pi R^3 \int_0^1 \sigma_{21}^+(t, \dot{\gamma}_R)\, x^3\, dx \tag{7.9}$$

The $\sigma_{21}^+(t, \dot{\gamma}_R)$ term now comes out and the integral evaluates to 1/4, giving the expression

$$\sigma_{21}^{+}(t, \dot{\gamma}_R) \cong \frac{2M(t)}{\pi R^3} \tag{7.10}$$

On the other hand if the relationship between σ_{21}^{+} and $\dot{\gamma}_R$ is nonlinear, then something else must be done [11]. In particular, we can regard Equation 7.8 as a second-moment integration. The first-order approximation to this moment integration [12] is

$$M(t) \cong 2\pi R^3\left[\frac{1}{3}\,\sigma_{21}^{+}\left(t, \frac{3}{4}\dot{\gamma}_R\right)\right] \tag{7.11}$$

or

$$\sigma_{21}^{+}\left(t, \frac{3}{4}\dot{\gamma}_R\right) \cong \frac{3M(t)}{2\pi R^3} \tag{7.12}$$

Simulation of the stress at various radii during start-up is illustrated in Figure 7.4 for a realistic nonlinear constitutive equation. The curvature is very mild, so it is no surprise that the first-order Gaussian integration of Equation 7.8 provides an excellent estimate. Thus, the *magic* of 3/4 in the Cross-Kaye approximation is equivalent to that of the Gaussian quadrature: pick the abscissa values wisely and the integration of curved functions can be accomplished with very few terms.

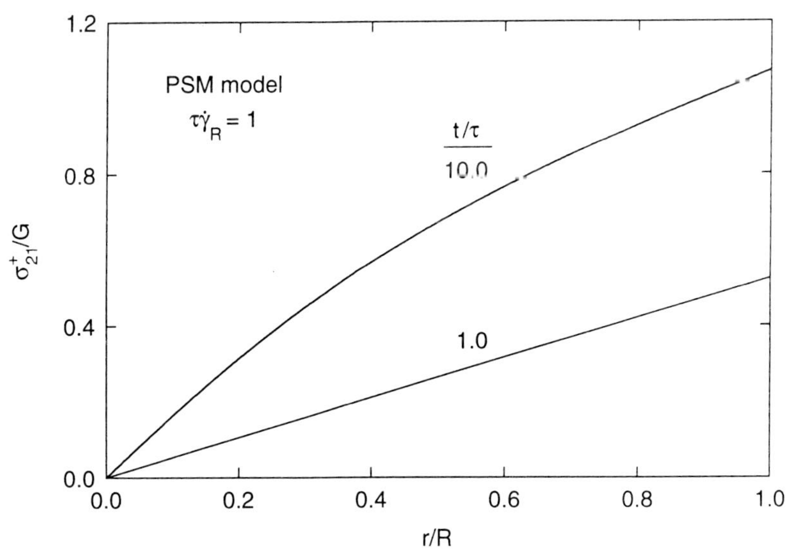

Figure 7.4 Illustration of the curvature of the stress in the radial direction during startup for a transient flow of a nonlinear viscoelastic fluid.

If the isochronic stress-strain relationship is highly nonlinear, things could get more difficult and one is forced back to derivatives or full double integration. In `TRANST` we offer a compromise – up to fifth-order Gaussian moment integration over radius – to fit the apparent stress. In Section 7.4, we will explore in more detail the error involved with this approximation.

What this all implies is that the parallel-plate geometry for shear-sensitive fluids does not provide sufficient information on what is happening at the outer radius. Finding $\sigma_{21}(\dot{\gamma}_R)$ always requires some approximation which may be accomplished by assuming a particular flow model for the data or by calculating derivatives up to the endpoint of a set of data and using Equation 7.2 (see Section 7.1.3). The other choice is to give up 25% of the shear-rate range by using the Cross-Kaye approximation. While 25% sounds large, it represents only about an eighth of a decade. In general, one cannot correspondingly increase the shear rate by 25% to recover the range because of the onset of elastic or inertial stabilities. The range of shear rates depicted in Figures 7.2 and 7.3 is included for completeness; it is not our intention to imply that torsional flow can be used for this entire range.

While Equation 7.5 holds out the promise of getting information about the material at the outer radius, it is a hollow promise because the derivative must be evaluated at the outer radius. No problem? Note that the plots of ln M versus ln $\dot{\gamma}_R$ in Figure 7.5 are not at all linear. To know what these curves are going to do at higher shear rates one needs data at higher shear rates – and so it goes.

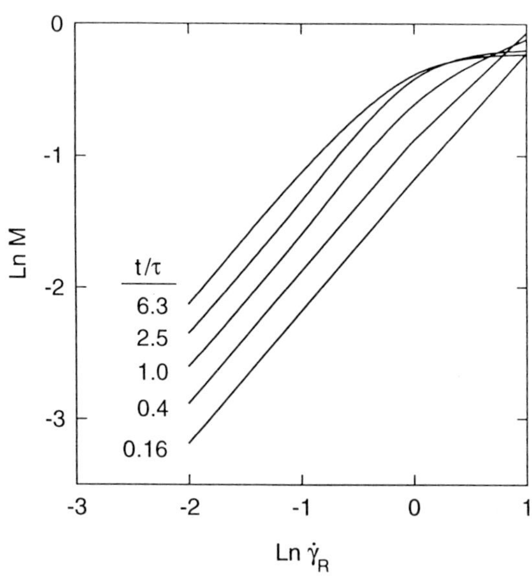

Figure 7.5 Plot required to apply traditional correction for curvature in the stress profile. The slope is needed at each time and at each shear rate ($\dot{\gamma}_R$).

Here, two choices are offered: a derivative-finding routine for steady flows and settle for three-fourths the shear rate for transient flows. However, we have a twist (!) for the latter; the procedure goes like this:

- Take the transient torque data at fixed angular velocity.
- Assign the values $3M/2\pi R^3$ to the transient shear stress at $3\dot{\gamma}_R/4$.
- Take these values and model them with a nonlinear constitutive equation, using the program `FITSTR` in Chapter 10.
- Predict what the torque response will look like, using the program `TRANST` in this chapter.
- If the predicted torques are in satisfactory agreement with the experimental values, use the model to predict the stress growth at $\dot{\gamma}_R$, or any other shear rate for that matter.

7.1.3 Taking Derivatives

Articles have been written entirely on the subject of finding accurate derivatives given discrete points having random error [13]. As it turns out, the best advice many of these articles give is that one should know the data and the phenomena behind their variation. As with the error of the predicted values, the error in the predicted slopes will increase as one attempts to find slopes near the ends of the data. It is not difficult to show that this error may also be systematic; that is, the estimated slopes at the ends may be consistently high or low depending upon the treatment. Surely one of the worst errors is to throw a polynomial through the stress-rate data. While the fit may appear to be excellent, the slopes near the ends are very likely to be badly off. From laboratory courses, we have observed instances in particularly bad cases where the slopes are even of the wrong sign near the ends. The polynomial function can be regarded as a Taylor expansion around the center of the data that is to be fit. It is very good at the center and very poor far away, even if it looks like it is fitting the data well. Curvature-minimizing cubic spline algorithms are convenient, but are likely to give strange slopes at abscissa values that are very close to one another. Such situations are likely to occur when repeat experimental runs are combined for analysis.

While there is no way of being certain that the slope is correct, it is good practice to at least have an estimate of the error. Thus, it is best to describe the flow curve with realistic functions and carry through the error propagation analysis to arrive at the final answer. This error must be added to any that might come from many other sources, some of which have been mentioned. (See Walters [14] for a more complete listing of these errors.) One type of error that rheologists tend to ignore – and they are not helped by the instrumentation manufacturers – arises from the steady-shear data gathered with progressively increasing (or decreasing) shear rate.

Any time effects, such as degradation, become irreparably confounded with shear rate. While one can sweep up then down in shear rate, this route takes twice as long. Randomly chosen shear rates is a better approach, and examination of the time dependence of the residuals can give some information about degradation or other time-dependent effects.

We pointed out that it is helpful to know something about the phenomena behind the data before attempting differentiation. What do we know about the phenomena? In Equation 7.2, the slope needed is that of apparent shear stress ($2M/\pi R^3$) versus shear rate at the outer radius on a logarithmic plot. Generally, this curve will be concave downward due to the pseudoplastic nature of polymer melts. The relationship will not be the same as the stress-rate flow curve for the melt, but it is not difficult to show that there is a strong and unique relationship (see Chapter 8 on capillary flow). Most important, the limiting slopes at high and low shear rates will be the same as the flow curve, and we expect the second derivatives to fade away. In view of the Cross-Kaye arguments, we expect the curve to be almost indistinguishable from the true stress-rate curve if the rate is shifted by 0.1249 decade. Thus, it is reasonable to use the same functional forms. In between, we expect some wiggles, perhaps, if the molecular weight distribution is multimodal, and it would be a shame to lose these details by using a global function with no inflections. A reasonable approach is to use a local derivative method with some help at the ends. Possibilities include moving strip methods and modified spline functions with starting values determined by the slopes of a global function. Unfortunately, we do not know in advance what this function should be.

Suppose we fit some arbitrary, but well-behaved, global function to the data; it might be the Carreau [11] or Cross viscosity model, for example (see Chapter 9 for details). Using a global function implies we know something about the behavior of the data even before we are in possession of the data. Is this disagreeable? Perhaps, but the behavior is often based on much previous empirical knowledge. Another aspect is that the data at the ends may have systematic errors: viscous heating or secondary flows at the high shear rates or degradation at the low ends. The global function in effect uses the inner points to reduce the influence of these problems.

The method we present here is a compromise. We use the global function to furnish most of the slope. The residuals are subsequently treated locally to correct the slope. The advantage is that the curvature in the residuals is reduced over that in the original data, and, unlike a transform, the statistical aspects remain unchanged.

7.2 Description of the Algorithms

Three programs are supplied. The principal program `TWIST` takes steady-state data and corrects for non-Newtonian behavior. `STEADY` calculates torques for several viscosity models, while TRANST does the same thing for transient shear flows but for a nonlinear integral model.

7.2.1 TWIST

The aim of program TWIST is to provide accurate corrections for steady-flow data, even when faced with moderate noise. The data are read in from a file in groups where the geometry (plate radius and gap) are constant within the group. As coded in Section 7.5.1, up to 30 rotational rates can be accommodated in each group. The groups are combined after calculating the shear rate at the plate edge and the apparent shear stress at the same position ($2M/\pi R^3$). These data are sorted according to shear rate and loaded into a single array. At this point, the operator can choose a two-parameter viscosity model for rough fitting of the data. We restricted the program to two-parameter functions because they have the best chance of working with scant, almost-Newtonian data. These models should account for most, but not all, of the curvature in the flow curve. The models, listed in Table 9.1, are the Bueche-Harding (BU) and Eyring (EY) functions and are imbedded in the program VISFIT (Section 9.5) as the subroutines FXBU and FXEY, respectively. The fitting of the data is accomplished with the Gauss-Newton method via the subroutine GAUSSN, and is explained in more detail in Chapter 1.

The curvature-reducing aspect is illustrated in Figure 7.6, which shows the normalized stress versus rim shear rate that would be produced by an inverse exponential (IE) model fluid if it were subjected to torsional flow between parallel plates. The IE model is sharply curved – ridiculously curved. It turns from slope 1 to slope 0 in about a decade of shear rate. The best fit of the Bueche-Harding model to these data is shown as the dashed line; it decidedly has grave problems. Yet, when subtracted from the data, the curvature in the residuals is still reduced by a significant amount, as illustrated in the figure. This makes it much easier to get accurate slopes by using the residuals.

Because most of the curvature in realistic flow curves can be removed by subtracting the global function, straight-line segments through a strip of three points are usually suitable for measuring the slopes in the residuals, and TWIST attempts this first. If the noise is high, more points are included in the strip until the error in the slope is brought down to a level specified by the operator. Before the slope is fixed, a final check is done with a linearized Cross function (Chapter 9), with the parameter a = 1, on that strip alone. This guards against the possibility of variance introduced merely by high residual curvature. The slope with the lowest error is chosen for the stress correction. Should these steps be unable to bring the estimated error in the corrected stress below the desired level, a warning is written into the results file. All these steps are repeated at every data point.

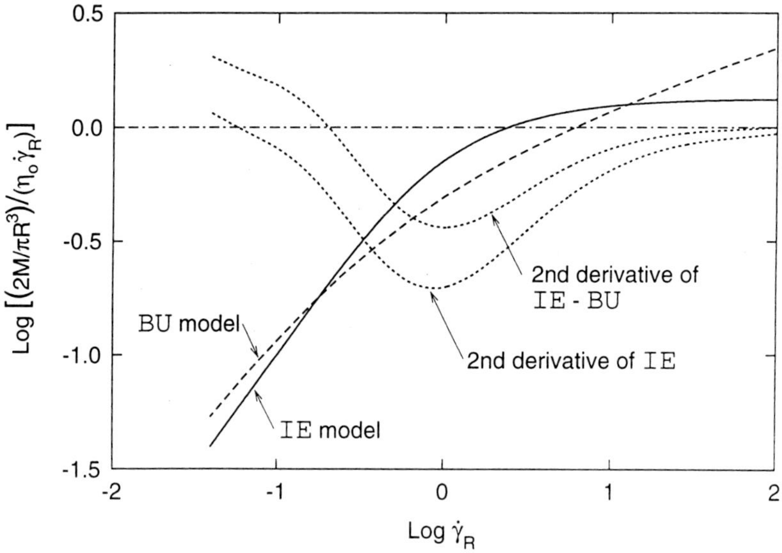

Figure 7.6 Illustration of the reduction in curvature on subtraction of a global viscosity function (the Bueche-Harding model BU) from the normalized stress data (generated with the inverse exponential model IE). The dotted lines are the values of the second derivatives.

When the slopes are known, the viscosities or the differences between the first and second normal stress coefficients are calculated according to Equation 7.2 or Equation 7.3, respectively, and written to the results file. The corrections according to the Cross-Kaye approximation are also recorded.

7.2.2 STEADY

STEADY can calculate highly accurate torques given a model for the viscosity of the fluid. Amazingly enough, the program will do the same for the axial thrust on the plates given a model for the first normal stress coefficient, Ψ_1, if one is willing to neglect the second normal stress. The reason STEADY can handle η or Ψ_1 is that both are third moments of the limiting values at zero shear rate of the coefficients of these functions (η_o and Ψ_1^o, respectively), modified by a function that depends only on shear rate. Thus,

$$M = \frac{2\pi R^3}{\dot{\gamma}_R^3} \int_0^{\dot{\gamma}_R} \eta \dot{\gamma}^3 \, d\dot{\gamma} = \frac{2\pi R^3}{\dot{\gamma}_R^3} \eta_o \int_0^{\dot{\gamma}_R} \frac{\eta}{\eta_o} \dot{\gamma}^3 \, d\dot{\gamma} \tag{7.13}$$

and

$$F \cong \frac{\pi R^2}{\dot{\gamma}_R^2} \int_0^{\dot{\gamma}_R} \Psi_1 \dot{\gamma}^3 \, d\dot{\gamma} = \frac{\pi R^2}{\dot{\gamma}_R^2} \Psi_1^{\mathrm{o}} \int_0^{\dot{\gamma}_R} \frac{\Psi_1}{\Psi_1^{\mathrm{o}}} \dot{\gamma}^3 \, d\dot{\gamma} \tag{7.14}$$

These expressions can be transformed into nondimensional forms by substituting $x = \dot{\gamma}/\dot{\gamma}_R$ to arrive at

$$\frac{2M/\pi R^3}{\eta_{\mathrm{o}} \dot{\gamma}_R} = 4 \int_0^1 f(x\bar{\tau}; P_i)\, x^3 \, dx \tag{7.15.1}$$

and

$$\frac{4F/\pi R^2}{\Psi_1^{\mathrm{o}} \dot{\gamma}_R^2} = 4 \int_0^1 f(x\bar{\tau}; P_i)\, x^3 \, dx \tag{7.15.2}$$

where f is either η/η_{o} or $\Psi_1/\Psi_1^{\mathrm{o}}$, and $x\bar{\tau} = x\dot{\gamma}_R\tau$ and P_i are any nondimensional parameters required to describe f.

STEADY does these 0-to-1 integrations by using the Romberg extrapolation technique, which is described thoroughly in *Numerical Recipes* [15]. With a few minor modifications to the *Numerical Recipes* routine QROMB, the Romberg technique, implemented in here as QROMB2, finds the value of the integral by extrapolating to zero spacing of the ordinate values where the integrand is evaluated. An error can be estimated and thus controlled. For this problem, the integrand is the product of the reduced viscosity function and x^3. To provide for different shear rates, the viscosity functions are fed not different values of shear rate but different values of the time constant. It is really the same thing; for every shear rate there is a multiplicative time constant. The shear rate then is always in the range of 0 to 1 to accommodate the moment integration.

We have provided a selection of viscosity models, most of which are tabulated in Table 9.1. As coded in the function subprogram VIS in Section 7.5.2, the models from Table 9.1 include the two-parameter Bueche-Harding (BU) and Eyring (EY), the three-parameter Carreau (CA), Cross (CR), and Sutterby (SU), and the four-parameter Vinogradov (VN). Two additional functions are available: the double Cross (DC) and the inverse exponential (IE) models. The DC model, a sum of two Cross functions with the parameter $a = 1$, would be appropriate for viscosity data exhibiting wiggles

$$\frac{\eta}{\eta_{\mathrm{o}}} = \frac{1 - \beta}{1 + \tau_1 \dot{\gamma}} + \frac{\beta}{1 + \tau_2 \dot{\gamma}} \tag{7.16}$$

where β is a nondimensional parameter and τ_i are the characteristic time constants. The inverse exponential model, which turns a very sharp corner, is for testing purposes. The inverse exponential model corresponds to the steady viscosity for the modified Tanner-Simmons model (Model 6, Chapter 10), but only if the relaxation modulus is a single exponential:

$$\frac{\eta}{\eta_o} = 1 - \exp\left(-\frac{1}{\tau\dot{\gamma}}\right) \tag{7.17}$$

For normal-stress data, the expression for the inverse exponential is

$$\frac{\Psi_1}{\Psi_1^o} = 1 - \exp\left(-\frac{1}{\tau\dot{\gamma}}\right)\left(1 + \frac{1}{\tau\dot{\gamma}}\right) \tag{7.18}$$

and is given the code `IN`.

7.2.3 TRANST

`TRANST` is similar to `STEADY` except it is intended for transient flows starting from rest. `TRANST` is restricted to the Papanastasiou-Scriven-Macosko (PSM) [16] and the modified Tanner-Simmons (MTS) models, but other functions may be added by following the guidelines outlined in Chapter 10.

The problem of transients is one of double integration, so the computation can take up much time without any difficulty. To control this, `TRANST` asks for specific time windows and point spacing. The user can also force the integration over radius to be of low order (even down to 1). With order 1, only one shear rate ($3\dot{\gamma}_R/4$) is being calculated. The integration over radius is done with the third-moment Gaussian method previously described (Equation 7.9).

The program `TRANST` allows the operator to choose the order of the integration; higher orders will be more accurate but take proportionally longer. The errors decrease rapidly with order; Table 7.1 shows the results for the third moment. However, in practice, the values of the higher derivatives may not be all that small compared to the lower ones. This means that one may get considerably less error than prescribed at low orders but lose this advantage at high orders.

The incorporation of a single-integral constitutive model into the integration over radius is illustrated below with a Lodge rubberlike liquid model for unidirectional shearing at a constant shear rate:

$$\sigma_{21}(t) = \int_{-\infty}^{t} m(t - s)\,[\gamma(t) - \gamma(s)]\,ds \tag{7.19}$$

where $m(t-s)$ is the memory function. Recognizing that, for flows starting from rest at $t = 0$, the strain remains constant from $-\infty$ to 0 results in the expression

Table 7.1 Errors in Third-Moment Gaussian Quadrature

Order, n	Error/$f(n)$
1	3.33×10^{-3}
2	1.88×10^{-5}
3	1.97×10^{-8}
4	1.89×10^{-11}
5	1.19×10^{-14}

$f(n)$ is the maximum value of the nth derivative of the function somewhere in the interval (in this case, 0 to 1).

$$\sigma_{21}(t) = \dot{\gamma}\, t\, G(t) + \int_0^t m(t-s)\, \dot{\gamma}\,(t-s)\, ds \tag{7.20}$$

This can be written for any radius r, using the variable $x = r/R = \dot{\gamma}/\dot{\gamma}_R$,

$$\sigma_{21}(t) = x\, \dot{\gamma}_R\, t\, G(t) + \int_0^t m(t-s)\, \dot{\gamma}_R x\,(t-s)\, ds \tag{7.21}$$

where $\dot{\gamma}_R$ and x are independent of time and, therefore, can be factored out.

The moment balance for torque is

$$M(t) = 2\pi \int_0^R \sigma_{21}^{+}(t, \dot{\gamma})\, r^2\, dr \tag{7.22}$$

which, upon substitution of $x = r/R = \dot{\gamma}/\dot{\gamma}_R$, becomes

$$M(t) = 2\pi R^3 \int_0^1 \frac{\sigma_{21}^{+}(t, \dot{\gamma}_R x)}{x}\, x^3\, dx \tag{7.9}$$

Inserting into the preceding equation the expression for stress, the final result is

$$M(t) = 2\pi R^3 \int_0^1 x^3 \left[\dot{\gamma}_R\, t\, G(t) + \int_0^t m(t-s)\, \dot{\gamma}_R\,(t-s)\, ds \right] dx \tag{7.23}$$

Clearly, this is a trivial case because the time integral does not depend on x and can be factored out. However, nonlinear models will retain a dependence on x and thus must be integrated to arrive at the torque.

7.3 Execution of the Programs

7.3.1 TWIST

The execution of TWIST requires the subroutines ALINR, AMAX, AMIN, GAUSSN, and SORT2 from Appendix B, and the subprograms FXBU and FXEY for the viscosity models imbedded in VISFIT (Section 9.5). Refer to Appendix A for the program dependencies. An ASCII data file must be assembled that consists of the rotational velocities of the plate and the torques. The normal forces can also be included, if available. All units must be consistent; that is, for viscosities to be expressed in Pa•s, the torques must be in N•m, and the plate gap and radius in m.

The first record in the data file is a title; insert anything of up to 79 characters to help identify the output, since it will be copied directly to the results file. The next record has three entries: the radius of the plates, the gap separating them, and the number of readings made with this geometry. If this number is greater than 30, all data over this number will be dumped. The next record is the first speed (radians per second) and the resulting torque and normal force. Ignore the third entry if normal-force data is unavailable. This format continues until the number of readings is equal to the number indicated on the second record. Speeds can be repeated as often as desired, but there should be at least four distinct speeds (in no particular order).

The data entry continues with other plate gaps or sizes, up to a maximum of five different geometries. As mentioned, it is good experimental policy to vary the geometry to check for edge or surface effects. Because TWIST keeps track of the data, the results can be checked for these boundary effects after the corrections are made. For example, if the small gaps are producing consistently lower viscosities, one might suspect alignment problems or slip.

The program starts with the usual questions concerning the identity of the data and results files. Checks are made on the status of these; if you want to write results to an old file, this will be done at the bottom of the file and all the old contents will be preserved. If no data file is found, enter the data via the keyboard, but it is very easy to make mistakes this way. At this point, TWIST will require operator interactions. First, TWIST must be informed whether the analysis will be performed on the shear or normal-stress data. The second interaction sets the error level. Choose something consistent with the level of noise in the data. While entering 0 is allowed, the performance of the algorithm may not be optimal because too many points will be used in finding the slopes; this has a corner-rounding effect. Initial values for the parameters log η_o, or $\log(\Psi_1 - \Psi_2)$ for normal-stress data, and log τ are displayed; these estimates are based on the range of data found in the file and probably will suffice. However, if there are problems with convergence, this would be the place to experiment. Choose N to proceed with the default parameters, Y otherwise.

The program will next ask for the viscosity model; enter BU for the Bueche-Harding model or EY for the Eyring model. The results should be independent of this choice, but this will be a good thing to check.

At this point, the subroutine GAUSSN is called to fit the model to the data. It is characteristic of the Gauss-Newton iterative method that it either converges (quadratically) or gets into trouble very rapidly. If problems exist, check the data for mistakes and experiment with new starting values. If the problem is a Divide by zero error, the covariance matrix may be close to singular, which often indicates that the data do not have enough curvature to be described by the model. In this case, use a power-law analytical solution. Sample input/output files are listed in Sections 7.6.1 through 7.6.3.

7.3.2 STEADY

The execution of STEADY requires the subroutine QROMB2 found in Appendix B and the function subprogram ASINH imbedded in the Chapter 9 program VISFIT. No data files are needed to operate STEADY. However, the subroutine that handles the file opening and status checks will inquire anyway. Just type in N to proceed. The next query will address the name of the output file; enter any valid file name.

A viscosity model must be specified. Refer to Section 7.2.2 for the available models. Descriptions of these models are provided in Table 9.1 with two additional models, the double Cross and the inverse exponential model, available for reasons discussed in Section 7.2.2. Tests with some models will be described in Section 7.4.2.

The output in the results file consists of nondimensional quantities. Refer to Section 7.6.5 for sample listings. In the first column is the time constant multiplied by shear rate, $\tau\dot{\gamma}$. The second column is the reduced viscosity; that is, the viscosity η, predicted by the model, divided by its zero-shear-rate viscosity η_o. Both the first and second columns refer to the outside edge of the plates. The third column is the torque, but again in reduced form, specifically

$$\text{Reduced torque} = \frac{2M/\pi R^3}{\eta_o \dot{\gamma}_R} \tag{7.24}$$

To predict the actual torque on running a real fluid which has been characterized by a viscosity model (e.g., by using VISFIT, Chapter 9), multiply the third column by $\eta_o \dot{\gamma}_R \pi R^3/2$.

On running STEADY, note that the data under the second and third columns are close, with the torque always being higher. This is because the torque is being increased by the fluid near the inside, which is not being sheared as rapidly and thus has a higher viscosity. These columns would be identical if the fluid were Newtonian.

7.3.3 TRANST

TRANST operates in much the same fashion as STEADY in that calls the subroutine QROMB2 (via the function subprograms MTS and PSM) and does not require a data file. In addition, TRANST uses the function subprograms XMEM1 and XMOD1 found in Sections 10.6.7 and 10.6.8, respectively. To run the program, choose either the MTS or the PSM model. The next requirement is to enter the parameters for these models: the characteristic modulus G, the breadth parameter B, the time constant τ, and the nonlinear parameter x_{NL}. Refer to Chapter 10 for a full description of the models and how the parameters influence the stress. Next, the program will request the order for the integration. It would be prudent to start with 1 to provide the quickest results. Higher accuracy and longer computational time come with higher orders. The final question refers to the span and spacing of the calculated points. To be cautious, choose a starting time that is, for example, a tenth of the time constant for the fluid. A finishing time of 10τ at a strain of around 10 will probably give enough information to see what is going on. The multiplier must be greater than 1; if you enter 10 it will perform the calculation every decade. A sample dialog for the execution of TRANST is provided in Section 7.6.6.

At this point the integrations will begin. Remember that both time and radius are involved, so things might be slow. If a high order for the integration over radius is chosen, and the output is slow in coming, break off the calculation and restart with a lower integration order.

The output of TRANST appears in several columns. A sample listing of an output file from TRANST is given in Section 7.6.7. The first column corresponds to time. The second column, labeled STRR, carries the true shear stress at the outer edge of the plate. The third column carries the apparent shear stress at the outer edge; it is labeled STRA. STRA will be higher than STRR if the fluid is shear thinning and is simply $2M/\pi R^3$. The fourth column, labeled STR34, contains the actual stress at $3R/4$. The last column contains the apparent stress at $3R/4$ or $3M/2\pi R^3$. The last two columns should be nearly identical for the reasons explained in the introduction. They will be identical if the order of integration is 1.

TRANST is closely allied with STRESS in Chapter 10. The output under the column heading STRR should be identical to the output from STRESS if all the kinematic and material variables are the same. Given real data from the start-up of torsional flow, one can assume that the observed $3M/2\pi R^3$ is assignable to a rate equal to $3\dot{\gamma}_R/4$. Taking this data, one can proceed to Chapter 10 and use FITSTR to fit a nonlinear constitutive model. The agreement of the experimental torque with that which would be produced by such a fluid can subsequently be checked with TRANST. Iterating on this is possible but probably unwarranted.

7.4 Application of the Programs

7.4.1 TWIST

For TWIST, we are concerned with several issues, including accuracy and behavior with noisy data. For accuracy, we can generate torque and axial force data with STEADY and attempt to reproduce the viscosity with TWIST. The worst problems for TWIST will be high curvature in the data; much more than can be removed by the BU or EY models. For this reason, two inverse exponential functions were inserted into STEADY; they are useless for anything but giving high curvature – far too curved to fit even the narrowest fraction of a real polymer. One function is a viscosity model (Equation 7.17), and the other function (Equation 7.18) is for normal stress.

The application of TWIST to the torque and axial force produced by the inverse-exponential functions is shown in Figure 7.7. With the viscosity there is no problem; the data mimic the known viscosity function in spite of the very high curvature. For the normal stress coefficient, the derivative technique has trouble

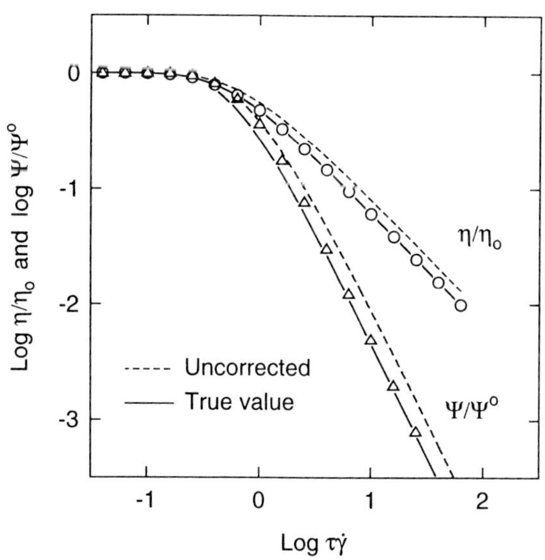

Figure 7.7 Application of TWIST to the correction of viscosity (η) and the normal stress coefficients (Ψ_1 - Ψ_2, and denoted as Ψ). The symbols are the corrected values. With the extreme curvature in Ψ, TWIST undercorrects.

turning the corner fast enough. Upon seeing this problem, we were tempted to go to higher-order local derivative techniques, but there is always a penalty. Here, the penalty would be the handling of noise. Again, these high curvatures are never expected for real materials.

With more realistic functions the ability to reproduce the data is very good. One feature we were particularly concerned about was the preservation of wiggles in the viscosity function, as a result of blending, for example. In Figure 7.8, such behavior is generated using the DC model (Equation 7.16); the model consists of a sum of two Cross functions of equal amplitudes but with characteristic times spaced by two decades. The apparent viscosity is the upper curve obtained by assuming the stress at the outer radius is the Newtonian value, $2M/\pi R^3$. The points on the plot are the corrected values of viscosity using TWIST. The bump is faithfully preserved. Also shown in Figure 7.8 is the corrected viscosity using the Cross-Kaye approximation; it tends to overcorrect slightly (see inset) where the slope of the viscosity function is high. (This effect is maximum where the slope is -0.5.) The printout of TWIST carries the results of the application of the Cross-Kaye approximation for comparison with the conventional method.

In Figure 7.9, the exercise has been repeated but with 2% noise added to the torque data. The correction for noisy data lie substantially along the true viscosity curve that was used to generate the torque data with STEADY. While the noise practically wipes out the wiggle in the curve, there are no systematic deviations of

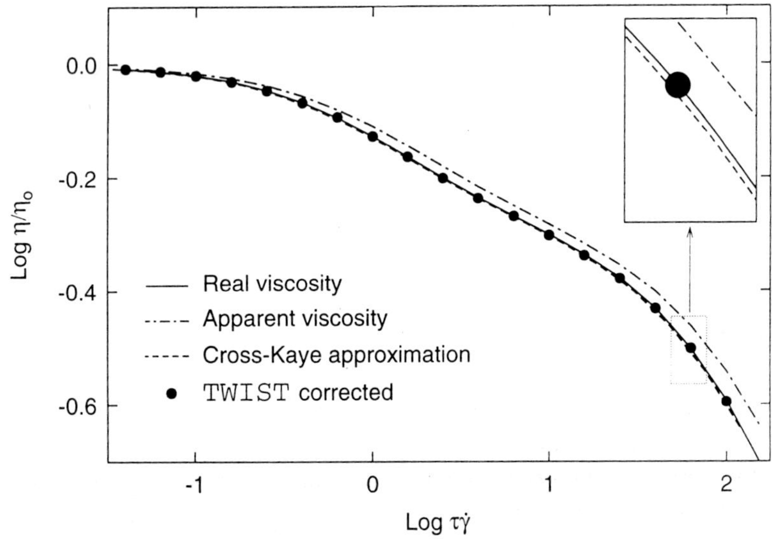

Figure 7.8 Illustration of the correction of viscosity data with TWIST. The viscosity function contains a slight inflection. The inset shows the relative placement of the corrections using TWIST and the Cross-Kaye approximation in more detail.

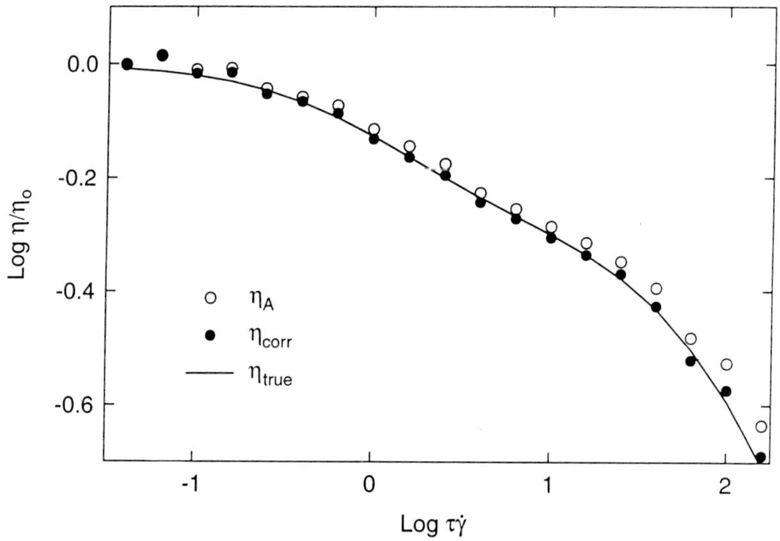

Figure 7.9 Behavior of TWIST with noisy data. Random noise with a standard deviation of 2% was added to the torque values.

the points from the true viscosity curve. However, the message concerning wiggles is clear; if the experiment is to detect them, the data need to be both numerous and precise!

The residuals from Figure 7.9 are plotted in Figure 7.10. While there is slight suggestion of positive serial correlation at low shear rates – the points seem undercorrected – it could have gone the other way depending upon the errors in the points. It does illustrate the difficulties at the ends of the set, and for this there is no apparent cure.

7.4.2 STEADY

The practical usefulness of the program STEADY is for checking the results of TWIST. Given that the parallel-plate data have been corrected and fitted to a viscosity function, the validity of this function can be checked by using STEADY to regenerate the original data. (In this discussion, the same remarks apply to $\Psi_1 - \Psi_2$.) STEADY must, therefore, introduce negligible error of its own.

To test the accuracy of STEADY it is necessary to have analytic solutions. An easy one is for a Cross model with the nondimensional parameter a = 2. (Refer to a Chapter 9 for a detailed description of this model.) A value of 2 for the parameter

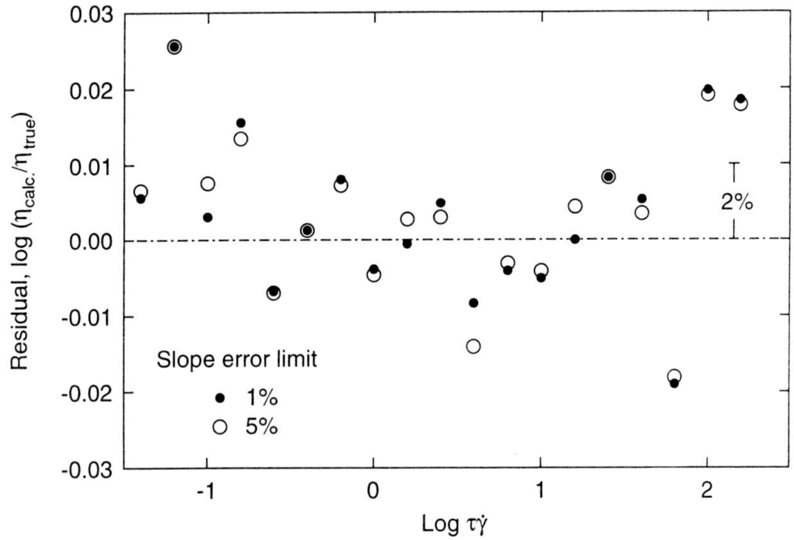

Figure 7.10 The residuals from Figure 7.9 showing the influence of attempts to limit the error while using TWIST.

a in the Cross model gives a very highly curved *knee* in the viscosity function; thus, it represents a good test. With this function, Equation 7.15.1 can be integrated to give the following expression for torque:

$$\frac{2M/\pi R^3}{\eta_o \dot{\gamma}_R} = \frac{2}{\dot{\gamma}_R^4}\left[\dot{\gamma}_R^2 - \ln(1 + \dot{\gamma}_R)\right] \tag{7.25}$$

The results are given in Table 7.2. Anything works until the knee ($\tau\dot{\gamma}_R = 1$) gets well into the inner part of the plates. Here, the advantages of an adaptive integration technique, such as the Romberg extrapolation available in *Numerical Recipes* [15], become evident. Although a fifth-order Gaussian quadrature performs well, considering it involves only five evaluations of the integrand, STEADY is coded with QROMB2 because of its controllable accuracy. With TRANST, it is clear why Gaussian quadrature is appropriate when time is of the essence.

The results using the Cross model with $a = 4$ are similar; the torque is given by the very simple expression

$$\frac{2M/\pi R^3}{\eta_o \dot{\gamma}_R} = \frac{\ln\left(1 + \dot{\gamma}_R^4\right)}{\dot{\gamma}_R^4} \tag{7.26}$$

This is considerably more severe than the former example, yet QROMB2 shows four-

Table 7.2 Precision Test for STEADY in Calculating Reduced Torque

			Gaussian Quadrature Order	
$\tau\dot{\gamma}_R$	Analytic	QROMB2	Fourth	Fifth
0.01	0.999934	0.999933	0.999933	0.999933
0.1	0.993383	0.993383	0.993383	0.993383
1.0	0.613705	0.613706	0.613706	0.613706
10.0	0.019077	0.019077	0.019021	0.019070
100.0	0.0001998	0.0001998	0.0001971	0.0001984

figure precision up to $\tau\dot{\gamma}_R = 10$.

7.4.3 TRANST

TRANST is easy to check against STRESS (Chapter 10) to confirm that the stress-versus-time aspects are being handled correctly. Checking the accuracy of the torque is less easily done directly, although it uses the same Gaussian quadrature that showed good results in the tests of STEADY discussed previously.

To work an example through the full cycle, we examined the NBS Fluid 40 data of Ganani and Powell [7]. For this fluid, the transient torque curve at 17 s^{-1} was compared with transient results from a cone-and-plate geometry at the same shear rate. The authors were able to correct the torque by examining a range of shear rates and finding slopes for the application of Equation 7.5. Our rendition of their results is shown in Figure 7.11. The peak of the torque overshoot occurs at a later time than the peak in the corrected stress. The torques in this figure have been multiplied by an unknown constant, so we cannot get the apparent stress. However, we took the corrected stresses and applied FITSTR using the PSM model. From Figure 7.12, it is evident that the fit was not perfect – the PSM model has difficulty describing such a sharp stress maximum – but good enough for a qualitative illustration.

With the stress model in hand, TRANST was invoked to predict the torque that would result if such a fluid were sheared in parallel-plate torsional flow. The result is also shown in Figure 7.12. In particular, note that the shift in the peak stress is properly accounted for.

Shown in Figure 7.13 is the application of TRANST to illustrate the usefulness of the Cross-Kaye approximation for transient flows. Here, the transients were calculated by using the PSM model for a shear rate at the outer radius of $\dot{\gamma}_R = 10/\tau$ Four quantities are plotted: the apparent stress at the outer radius $2M/\pi R^3$, the actual

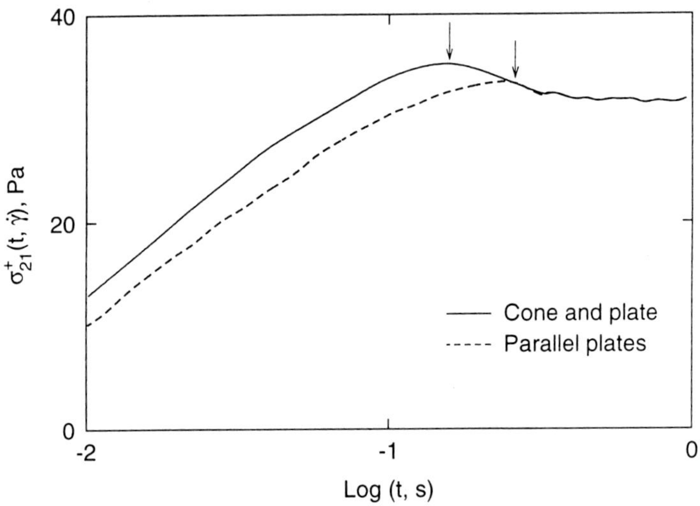

Figure 7.11 Reproduction of the transient experiments of Ganani and Powell on NBS Fluid 40 [7]. The arrows show the peaks in the transients. The curve for the parallel plates is the apparent shear stress normalized to coincide with the cone-and-plate curve at long times. Adapted from Ref. 7. Copyright © 1985 by The Society of Rheology, Inc. Reprinted by permission of the American Institute of Physics.

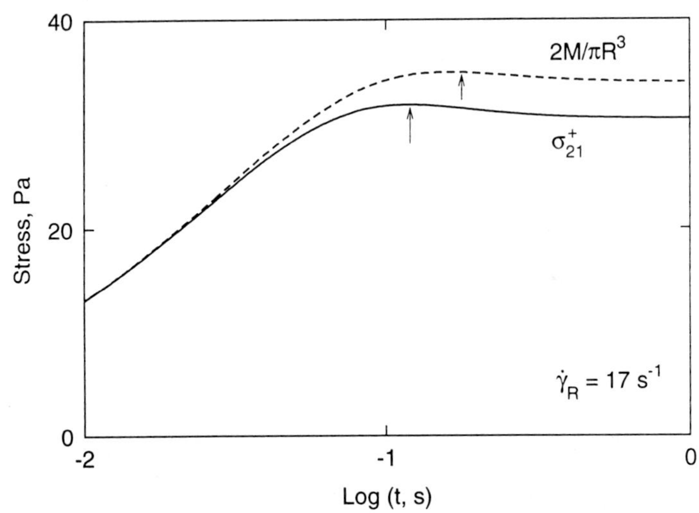

Figure 7.12 Depiction of the predictions from TRANST for the experiment in Figure 7.11. Note that the displacement of the peaks is correctly predicted. The dotted curve would have to be multiplied by about 0.87 to give the same type of plot as Figure 7.11.

stress at the outer radius, $\sigma_{21}^{+}(t, \dot{\gamma}_R)$, three-fourths of the apparent stress ($3M/2\pi R^3$), and the actual stress at $3R/4$. The outer radius quantities do not compare well, as noted by Ganani and Powell [7], but the quantities at $3R/4$ are in good agreement.

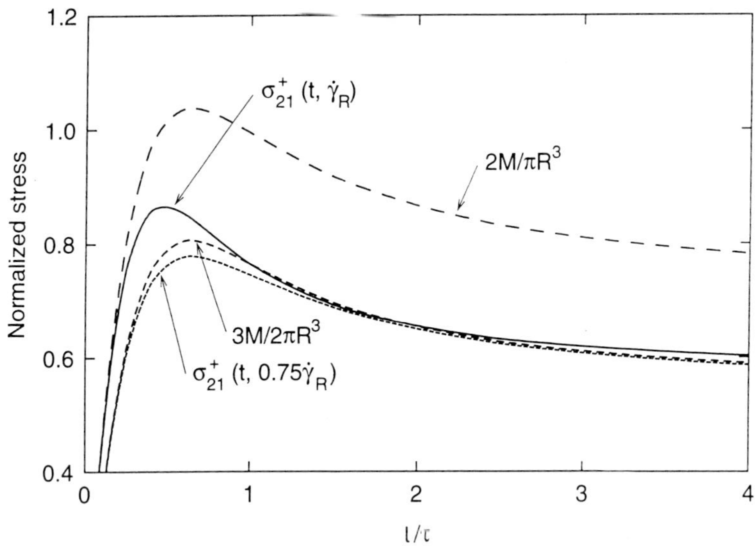

Figure 7.13 Application of the Cross-Kaye approximation to transients. From the top down the curves, normalized with respect to the characteristic modulus G from the PSM model in TRANST, are the apparent stress, the actual stress at the outside radius R, the apparent stress at $3R/4$, and the actual stress at $3R/4$. Note the near coincidence of the lower two curves.

7.5 Program Listings

7.5.1 TWIST

Imbedded into the code is the function subprogram FSLOPE inherent to the execution of TWIST.

```
      PROGRAM TWIST
C Corrects steady-state data from the parallel-plate torsional flow using
C  the Bueche-Harding or Eyring viscosity model.
C Requires the routines ALINR, AMAX, AMIN, GAUSSN, and SORT2
C  and the subroutines FXBU and FXEY from program VISFIT (Chapter 9).
C Data file format:
C         1st record - Title, up to 79 characters
C         2nd record - plate radius, plate gap, number of readings for
C                      this geometry. Up to five different geometries may
C                      be entered.
C subsequent records - rotational speed (rad/s), torque, normal force
```

```
      IMPLICIT DOUBLE PRECISION (A-H,O-Z)
      PARAMETER (TWO=2.D0, PI=3.1415926536)
      PARAMETER (NPAR=2,NRUNS=5,NWS=30,NPT=NRUNS*NWS,NPD=10)
      CHARACTER TITLE*80,JUNK*80,VM*2,ANS*1
      DIMENSION C(11),R(NRUNS),NW(NRUNS),H(NRUNS),RT(NWS,NRUNS)
      DIMENSION ST(NWS,NRUNS),U(NPT),V(NPT),P(NPT/2),Q(NPT/2)
C Variables for GAUSSN.
      DIMENSION A(NPT,NPD),AT(NPD,NPT),B(NPD),ERR(NPD),DER(NPD)
      DIMENSION RR(NPT,1),X(NPT),Y(NPT)
      EXTERNAL FXBU,FXEY
C t-statistic for 95% confidence interval with N degrees of freedom.
      T95(N)=1.96+2.3539/N+3.1902/N**2+0.7004/N**3+4.5015/N**4
10    CALL MSHEAD
      CALL UNITS(NTER,NINT,NOUT)
      WRITE(NOUT,'(21X,A)')'Parallel-Plate Torsional Flow'
      WRITE(NOUT,'(21X,30(''-'')/)')
      READ(NINT,'(A80)')TITLE
      WRITE(NOUT,'(A80)')TITLE
C Read in data.
      IFLAG=0
      WRITE(NTER,*)' Shear or normal stresses (S/N)?'
      READ(NTER,'(A1)')ANS
      IF(ANS.EQ.'N'.OR.ANS.EQ.'n')IFLAG=1
      K=0
      DO 30 J=1,NRUNS
        IALARM=0
        READ(NINT,*,END=40)R(J),H(J),NW(J)
        IF(NW(J).GT.NWS)THEN
          WRITE(NTER,*)' Number of rates too high.'
          IALARM=1
        ENDIF
        DO 20 I=1,NW(J)
          IF(I.GT.NWS)THEN
            READ(NINT,'(A80)')JUNK
            WRITE(NTER,*)' Reading discarded'
          ELSE
            IF(IFLAG.EQ.0)THEN
              READ(NINT,*)RT(I,J),ST(I,J)
            ELSE
              READ(NINT,*)RT(I,J),ST(I,J),ST(I,J)
            ENDIF
            K=K+1
C Compute apparent rates and stresses.
            RT(I,J)=RT(I,J)*R(J)/H(J)
            X(K)=LOG10(RT(I,J))
            IF(IFLAG.EQ.0)THEN
              ST(I,J)=TWO*ST(I,J)/PI/R(J)**3
            ELSE
              ST(I,J)=4.D0*ST(I,J)/PI/R(J)**2
            ENDIF
            Y(K)=LOG10(ST(I,J))
          ENDIF
20      CONTINUE
        IF(IALARM.EQ.1)NW(J)=NWS
30    CONTINUE
40    NR=J-1
      NP=K
C Maximum and minimum rates and stresses to be used later.
      XMIN=AMIN(NP,X)
      XMAX=AMAX(NP,X)
      YMIN=AMIN(NP,Y)
      YMAX=AMAX(NP,Y)
      WRITE(NTER,*)' Enter fractional error wanted in corrected stress.'
      READ(NTER,*)CERRMX
C Get fit of flow curve to selected function: Provide initial guesses for
C  zero-shear-rate viscosity, B(1), and time constant, B(2).
      B(2)=-(XMAX+XMIN)/2.D0
      B(1)=YMIN-XMIN
```

```
      IF(IFLAG.EQ.0)THEN
        WRITE(NTER,50)'viscosity',B(1),B(2)
      ELSE
        WRITE(NTER,50)'Psi1-Psi2',B(1),B(2)
      ENDIF
50      FORMAT(/7X,'Initial values:'//14X,'Log ',A,' =',F7.2
     &          /10X,'Log time constant =',F7.2/)
      WRITE(NTER,*)' Revise these initial values? (Y/N)'
      READ(NTER,'(A1)')ANS
      IF(ANS.EQ.'Y'.OR.ANS.EQ.'y')THEN
        WRITE(NTER,*)' Enter new initial values.'
        READ(NTER,*)(B(I),I=1,NPAR)
      ENDIF
C Choose the Bueche-Harding (BU) or Eyring (EY) viscosity model.
      WRITE(NTER,*)' Choose viscosity model (BU or EY).'
      READ(NTER,'(A2)')VM
C Specify convergence criterion.
      CC=1.D-4
C Fit the viscosity model chosen.
      IF(VM.EQ.'BU'.OR.VM.EQ.'bu'.OR.VM.EQ.'Bu')
     &  CALL GAUSSN(FXBU,CC,NPAR,NP,X,Y,B,A,AT,RR,ERR,SEE)
      IF(VM.EQ.'EY'.OR.VM.EQ.'ey'.OR.VM.EQ.'Ey')
     &  CALL GAUSSN(FXEY,CC,NPAR,NP,X,Y,B,A,AT,RR,ERR,SEE)
      WRITE(NTER,60)VM
      WRITE(NOUT,60)VM
60    FORMAT(/3X,'Final values for ',A2,' model:'/)
      IF(IFLAG.EQ.0)THEN
        WRITE(NTER,70)'viscosity',B(1),T95(NP-NPAR)*SEE*ERR(1)
        WRITE(NOUT,70)'viscosity',B(1),T95(NP-NPAR)*SEE*ERR(1)
      ELSE
        WRITE(NTER,70)'Psi1-Psi2',B(1),T95(NP-NPAR)*SEE*ERR(1)
        WRITE(NOUT,70)'Psi1-Psi2',B(1),T95(NP-NPAR)*SEE*ERR(1)
      ENDIF
70    FORMAT(14X,'Log ',A,' =',F8.3,' +/-',F8.3)
      WRITE(NTER,80)B(2),T95(NP-NPAR)*SEE*ERR(2)
      WRITE(NOUT,80)B(2),T95(NP-NPAR)*SEE*ERR(2)
80    FORMAT(10X,'Log time constant =',F8.3,' +/-',F8.3)
      WRITE(NTER,90)SEE
      WRITE(NOUT,90)SEE
90    FORMAT(/1X,'Standard error of estimate =',1P,E10.3)
      WRITE(NOUT,100)
100   FORMAT(/T61,'3R/4'/T6,'Shear',T17,'Apparent',T30,'Shear',
     &          T52,21('-'))
      IF(IFLAG.EQ.0)THEN
        WRITE(NOUT,110)
110     FORMAT(T7,'Rate',T18,'Stress',T30,'Stress',T40,'Viscosity',
     &          T54,'Rate',T64,'Viscosity'/6(3X,9('-')))
      ELSE
        WRITE(NOUT,120)
120     FORMAT(T7,'Rate',T18,'Stress',T30,'Stress',T40,'Psi1-Psi2',
     &          T54,'Rate',T64,'Psi1-Psi2'/6(3X,9('-')))
      ENDIF
C Do corrections.
C Place X in U and Y in V and sort U,V with respect to U.
C Keep track of K via X.
      CALL SORT2(NP,X,Y,U,V)
C Subtract off fitted curve to get residuals and reload in V.
      DO 130 K=1,NP
        IF(VM.EQ.'BU'.OR.VM.EQ.'bu'.OR.VM.EQ.'Bu')
     &    CALL FXBU(NPAR,U(K),B,VK,DER)
        IF(VM.EQ.'EY'.OR.VM.EQ.'ey'.OR.VM.EQ.'Ey')
     &    CALL FXEY(NPAR,U(K),B,VK,DER)
        V(K)=V(K)-VK
130   CONTINUE
      K=0
      NMAX=NP/2
      NMIN=3
```

```
      DO 210 J=1,NR
        WRITE(NOUT,'(/A,I3/)')' Run number:',J
        NPRINT=NW(J)/8+1
        DO 200 I=1,NW(J)
          K=K+1
          DMDW=FSLOPE(VM,B,X(K))
C Adjust slopes by looking at residuals.
C Find correct U value given X(K).
          DO 140 L=1,NP-1
            IF(U(L).GE.X(K))GOTO 150
140       CONTINUE
C Find slope in residuals and check for error.
150       DO 160 NST=NMIN,NMAX
            NSTART=MIN(NP-NST+1,MAX(1,L-NST/2))
            CALL ALINR(NST,U(NSTART),V(NSTART),C)
C If error in slope is acceptable, proceed.
            IF(ABS(C(10)*T95(NST-NPAR)).LE.CERRMX*(3.D0+DMDW).OR.
     &        (IFLAG.EQ.1.AND.ABS(C(10)*T95(NST-NPAR)).LE.
     &        CERRMX*(1.D0+DMDW)))GOTO 170
160       CONTINUE
170       DMDW=DMDW+C(2)
          ERR1=C(10)*T95(NST-NPAR)
          DO 180 M=NSTART,NSTART+NST
            P(M-NSTART+1)=1.D1**(-U(M))
            IF(VM.EQ.'BU'.OR.VM.EQ.'bu'.OR.VM.EQ.'Bu')
     &        CALL FXBU(N,U(M),B,VK,DER)
            IF(VM.EQ.'EY'.OR.VM.EQ.'ey'.OR.VM.EQ.'Ey')
     &        CALL FXEY(N,U(M),B,VK,DER)
            Q(M-NSTART+1)=1.D1**(-V(M)-VK)
180       CONTINUE
          CALL ALINR(NST,P,Q,C)
          DMDW2=C(2)/(C(2)+C(1)*RT(I,J))
          ERR2=DMDW2*ABS(C(1))*ST(I,J)*(ABS(C(9)*T95(NST-NPAR)/C(1))+
     &         ABS(C(10)*T95(NST-NPAR)/C(2)))
          ERRMN=ERR1
          IF(ERR2.LT.ERR1)THEN
            DMDW=DMDW2
            ERRMN=ERR2
          ENDIF
C Final check on error.
          IF(ERRMN.GT.CERRMX*(3.D0+DMDW).OR.
     &      (IFLAG.EQ.1.AND.ERRMN.GT.CERRMX*(1.D0+DMDW)))
     &      WRITE(NOUT,*)' Warning: Error high on point below.'
C Guard against out-of-range values.
          IF(DMDW.GT.1.D0)DMDW=1.D0
          IF(DMDW.LT.0.D0)DMDW=0.D0
C Finally, the correction.
          IF(IFLAG.EQ.0)THEN
            STRESS=ST(I,J)*(.75D0+.25D0*DMDW)
            VIS=STRESS/RT(I,J)
            VIS34=ST(I,J)/RT(I,J)
            RT34=.75D0*(.75D0**DMDW*(1.D0+DMDW/3.D0))*RT(I,J)
            WRITE(NOUT,190)RT(I,J),ST(I,J),STRESS,VIS,RT34,VIS34
          ELSE
            STRESS=ST(I,J)*(.5D0+.5D0*DMDW)
            PSI=STRESS/RT(I,J)**2
            PSI34=ST(I,J)/RT(I,J)**2
            RT34=.75D0*RT(I,J)
            WRITE(NOUT,190)RT(I,J),ST(I,J),STRESS,PSI,RT34,PSI34
          ENDIF
190       FORMAT(1P,6E12.3)
200     CONTINUE
210   CONTINUE
      CALL MSTAIL(*10)
      STOP
      END
```

```
      DOUBLE PRECISION FUNCTION FSLOPE(VM,B,X)
C Returns log-log slope of stress versus rate given viscosity model VM,
C  parameters B, and the log(base 10) of shear rate X.
C  Note that B(2) is the log(base 10) of the time constant.
      IMPLICIT DOUBLE PRECISION (A-H,O-Z)
      CHARACTER VM*2
      DIMENSION B(4)
      IF(VM.EQ.'BU'.OR.VM.EQ.'bu'.OR.VM.EQ.'Bu')THEN
        RR=1.D1**(.75D0*(B(2)+X))
        FSLOPE=1.D0-.75D0*RR/(1.D0+RR)
      ELSE IF(VM.EQ.'EY'.OR.VM.EQ.'ey'.OR.VM.EQ.'Ey')THEN
        XR=1.D1**(X+B(2))
        Y=XR+SQRT((XR**2+1.D0))
        FSLOPE=XR*(1.D0+XR/(XR**2+1.D0))/Y/LOG(Y)
      ENDIF
      RETURN
      END
```

7.5.2 STEADY

Imbedded into the code are the function subprograms `VIS` and `CSTEDY` inherent to the execution of `STEADY`.

```
      PROGRAM STEADY
C Calculates steady torque given viscosity model. Models should be written
C  with B(1) as zero-shear-rate viscosity and B(2) as time constant.
C The result is    2M/πR^3/[GDR*B(1)] = f[GDR*B(2)],
C  where GDR, listed as reduced torque, is the shear rate at rim (r = R)
C  and GDR*B(2) is listed as reduced rate.
C Requires the subroutine QROMB2 (modified version of the Numerical Recipes
C  code QROMB by Press et al., Cambridge University Press, 1986).
      IMPLICIT DOUBLE PRECISION (A-H,O-Z)
      PARAMETER (EPS=1.D-6,NPAR=4,NPPD=5,NDEC=6,TINY=1.D-9)
      CHARACTER VM*2
      DIMENSION RTOR(NDEC*NPPD+1),RVIS(NDEC*NPPD+1),GDR(NDEC*NPPD+1)
      COMMON /BLOC2/ B(NPAR),VM
      EXTERNAL CSTEDY
10    CALL MSHEAD
      CALL UNITS(NTER,NINT,NOUT)
      NPRM=2
C Refer to Table 9.1, Chapter 9 for entry codes to viscosity models.
C The DC function is a sum of two Cross viscosity models (Eq. 7.16).
C The IE function is the inverse exponential viscosity model (Eq. 7.17).
C The IN function is the inverse exponential normal-stress-coefficient
C   model (Eq. 7.18).
      WRITE(NTER,*)' Enter viscosity model.'
      READ(NTER,'(A2)')VM
      IF( VM.EQ.'CA'.OR.VM.EQ.'ca'.OR.VM.EQ.'Ca'
     &.OR.VM.EQ.'CR'.OR.VM.EQ.'cr'.OR.VM.EQ.'Cr'
     &.OR.VM.EQ.'SU'.OR.VM.EQ.'su'.OR.VM.EQ.'Su')THEN
        NPRM=3
        WRITE(NTER,*)' Enter nondimensional parameter value.'
        READ(NTER,*)B(3)
      ENDIF
      IF( VM.EQ.'VN'.OR.VM.EQ.'vn'.OR.VM.EQ.'Vn'
     &.OR.VM.EQ.'DC'.OR.VM.EQ.'dc'.OR.VM.EQ.'Dc')THEN
        NPRM=4
        WRITE(NTER,*)' Enter two nondimensional parameter values.'
        READ(NTER,*)B(3),B(4)
      ENDIF
      A=TINY
      C=1.D0
      B(1)=1.D0
```

```
      X=1.D0
      NP2=(NDEC*NPPD+1)/2+1
      WRITE(NTER,20)VM
      WRITE(NOUT,20)VM
20    FORMAT(/2X,'Viscosity model is: ',A2/)
      WRITE(NTER,30)(I,B(I),I=1,NPRM)
      WRITE(NOUT,30)(I,B(I),I=1,NPRM)
30    FORMAT(15X,'B(',I1,') =',F7.2)
      WRITE(NTER,40)
      WRITE(NOUT,40)
40    FORMAT(/T7,'Reduced',T22,'Reduced',T37,'Reduced'
     &       /T9,'rate',T21,'viscosity',T38,'torque'
     &       /T1,3(4X,(11('-'))))
      DO 60 I=1,NDEC*NPPD+1
C Scan a range of reduced shear rates, P(2), five points per decade.
        B(2)=10.D0**(REAL(I-NP2)/REAL(NPPD))
        CALL QROMB2(CSTEDY,A,C,SS,EPS)
        RTOR(I)=4.D0*SS
        RVIS(I)=VIS(VM,X,B)
        GDR(I)=B(2)
        WRITE(NOUT,50)GDR(I),RVIS(I),RTOR(I)
C Display partial results to terminal.
        IF((I-1)/NPPD*NPPD.EQ.I-1)WRITE(NTER,50)GDR(I),RVIS(I),RTOR(I)
50      FORMAT(1P,3E15.5)
60    CONTINUE
      WRITE(NOUT,70)
70    FORMAT(/T13,'Logs of above numbers'/)
      WRITE(NOUT,'(3F15.4)')
     &      (LOG10(GDR(I)),LOG10(RVIS(I)),LOG10(RTOR(I)),I=1,NDEC*NPPD+1)
      CALL MSTAIL(*10)
      STOP
      END
C
      DOUBLE PRECISION FUNCTION VIS(VM,SR,B)
C Returns reduced viscosity given model (VM), shear rate (SR) and
C  parameters (B) for the model. B(1) is the reduced zero-shear-rate
C  viscosity. B(2) is the reduced time constant.
C Requires the function subprogram ASINH from Chapter 9.
      IMPLICIT DOUBLE PRECISION (A-H,O-Z)
      PARAMETER (NPAR=4)
      CHARACTER VM*2
      DIMENSION B(NPAR)
      X=B(2)*SR
      IF(VM.EQ.'BU'.OR.VM.EQ.'bu'.OR.VM.EQ.'Bu')THEN
        VIS=B(1)/(1.D0+X**.75D0)
      ELSE IF(VM.EQ.'EY'.OR.VM.EQ.'ey'.OR.VM.EQ.'Ey')THEN
        VIS=B(1)*ASINH(X)/X
      ELSE IF(VM.EQ.'IE'.OR.VM.EQ.'ie'.OR.VM.EQ.'Ie')THEN
        VIS=B(1)*(1.D0-EXP(-1.D0/X))
      ELSE IF(VM.EQ.'IN'.OR.VM.EQ.'in'.OR.VM.EQ.'In')THEN
        VIS=B(1)*(1.D0-EXP(-1.D0/X)*(1.D0/X+1.D0))
      ELSE IF(VM.EQ.'CA'.OR.VM.EQ.'ca'.OR.VM.EQ.'Ca')THEN
        VIS=B(1)/(1.D0+X)**B(3)
      ELSE IF(VM.EQ.'CR'.OR.VM.EQ.'cr'.OR.VM.EQ.'Cr')THEN
        VIS=B(1)/(1.D0+X**B(3))
      ELSE IF(VM.EQ.'SU'.OR.VM.EQ.'su'.OR.VM.EQ.'Su')THEN
        VIS=B(1)*(ASINH(X)/X)**(1.D0-B(3))
      ELSE IF(VM.EQ.'VN'.OR.VM.EQ.'vn'.OR.VM.EQ.'Vn')THEN
        X2=X**B(3)
        VIS=B(1)/(1.D0+B(4)*SQRT(X2)+X2)
      ELSE IF(VM.EQ.'DC'.OR.VM.EQ.'dc'.OR.VM.EQ.'Dc')THEN
        VIS=B(1)*((1.D0-B(3))/(1.D0+B(4)*X)+B(3)/(1.D0+X))
      ENDIF
      RETURN
      END
```

```
      DOUBLE PRECISION FUNCTION CSTEDY(X)
C Integrand for the calculation of reduced torque or axial thrust.
      PARAMETER (NPAR=4)
      IMPLICIT DOUBLE PRECISION (A-H,O-Z)
      CHARACTER VM*2
      COMMON /BLOC2/ B(NPAR),VM
      CSTEDY=X**3*VIS(VM,X,B)
      RETURN
      END
```

7.5.3 TRANST

Imbedded into the code are the subprograms CERN5A, CERN6A, GAUSSI, MTS, and PSM inherent to the execution of TRANST.

```
      PROGRAM TRANST
C Calculates transient torque for the nonlinear integral models.
      IMPLICIT DOUBLE PRECISION (A-H,O-Z)
      PARAMETER (NMAX=100,NPAR=4,TINY=1.D-7)
      CHARACTER CMOD*3
      DOUBLE PRECISION MTS
      DIMENSION T(NMAX),STRR(NMAX),STRA(NMAX),STR34(NMAX),STRA34(NMAX)
      COMMON /BLOC1/ TF,SRR,ALFA,XX
      COMMON /BLOC2/ B(NPAR)
      EXTERNAL MTS,PSM
10    CALL MSHEAD
      CALL UNITS(NTER,NINT,NOUT)
      WRITE(NTER,'(2(/4X,A)/4(/4X,A)/1(/4X,A)/2(/4X,A)/5(/4X,A))')
     &'Calculates the following quantities for transient,              ',
     &'parallel-plate torsional flow:                                  ',
     &'  STRR: stress at the outer radius (R)                          ',
     &'  STRA: apparent stress at R; 2M/pi R**3 where M is the torque.',
     &' STR34: the stress at 3R/4                                      ',
     &'STRA34: apparent stress at 3R/4; 3M/2 pi R**3                   ',
     &'The models provided are:                                        ',
     &'   PSM:  Papanastasiou, Scriven and Macosko                     ',
     &'   MTS:  Modified Tanner Simmons                                ',
     &'  The stress is calculated at specified times and fixed radii  ',
     &'(corresponding to fixed shear rates). The values and number of ',
     &'radii are fixed by Order of Integration for the Gaussian moment',
     &'integration over radius.  Must supply the G, B, TAU and XNL    ',
     &'terms for the model; see Chapter 10 for details.               '
      WRITE(NTER,'(/2X,A)')'Enter name of model (PSM or MTS).'
      READ(NTER,'(A3)')CMOD
      IF(CMOD.EQ.'psm')CMOD='PSM'
      IF(CMOD.EQ.'mts')CMOD='MTS'
20    WRITE(NTER,*)' Enter G, B, TAU, XNL;  0.,0.,0.,0. to quit.'
      READ(NTER,*)(B(I),I=1,NPAR)
      IF(B(1).LT.TINY)GOTO 120
      WRITE(NOUT,'(A,1P,E14.3,0P,F12.2,1P,E14.3,0P,F12.2/)')
     &      ' G, B, TAU, XNL:',(B(I),I=1,4)
30    WRITE(NTER,*)' Enter shear rate at outside radius; 0. to stop.'
      READ(NTER,*)SRR
      IF(SRR.LT.TINY)GOTO 20
      WRITE(NOUT,'(A,F9.2)')' Shear rate at r = R:',SRR
40    WRITE(NTER,'(/2X,A)')
     &      'Enter order (1 to 4) of integration over radius; 0 to stop.'
      READ(NTER,*)XNO
      NO=INT(XNO+TINY)
      IF(NO.LT.1)GOTO 30
      IF(NO.GT.4)GOTO 40
```

```
      WRITE(NOUT,50)NO
50    FORMAT(/1X,'Order of integration =',I2)
60    WRITE(NTER,*)' Enter starting time (t > 0.), ending time, and time
     & multiplier ( > 1).'
      READ(NTER,*)DT,ET,FT
      IF(DT.LE.0.D0.OR.FT.LE.1.D0)GOTO 60
      WRITE(NTER,70)
      WRITE(NOUT,70)
70    FORMAT(/T8,'Time',T22,'STRR',T36,'STRA',T49,'STR34',T63,'STRA34'/
     &       5(3X,11('-')))
      IFLAG=0
      DO 110 I=1,NMAX
        TF=DT*FT**(I-1)
        IF(TF.GE.ET-DT)THEN
          IFLAG=1
          TF=ET
        ENDIF
        T(I)=TF
C Do integration over reduced radius using Gaussian quadrature (GAUSSI).
        IF(CMOD.EQ.'PSM')THEN
          STRA(I)=4.D0*SRR*GAUSSI(NO,PSM)
          STRR(I)=SRR*PSM(1.D0)
          STR34(I)=.75D0*SRR*PSM(.75D0)
        ENDIF
        IF(CMOD.EQ.'MTS')THEN
          STRA(I)=4.D0*SRR*GAUSSI(NO,MTS)
          STRR(I)=SRR*MTS(1.D0)
          STR34(I)=.75D0*SRR*MTS(.75D0)
        ENDIF
        STRA34(I)=.75D0*STRA(I)
        WRITE(NTER,80)TF,STRR(I),STRA(I),STR34(I),STRA34(I)
        WRITE(NOUT,80)TF,STRR(I),STRA(I),STR34(I),STRA34(I)
80      FORMAT(1P,5E14.5)
        IF(IFLAG.EQ.1)THEN
          WRITE(NOUT,90)
90        FORMAT(/26X,'Logs of above numbers'/)
          WRITE(NOUT,100)(LOG10(T(J)),LOG10(STRR(J)),LOG10(STRA(J)),
     &                    LOG10(STR34(J)),LOG10(STRA34(J)),J=1,I)
100       FORMAT(5(F14.4))
          GOTO 40
        ENDIF
110   CONTINUE
120   CONTINUE
      CALL MSTAIL(*10)
      STOP
      END
C
      DOUBLE PRECISION FUNCTION CERN5A(S)
C Total integrand for PSM model.
C Requires the function subprogram XMEM1 from Section 10.6.7.
      IMPLICIT DOUBLE PRECISION (A-H,O-Z)
      COMMON /BLOC1/ TF,SRR,ALFA,XX
      COMMON /BLOC2/ G,B,TAU,XNL
      GF=0.D0
      CERN5A=XMEM1(TF-S,G,B,TAU,GF)*(TF-S)/
     &        (1.D0+ALFA*(SRR*XX*(TF-S))**2)
      RETURN
      END
C
      DOUBLE PRECISION FUNCTION CERN6A(S)
C Total integrand for MTS model.
C Requires the function subprogram XMEM1 from Section 10.6.7.
      IMPLICIT DOUBLE PRECISION (A-H,O-Z)
      COMMON /BLOC1/ TF,SRR,BZ,XX
      COMMON /BLOC2/ G,B,TAU,XNL
      GF=0.D0
      CERN6A=XMEM1(TF-S,G,B,TAU,GF)*(TF-S)
      RETURN
      END
```

```
      DOUBLE PRECISION FUNCTION GAUSSI(NO,FXT)
C Returns integral from 0 to 1 of third moment of FXT using Gaussian moment
C  integration. Constants from "Handbook of Mathematical Functions" p. 922.
      DOUBLE PRECISION FXT,SS,X(5,5),W(5,5)
      EXTERNAL FXT
      DATA(X(1,I),I=1,1)/.8D0/
      DATA(W(1,I),I=1,1)/.25D0/
      DATA(X(2,I),I=1,2)/.5298579359,.8987134927/
      DATA(W(2,I),I=1,2)/.0669052498,.1830947502/
      DATA(X(3,I),I=1,3)/.3632646302,.6988112692,.9379241006/
      DATA(W(3,I),I=1,3)/.0164790593,.1045998976,.1289210432/
      DATA(X(4,I),I=1,4)/.2614777888,.5358464461,.7902832300,
     &                   .9578470806/
      DATA(W(4,I),I=1,4)/.0046583671,.0425417241,.1090043689,
     &                   .0937955399/
      DATA(X(5,I),I=1,5)/.1962120074,.4171002118,.6485700042,
     &                   .8456051500,.9694357035/
      DATA(W(5,I),I=1,5)/.0015206894,.0169573249,.0604449532,
     &                   .1003165045,.0707605281/
      SS=0.D0
      DO 10 I=1,NO
        SS=SS+W(NO,I)*FXT(X(NO,I))
10    CONTINUE
      GAUSSI=SS
      RETURN
      END
C
      DOUBLE PRECISION FUNCTION MTS(X)
C Provides integrand for Gaussian integration using the MTS model.
C Requires the subroutine QROMB2 and the function subprogram XMOD1.
      IMPLICIT DOUBLE PRECISION (A-H,O-Z)
      PARAMETER (EPS=1.D-6,TINY=1.D-9,UNDER=158)
      EXTERNAL CERN6A
      COMMON /BLOC1/ TF,SRR,BZ,XX
      COMMON /BLOC2/ G,B,TAU,XNL
      XX=X
      GF=0.D0
      TZ=0.D0
      BZ=10.D0**XNL
      RATE=SRR*XX+TINY
C If rate is very high, TZ will be close to TF.
      IF(TF.GT.TINY.AND.BZ/TF/RATE.LT.UNDER)TZ=TF*EXP(-BZ/TF/RATE)
C Must use modified TZ here because first line is uv portion of
C integration-by-parts procedure of the integral from TZ to TF.
      CALL QROMB2(CERN6A,TZ,TF,SS,EPS)
      MTS=(TF-TZ)*XMOD1(TF-TZ,G,B,TAU,GF)
C Add in integral over time TZ to TF.
     & +SS
      RETURN
      END
C
      DOUBLE PRECISION FUNCTION PSM(X)
C Provides integrand for Gaussian integration using the PSM model.
C Requires the subroutine QROMB2 and the function subprogram XMOD1.
      IMPLICIT DOUBLE PRECISION (A-H,O-Z)
      PARAMETER (EPS=1.D-6)
      COMMON /BLOC1/ TF,SRR,ALFA,XX
      COMMON /BLOC2/ G,B,TAU,XNL
      EXTERNAL CERN5A
      TZ=0.D0
      GF=0.D0
      XX=X
      ALFA=EXP(-XNL)/3.D0
      CALL QROMB2(CERN5A,TZ,TF,SS,EPS)
      PSM=TF*XMOD1(TF,G,B,TAU,GF)
     &    *1.D0/(1.D0+ALFA*(SRR*XX*TF)**2)
     &    +SS
      RETURN
      END
```

7.6 Sample Listings of INPUT/OUTPUT Files

7.6.1 Data File for TWIST

Shown here is a data file in the format required for running TWIST.

```
Noisy shear rate and torque data generated with STEADY, Double Cross
       1. 1.   11
  .1000E-02  .1610E-02
  .3981E-02  .4930E-02
  .6310E-02  .7212E-02
  .1000E-01  .9074E-02
  .1585E-01  .1375E-01
  .1000E+00  .3262E-01
  .1585E+00  .4322E-01
  .2512E+00  .5438E-01
  .3981E+00  .6338E-01
  .6310E+00  .7755E-01
  .2512E+01  .1506E+00
      1.   1.   10
  .1000E+01  .1285E+00
  .1585E+01  .1439E+00
  .1585E-02  .2514E-02
  .2512E-02  .3480E-02
  .2512E-01  .1485E-01
  .3981E-01  .1796E-01
  .6310E-01  .2614E-01
  .3981E+01  .1777E+00
  .6310E+01  .1987E+00
  .1000E+02  .1895E+00
```

7.6.2 Terminal Dialog with TWIST

Reproduced here is the terminal dialog for running TWIST with the data file TWIST.DAT listed in Section 7.6.1.

```
>TWIST
 Enter name of data file (N, if not applicable).
TWIST.DAT
 Results to be written to NEW file: TWIST.RES
 OK (Y/N)?
Y
 Shear or normal stresses (S/N)?
S
 Enter fractional error wanted in corrected stress.
0.05

      Initial values:

             Log viscosity =    .01
         Log time constant =   1.00

 Revise these initial values? (Y/N)
N
 Choose viscosity model (BU or EY).
BU
```

```
  Final values for BU model:

              Log viscosity =   -.047 +/-    .082
         Log time constant =   1.501 +/-    .158

Standard error of estimate = 8.575E-02

 The results are in TWIST.RES
 Run another set of data? (Y/N)
N
Stop - Program terminated.
```

7.6.3 Output File from **TWIST**

Reproduced here is the output file from `TWIST` using the data and terminal dialog listed in Sections 7.6.1 and 7.6.2, respectively.

```
                 Parallel-Plate Torsional Flow
                 ------------------------------

Noisy shear rate and torque data generated with STEADY, Double Cross

   Final values for BU model:

              Log viscosity =   -.047 +/-    .082
         Log time constant =   1.501 +/-    .158

 Standard error of estimate = 8.575E-02
```

Shear Rate	Apparent Stress	Shear Stress	Viscosity	3R/4 Rate	3R/4 Viscosity
Run number: 1					
1.000E-03	1.025E-03	9.994E-04	9.994E-01	7.526E-04	1.025E+00
3.981E-03	3.139E-03	2.940E-03	7.386E-01	3.008E-03	7.884E-01
6.310E-03	4.591E-03	4.210E-03	6.672E-01	4.775E-03	7.276E-01
1.000E-02	5.777E-03	5.240E-03	5.240E-01	7.571E-03	5.777E-01
1.585E-02	8.754E-03	7.673E-03	4.841E-01	1.201E-02	5.523E-01
1.000E-01	2.077E-02	1.864E-02	1.864E-01	7.574E-02	2.077E-01
1.585E-01	2.751E-02	2.415E-02	1.524E-01	1.201E-01	1.736E-01
2.512E-01	3.462E-02	2.968E-02	1.181E-01	1.903E-01	1.378E-01
3.981E-01	4.035E-02	3.425E-02	8.604E-02	3.016E-01	1.014E-01
6.310E-01	4.937E-02	4.354E-02	6.901E-02	4.781E-01	7.824E-02
2.512E+00	9.587E-02	7.706E-02	3.068E-02	1.898E+00	3.817E-02
Run number: 2					
1.000E+00	8.181E-02	7.059E-02	7.059E-02	7.578E-01	8.181E-02
1.585E+00	9.161E-02	7.802E-02	4.922E-02	1.201E+00	5.780E-02
1.585E-03	1.600E-03	1.541E-03	9.720E-01	1.195E-03	1.010E+00
2.512E-03	2.215E-03	2.098E-03	8.353E-01	1.896E-03	8.819E-01
2.512E-02	9.454E-03	8.278E-03	3.295E-01	1.904E-02	3.763E-01
3.981E-02	1.143E-02	9.993E-03	2.510E-01	3.017E-02	2.872E-01
6.310E-02	1.664E-02	1.494E-02	2.367E-01	4.779E-02	2.637E-01
3.981E+00	1.131E-01	9.168E-02	2.303E-02	3.010E+00	2.842E-02
6.310E+00	1.265E-01	9.839E-02	1.559E-02	4.753E+00	2.005E-02
1.000E+01	1.206E-01	9.269E-02	9.269E-03	7.523E+00	1.206E-02

7.6.4 Terminal Dialog with STEADY

Reproduced here is the terminal dialog used for testing the accuracy of STEADY, with the Cross (CR) viscosity model having the nondimensional parameter $a = 2$, shown in Table 7.2.

```
>STEADY
 Enter name of data file (N, if not applicable).
N
 Enter name for results file.
STEADY.RES
 Results to be written to NEW file: STEADY.RES
 OK (Y/N)?
Y
 Enter viscosity model.
CR
 Enter nondimensional parameter value.
2

 Viscosity model is:  CR

                B(1) =   1.00
                B(2) =    .00
                B(3) =   2.00

      Reduced          Reduced          Reduced
       rate           viscosity          torque
    -----------     -----------      -----------
    1.00000E-03     9.99999E-01      9.99999E-01
    1.00000E-02     9.99900E-01      9.99933E-01
    1.00000E-01     9.90099E-01      9.93383E-01
    1.00000E+00     5.00000E-01      6.13706E-01
    1.00000E+01     9.90099E-03      1.90770E-02
    1.00000E+02     9.99900E-05      1.99856E-04
    1.00000E+03     9.99999E-07      1.99999E-06

 The results are in STEADY.RES
 Run another set of data? (Y/N)
N
Stop - Program terminated.
```

7.6.5 Output File from STEADY

Reproduced here is the output file from STEADY using the terminal dialog shown in Section 7.6.4. Note that the six decades of reduced shear rate and the number of points per decade are defined in the code by the integers NDEC and NPPD, respectively.

```
Viscosity model is: CR

             B(1) =   1.00
             B(2) =    .00
             B(3) =   2.00
```

Reduced rate	Reduced viscosity	Reduced torque
1.00000E-03	9.99999E-01	9.99999E-01
1.58489E-03	9.99997E-01	9.99998E-01
2.51189E-03	9.99994E-01	9.99996E-01
3.98107E-03	9.99984E-01	9.99989E-01
6.30957E-03	9.99960E-01	9.99973E-01
1.00000E-02	9.99900E-01	9.99933E-01
1.58489E-02	9.99749E-01	9.99833E-01
2.51189E-02	9.99369E-01	9.99580E-01
3.98107E-02	9.98418E-01	9.98945E-01
6.30957E-02	9.96035E-01	9.97354E-01
1.00000E-01	9.90099E-01	9.93383E-01
1.58489E-01	9.75497E-01	9.83563E-01
2.51189E-01	9.40649E-01	9.59831E-01
3.98107E-01	8.63193E-01	9.05493E-01
6.30957E-01	7.15253E-01	7.94853E-01
1.00000E+00	5.00000E-01	6.13706E-01
1.58489E+00	2.84747E-01	3.98040E-01
2.51189E+00	1.36807E-01	2.17047E-01
3.98107E+00	5.93509E-02	1.03704E-01
6.30957E+00	2.45034E-02	4.55575E-02
1.00000E+01	9.90099E-03	1.90770E-02
1.58489E+01	3.96529E-03	7.78664E-03
2.51189E+01	1.58239E-03	3.13726E-03
3.98107E+01	6.30559E-04	1.25603E-03
6.30957E+01	2.51126E-04	5.01368E-04
1.00000E+02	9.99900E-05	1.99856E-04
1.58489E+02	3.98091E-05	7.95985E-05
2.51189E+02	1.58487E-05	3.16942E-05
3.98107E+02	6.30953E-06	1.26186E-05
6.30957E+02	2.51188E-06	5.02368E-06
1.00000E+03	9.99999E-07	1.99999E-06

Logs of above numbers

-3.0000	.0000	.0000
-2.8000	.0000	.0000
-2.6000	.0000	.0000
-2.4000	.0000	.0000
-2.2000	.0000	.0000
-2.0000	.0000	.0000
-1.8000	-.0001	-.0001
-1.6000	-.0003	-.0002
-1.4000	-.0007	-.0005
-1.2000	-.0017	-.0012
-1.0000	-.0043	-.0029
-.8000	-.0108	-.0072
-.6000	-.0266	-.0178
-.4000	-.0639	-.0431
-.2000	-.1455	-.0997
.0000	-.3010	-.2120
.2000	-.5455	-.4001
.4000	-.8639	-.6634
.6000	-1.2266	-.9842
.8000	-1.6108	-1.3414
1.0000	-2.0043	-1.7195
1.2000	-2.4017	-2.1086
1.4000	-2.8007	-2.5035
1.6000	-3.2003	-2.9010
1.8000	-3.6001	-3.2998
2.0000	-4.0000	-3.6993
2.2000	-4.4000	-4.0991
2.4000	-4.8000	-4.4990
2.6000	-5.2000	-4.8990
2.8000	-5.6000	-5.2990
3.0000	-6.0000	-5.6990

7.6.6 Terminal Dialog with **TRANST**

Reproduced here is the terminal dialog for running TRANST using the PSM model.

```
>TRANST
 Enter name of data file (N, if not applicable).
N
 Enter name for results file.
TRANST.RES
 Results to be written to NEW file: TRANST.RES
 OK (Y/N)?
Y
   Calculates the following quantities for transient,
   parallel-plate torsional flow:

     STRR:  stress at the outer radius (R)
     STRA:  apparent stress at R; 2M/pi R**3 where M is the torque.
    STR34:  the stress at 3R/4
   STRA34:  apparent stress at 3R/4; 3M/2 pi R**3

   The models provided are:

      PSM:  Papanastasiou, Scriven and Macosko
      MTS:  Modified Tanner Simmons model

     The stress is calculated at specified times and fixed radii
   (corresponding to fixed shear rates).  The values and number of
   radii are fixed by Order of Integration for the Gaussian moment
   integration over radius.  Must supply the G, B, TAU and XNL
   terms for the model; see Chapter 10 for details.

 Enter name of model (PSM or MTS)
PSM
 Enter G, B, TAU, XNL: 0.,0.,0.,0. to quit.
1.2E+4  2.30  0.04  1.00
 Enter shear rate at outside radius; 0. to stop.
1.

 Enter order (1 to 4) of integration over radius; 0. to stop.
1
 Enter starting time (t > 0.), ending time, and time multiplier ( > 1).
0.01 10. 3.1623

      Time          STRR          STRA          STR34         STRA34
   -----------   -----------   -----------   -----------   -----------
   1.00000E-02   6.57660E+01   6.57662E+01   4.93247E+01   4.93247E+01
   3.16230E-02   1.75693E+02   1.75699E+02   1.31776E+02   1.31775E+02
   1.00001E-01   4.47760E+02   4.47930E+02   3.35975E+02   3.35948E+02
   3.16234E-01   1.06808E+03   1.07191E+03   8.04553E+02   8.03931E+02
   1.00003E+00   2.21106E+03   2.27934E+03   1.72105E+03   1.70951E+03
   3.16239E+00   2.92228E+03   3.35687E+03   2.60750E+03   2.51765E+03
   1.00000E+01   2.51086E+03   2.99181E+03   2.35930E+03   2.24386E+03

 Enter order (1 to 4) of integration over radius; 0. to stop.
0
 Enter shear rate at outside radius; 0. to stop
0
 Enter G, B, TAU, XNL: 0.,0.,0.,0. to quit.
0 0 0 0

 The results are in TRANST.RES
 Run another set of data? (Y/N)
N
Stop - Program terminated.
```

7.6.7 Output File from TRANST

Reproduced here is the output file from TRANST using the terminal dialog shown in Section 7.6.6.

```
G, B, TAU, XNL:       1.200E+04          2.30       4.000E-02          1.00

Shear rate at r = R:       1.00

Order of integration = 1

      Time           STRR           STRA           STR34          STRA34
   -----------    -----------    -----------    -----------    -----------
   1.00000E-02    6.57660E+01    6.57662E+01    4.93247E+01    4.93247E+01
   3.16230E-02    1.75693E+02    1.75699E+02    1.31776E+02    1.31775E+02
   1.00001E-01    4.47760E+02    4.47930E+02    3.35975E+02    3.35948E+02
   3.16234E-01    1.06808E+03    1.07191E+03    8.04553E+02    8.03931E+02
   1.00003E+00    2.21106E+03    2.27934E+03    1.72105E+03    1.70951E+03
   3.16239E+00    2.92228E+03    3.35687E+03    2.60750E+03    2.51765E+03
   1.00000E+01    2.51086E+03    2.99181E+03    2.35930E+03    2.24386E+03

                          Logs of above numbers

      -2.0000         1.8180         1.8180         1.6931         1.6931
      -1.5000         2.2448         2.2448         2.1198         2.1198
      -1.0000         2.6510         2.6512         2.5263         2.5263
       -.5000         3.0286         3.0302         2.9056         2.9052
        .0000         3.3446         3.3578         3.2358         3.2329
        .5000         3.4657         3.5259         3.4162         3.4010
       1.0000         3.3998         3.4759         3.3728         3.3510
```

References

1. Katsaros, J. D.; Malone; M. F.; Winter, H. H.: "The Effects of Flow on Miscibility in a Blend of Polystyrene and Poly(vinyl Methyl Ether)," *Polym. Eng. Sci.* **29**, 1434-1445 (1989).
2. Hindawi, I. A.; Higgins; J. S.; Weiss, R. A.: "Flow-Induced Mixing and Demixing in Polymer Blends," *Polymer* **33**, 2522-2529 (1992).
3. Cheng, D. C.-H.; "Further Observations on the Rheological Behaviour of Dense Suspensions," *Powder Technol.* **37**, 255-273 (1984).
4. Metzner, A. B.: "Rheology of Suspensions in Polymeric Liquids," *J. Rheol.* **29**, 739-775 (1985).
5. Powell, R. L.: "Rheology of Suspensions of Rodlike Particles," *J. Stat. Phys.* **62**, 1073-1094 (1991).
6. Soskey, P. R.; Winter, H. H.: "Large Step Shear Strain Experiments with Parallel-Disk Rotational Rheometers," *J. Rheol.* **28**, 625-645 (1984).
7. Ganani, E.; Powell, R. L.: "Transient Flow Viscometry," *J. Rheol.* **29**, 913-941 (1985).

8. Cross, M. M.; Kaye, A.: "Simple Procedures for Obtaining Viscosity/Shear Rate Data From a Parallel Disc Viscometer," *Polymer* **28**, 435-440 (1987).
9. Cross, M. M.: "Rheology of Non-Newtonian Fluids: A New Flow Equation for Pseudoplastic Systems," *J. Colloid Sci.* **20**, 417-437 (1965).
10. Cross, M. M.: "Analysis of Flow Data on Molten Polymers," *Eur. Polym. J.* **2**, 299-307 (1966).
11. Geiger, K.: "Weissenberg-Rabinowitsch-Auswertung der mit dem Platte-Platte-Rotationsrheometer gemessenen Fließkurve mittels eines Fließgesetzes vom Carreauschen Typ," *Rheol. Acta* **27**, 209-211 (1988).
12. Abramowitz, M.; Stegun, I. A. (Eds.): *Handbook of Mathematical Functions with Formulas, Graphs and Mathematical Tables.* Dover, New York, 1965; p. 921.
13. Hershey, H. C.; Zakin; J. L.; Simha, R.: "Numerical Differentiation of Equally Spaced and Not Equally Spaced Experimental Data," *I&EC Fund.* **6**, 413-421 (1967).
14. Walters, K.: *Rheometry.* Chapman and Hall, London, 1975.
15. Press, W. H.; Flannery, B. P.; Teukolsky, S. A.; Vetterling, W. T.: *Numerical Recipes in* FORTRAN: *The Art of Scientific Computing.* Cambridge University Press, Cambridge, 1986.
16. Papanastasiou, A. C.; Scriven; L. E.; Macosko, C. W.: "An Integral Constitutive Equation for Mixed Flows: Viscoelastic Characterization," *J. Rheol.* **27**, 387-410 (1983).

Chapter 8

Capillary Data Analysis

8.1 Introduction

Capillary rheometry is a well-established method for determining the shear viscosity function of polymer melts. With readily available commercial equipment, capillary rheometry can generate viscosity data for melts having viscosities up to 1 MPa•s over a shear rate range of 0.1 to 10^4 s^{-1}. The procedures for running the experiment and analyzing the data are widely available [1-8]. In this chapter, aside from the conventional approach, a recently introduced alternative is offered to handle capillary data analysis [9]. Both analytical techniques apply to steady, isothermal flows in capillaries having circular cross sections. Figure 8.1 shows the geometry of the capillary experiment.

The objective of capillary data analysis is to transform the raw information – total flow rate and total pressure drop – to corresponding shear-rate and shear-stress data at the capillary wall. Despite the availability of such analyses, they are rarely applied in full. There are probably many reasons that account for this shortcoming. The primary reason is the time aspect; conventional procedures require data from an extensive set of capillaries. The minimum requirement would be a set of four capillaries, that is, two different radii each with two different aspect ratios (L/R). Even with the availability of the capillaries, the time to carry out the experiments may be prohibitive. Another reason is the likelihood that attempts to correct raw data produce unexpected results such as negative end corrections from the Bagley plots. As will be discussed further, an explanation for this can be the presence of significant slip velocities along the capillary wall. Other factors include the influences of viscous heating and pressure on the melt viscosity.

The normal procedure for the reduction of isothermal capillary data involves three separate steps. The first step is to correct for the pressure drop at the capillary ends using the Bagley procedure [10]:

$$\sigma_w = \frac{P}{2(L/R - e)} \tag{8.1}$$

where σ_w is the true shear stress at the wall of the capillary of radius R and length L, P is the total pressure drop, and e is the end correction dependent on the apparent, uncorrected shear rate. The next step attempts to correct for slip between the melt and the capillary wall using the Mooney slip equation [11]:

$$V_s = \left[\frac{\partial \dot{\gamma}_{a,u}}{\partial(1/R)} \right]_{\sigma_w} = \left[\frac{\partial(4Q/\pi R^3)}{\partial(1/R)} \right]_{\sigma_w} \tag{8.2}$$

where V_s is the slip velocity, $\dot{\gamma}_{a,u}$ is the apparent, uncorrected shear rate, and Q is the total volumetric flow rate. The contribution from slip velocity, multiplied by the capillary cross section, can be deducted from the total flow to obtain the flow due solely to the shearing of the melt. The last step is a velocity-profile correction using the Rabinowitsch-Weissenberg-Mooney equation [11,12],

$$\dot{\gamma}_w = \dot{\gamma}_a \left[\frac{3}{4} + \frac{1}{4} \frac{d\ln \dot{\gamma}_a}{d\ln \sigma_w} \right] \tag{8.3}$$

where $\dot{\gamma}_w$ is the shear rate at the wall and $\dot{\gamma}_a$ is the apparent shear rate corrected for slip.

In addition to the capillary end corrections, there are barrel corrections including the effects of barrel pressure drop and conflicting flow patterns as the piston approaches the capillary [2,3]. Pressure gradients in the barrel are negligible only if the capillary is long and narrow relative to the barrel. One method to handle the barrel corrections is to add the constraint of constant capillary radius to the Bagley procedure in Equation 8.1. While the barrel corrections are usually negligible compared with the three corrections discussed, we have added the constraint of constant R to the Bagley procedure in the program CAPSEQ.

When the corrections are done in order as prescribed by Equations 8.1 to 8.3, it implies the following assumptions: First, the entrance pressure loss, P_e, which is the principal end correction, is assumed to depend only on the apparent, uncorrected shear rate $\dot{\gamma}_{a,u}$, therefore allowing Equation 8.1 to be rewritten as

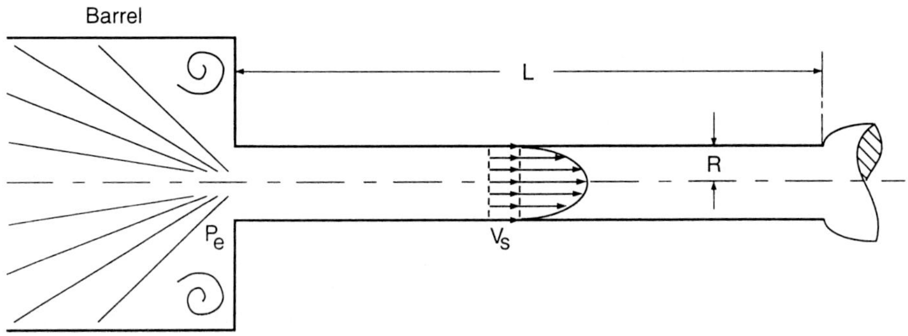

Figure 8.1 Schematic of the capillary flow experiment showing the velocity profile, slip velocity (V_s), and the flow patterns leading to entrance pressure losses (P_e).

$$\sigma_w = \frac{R[P - P_e(\dot{\gamma}_{a,u})]}{2L} \tag{8.4}$$

Second, the slip velocity is assumed to be a function of the true shear stress at the capillary wall, which is why the determination of the entrance pressure loss is a prerequisite to the calculation of the slip velocity. It has been recognized that small errors in wall shear stress can lead to large errors in slip velocity [13,14]. The apparent stress, based on the total pressure drop, could be much higher than that which is causing the melt to slip and is thus inappropriate. Given constant σ_w, Equation 8.2 can be integrated to yield the familiar form of the Mooney slip equation:

$$\dot{\gamma}_a = \dot{\gamma}_{a,u} - \frac{4V_s(\sigma_w)}{R} = \frac{4\left[Q - \pi R^2 V_s(\sigma_w)\right]}{\pi R^3} \tag{8.5}$$

The velocity-profile correction, which is performed last, depends on the assumption of a unique relationship between σ_w and $\dot{\gamma}_a$. Therefore, the slip velocity must be determined prior to any velocity-profile correction. The program `CAPSEQ` was specifically designed to handle these three steps sequentially.

A problem exists with the above correction sequence in that the argument for the entrance pressure loss function $P_e(\dot{\gamma}_{a,u})$ contains the total flow term Q, while the remainder of the pressure drop, $P - P_e$, which reflects the shear stress in the melt within the capillary, can only be a function of the apparent shear rate corrected for slip, $\dot{\gamma}_a$. Hence, the normal Bagley procedure is valid only if the slip velocities are low or if the radii of the capillaries are invariant. In the latter case, $\dot{\gamma}_{a,u}$ and $\dot{\gamma}_a$ will differ by a constant that will not vary with capillary length given the assumptions of the Mooney slip equation (Equation 8.2). A plot of P versus L/R at constant $\dot{\gamma}_{a,u}$ and R will then be linear with a slope of $2\sigma_w$ and an intercept of P_e.

The problems associated with the Bagley plot using variable capillary radii are illustrated in Figure 8.2, where the pressures and flow rates were calculated for realistic functions for $V_s(\sigma_w)$ and $P_e(\dot{\gamma}_{a,u})$. Slip depresses the slope; however, with constant-radius capillaries (R = 0.64 mm in Figure 8.2), the intercept (P_e) is correct. If capillaries of different radii were used (e.g., R = 0.64, 1.37, and 7.54 mm for aspect ratios L/R of 160, 80, and 40, respectively), the entrance pressure would be overestimated by 40%.

If severe slip is suspected, not only is a set of constant-radius capillaries required, but capillaries of different radii are needed to correct for slip via the Mooney equation. An even more controlled experiment would hold the total pressure constant as well, requiring a set of capillaries of various radii with constant aspect ratios.

In principle, a complete capillary matrix of constant radii and constant aspect ratios would be unnecessary if the functional forms for slip velocity $V_s(\sigma_w)$ and

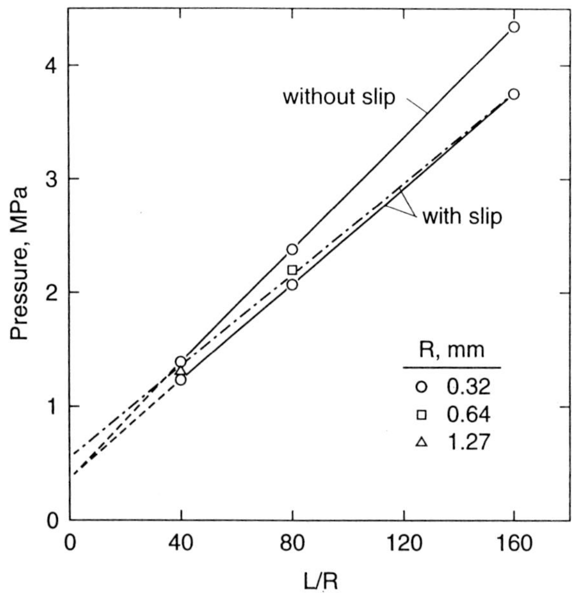

Figure 8.2 Simulation of errors possible with the Bagley procedure [10] using capillaries of variable radius when slip is present. The broken line, drawn through points obtained with capillaries of different radii, overestimates the entrance pressure loss by 40% [9].

entrance pressure $P_e(\dot{\gamma}_{a,u})$ were known beforehand. With these functions in hand, P-versus-Q data from an arbitrary collection of capillaries could be fully corrected. Furthermore, the corrections could be done with fewer capillaries. This concept was the basis for the development of an alternative method of analysis [9]. The features of the alternative program, CAPSIM, include the simultaneous correction, from an arbitrary set of capillaries, of P-versus-Q data to a viscosity function and the ability to cope with a minimum number of capillaries.

Aside from Equations 8.4 and 8.5, the last relation necessary for performing the simultaneous end and slip corrections is a relationship that describes the flow in the capillary as a function of the apparent shear rate:

$$\sigma_w = f(\dot{\gamma}_a) \tag{8.6.1}$$

$$\dot{\gamma}_a = g(\sigma_w) \tag{8.6.2}$$

where f is the flow function and g is the inverse flow function. Once the proper flow function is found, the inverse function g can be substituted into Equation 8.3 for the velocity-profile correction to yield

$$\dot{\gamma}_w = \frac{3}{4}\dot{\gamma}_a + \frac{1}{4}\sigma_w g'(\sigma_w) \tag{8.7}$$

where $g'(\sigma_w)$ is the first derivative of the inverse flow function. The use of the inverse flow function in the preceding equation is convenient but unnecessary. Alternatives for evaluating the derivative of the flow curve ($\dot{\gamma}_a$ versus σ_w) can be applied once the apparent shear rate is determined.

To satisfy Equations 8.4 to 8.6 simultaneously, we must choose forms for the slip velocity, entrance pressure, and flow functions. If complex forms for these functions were assigned, highly accurate corrections could be made, given sufficiently precise data. On the other hand, given typical experimental data, convergence problems could occur with such overparameterization. A compromise was made in the development of CAPSIM by limiting the number of parameters from four to six, the final count depending on whether correction for slip is warranted or feasible and on the flow function better suited to the data.

To obtain the forms for these functions, we examined several published sets of capillary data [10,14-34]. For entrance pressure loss, Han and Kim [19] reported a power-law dependence on shear rate. Subsequent review of the literature [10, 15-27] verified this relationship. Using the nomenclature introduced above, we can write the entrance-pressure function as

$$P_e(\dot{\gamma}_{a,u}) = \beta_1 \dot{\gamma}_{a,u}^{\beta_2} \tag{8.8}$$

where β_1 and β_2 are the entrance-pressure parameters. Similarly, a power-law relation was found to be adequate for the slip-velocity function [9,14]. The slip-velocity function is expressed as

$$V_s(\sigma_w) = \beta_3 \sigma_w^m \tag{8.9}$$

where β_3 is the slip-velocity coefficient and m is the slip-velocity exponent. Hatzikiriakos and Dealy [34] found a strong temperature dependence for β_3, whereas m is practically independent of temperature. In CAPSIM, m is fixed to reduce the number of search parameters in the minimization procedure. From published data [27-34], we found that the slip-velocity exponent m may be assumed to be equivalent to the inverse of the exponent in the power-law flow function in the region where slip is most likely to occur. Ramamurthy [32] and Hatzikiriakos and Dealy [14,34] report a critical wall shear stress region of between 90 and 180 kPa above which the assumption for the occurrence of slip is valid.

Fortunately, the form for the flow function for polymer melts has been thoroughly investigated, although in terms of the viscosity function. Several empirical expressions for the viscosity function are reviewed in Chapter 9. In CAPSIM, a choice between the two-parameter power-law and the three-parameter Cross [35,

36] models is available. In terms of a flow function, the power-law model takes the form

$$\sigma_w = \beta_4 \, \dot{\gamma}_a^{\beta_5} \tag{8.10}$$

where β_4 is a consistency index and β_5 is the power-law index. If the Cross viscosity model is used, the form of the flow function becomes

$$\sigma_w = \frac{\beta_4 \, \dot{\gamma}_a}{1 + (\beta_5 \, \dot{\gamma}_a)^{\beta_6}} \tag{8.11}$$

where β_4 is the zero-shear-rate viscosity, β_5 is the characteristic time constant, and β_6 is a nondimensional parameter. Whereas the application of Equation 8.10 is appropriate in the shear-thinning region, Equation 8.11 has a broad response extending to lower shear rates where Newtonian behavior is expected. While it might be argued that the form for the viscosity function is unimportant because it simply relays values of shear stress between Equations 8.5 and 8.6, it is conceivable that a slight distortion of the data may result if the choice is incorrect. For example, the Cross model may introduce curvature at both ends of an otherwise linear relationship between log η and log $\dot{\gamma}$.

8.2 Description of the Algorithms

The program `CAPSEQ` is best suited for constant-flow-rate data compiled from an adequate set of capillaries such that the sequential order for the correction of systematic errors can be followed. A minimum of two different aspect ratios is required for *each* capillary radius size. While `CAPSEQ` will run given just one radius size by skipping the slip correction, the statistical aspects of the end and velocity-profile corrections will not be as favorable. The alternative method `CAPSIM` is recommended if less than four, or arbitrary sets of, capillaries are used, although `CAPSIM` can be applied to extensive data sets as well.

8.2.1 CAPSEQ

In preparation for the constrained Bagley end-correction procedure, the algorithm starts by locating constant-radius capillaries of different lengths to address the problem associated with end corrections in the presence of slip. Recall, from Figure 8.2, that an analysis with nonconstant radii in the presence of significant slip can lead to poor end corrections with the Bagley procedure. For each radius size, data are sorted out by apparent, uncorrected shear rate, $\dot{\gamma}_{a,u}$. From Equation 8.4, the wall shear stress, σ_w, is calculated by using a first-order linear regression of P versus L/R via the subroutine ALINR. ALINR is programmed to return stress values given only two points.

The slip correction becomes necessary when stress-rate (σ_w versus $\dot{\gamma}_{a,u}$) data, obtained with different radius capillaries, do not collapse onto a single curve. Two initial conditions must be satisfied before proceeding with slip correction. At least two different radius sizes are required to avail of the Mooney slip equation (Equation 8.5). Second, the range of radius sizes in the data set must vary by at least a factor of 2. A wider R range increases the ability to extract more accurate values. If neither condition is met, the algorithm skips ahead to the velocity-profile correction procedure.

When correction for slip is warranted, Equation 8.5 dictates a linear plot of $\dot{\gamma}_{a,u}$ versus $1/R$ at constant σ_w with an intercept equal to the apparent shear rate corrected for slip, $\dot{\gamma}_a$. Prior to this, a cross plot must be performed to obtain constant-stress data; such a plot is illustrated in Figure 8.3. With the capillary radius size as a parameter, a flow function is fitted to plots of σ_w versus $\dot{\gamma}_{a,u}$. To retain consistency with the alternative method CAPSIM, the choice for the flow function is limited to those expressed in Equations 8.10 and 8.11 and is made by comparing the standard error of estimate for each model. To facilitate the fitting processes, both equations are rescaled in terms of the logarithmic values of shear stress and shear rate. The rescaling of the power-law function reduces Equation 8.10 to a simple linear regression analysis which is handled by ALINR. The fitting of the three-parameter Cross function (Equation 8.11) is accomplished by the nonlinear regression technique GAUSSN introduced in Chapter 1. The initial estimates for fitting the Cross model are found by procedures similar to those outlined in Chapter 9. Thus, the initial estimate for the zero-shear-rate viscosity parameter (β_4 in Equation 8.11) is obtained by sorting the flow data and using the viscosity value at the lowest shear rate. At high shear rates, the parameter β_6 is derivable from the power-law index parameter, β_5 in Equation 8.10, by the equation

$$\beta_6 = 1 - (\beta_5)_{\text{Equation 8.10}} \tag{8.12}$$

The parameter β_5 is available because the power-law flow function is fit prior to the Cross function.

As opposed to nonlinear flow functions, higher-order logarithmic polynomial functions are easier to program and provide a faster, surer solution. However, fitting a polynomial function to data that is inherently nonlinear becomes unsuitable when situations arise that call for extrapolation. Given the requirement for constant-stress data, situations will occur that require substantial extrapolation because the shear-rate range is typically a strong a function of capillary radius. Referring to Figure 8.3, the application of a third-order polynomial would underestimate the value of $4Q/\pi R^3$ compared to the extrapolation using the Cross model. Although not shown in the

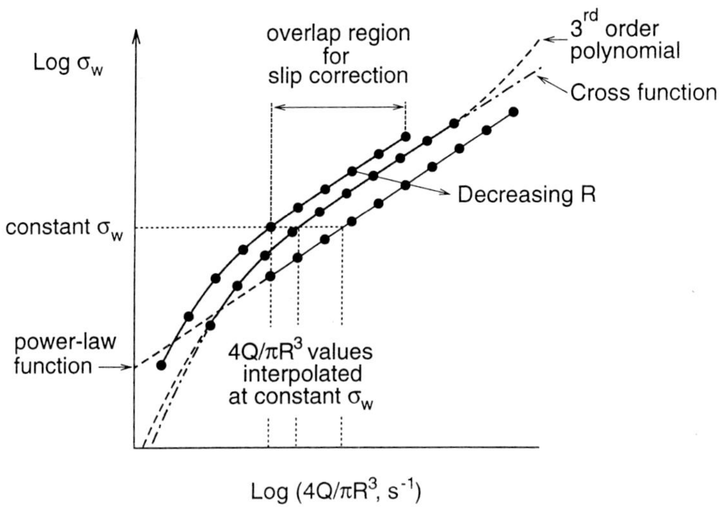

Figure 8.3 Schematic illustrations of the proper analysis and errors in slip velocity corrections. The flow curves are widely spaced on the ordinate for clarity.

figure, fitting a second-order polynomial exacerbates the situation. As seen in Figure 8.3, even an extrapolation via a power-law function (first-order polynomial) results in a crossover with data from larger capillaries at lower shear rates. To guard against such undesirable events, three statistical conditions are used, which are discussed in more detail later. Regardless of the type of empirical relationship, extrapolation is intrinsically dangerous. In fact, a more accurate analysis would limit the shear-rate window to that covered only by every capillary radius size, as illustrated in Figure 8.3. The alternative method, `CAPSIM`, avoids extrapolations altogether.

For each value of wall shear stress, σ_w, obtained from the constant-radius Bagley plot, interpolated values of the apparent, uncorrected shear rate ($\dot{\gamma}_{a,u}$) are found for all other capillary radii (see "constant σ_w" line in Figure 8.3). This step is repeated for each value of wall shear stress obtained from the Bagley plot. Interpolation to find $\dot{\gamma}_{a,u}$ is a simple process for the linearized power-law function. The subroutine `EQSOLN` is employed when the nonlinear Cross function is chosen. `EQSOLN` uses the Newton-Raphson iterative method [37].

The apparent shear rate is obtained by calling the subroutine `ALINR` to provide the solution to the Mooney slip equation (Equation 8.5). At this juncture, three final conditions must be satisfied before proceeding with any slip correction. No correction is attempted if a negative value for either the intercept or slope is obtained. While being physically meaningless, it is not a remote possibility in instances of viscous heating or, as discussed above, with extrapolated data from a constant-shear-stress set at low shear rates. Finally, for sets with at least three different radii, a statistical test for zero slope is performed. The correction for slip proceeds if the hypothesis that the slope is zero can be rejected at a 0.80 level of significance; otherwise, $\dot{\gamma}_a = 4Q/\pi R^3$.

To proceed with the velocity-profile correction, we calculate the term $d(\ln \dot{\gamma}_a)/d(\ln \sigma_w)$ in Equation 8.3 by using a derivative-finding method in the subroutine SLOPE. For capillary data, SLOPE fits a line to a moving strip of unevenly spaced points to determine the derivative. Prior to calling SLOPE, the subroutine SORT2 sorts the shear-stress data from all capillaries into ascending order while making the corresponding rearrangement of the shear-rate data.

8.2.2 CAPSIM

The program CAPSIM was designed to handle simultaneously the tedious corrections for systematic errors and to cope with a minimum number or an arbitrary set of capillaries by assuming forms for the $P_e(\dot{\gamma}_{a,u})$ and $V_s(\sigma_w)$ functions. Once the functional forms for the entrance pressure loss, slip velocity, and flow are established, such as those described in Equations 8.8 to 8.11, the optimization of the calculated pressure drops, with respect to the data, is performed by adjustment of the parameters defining these functions.

For constant-flow-rate viscometers, the pressure-drop data is expected to contain all the experimental error. Hence, the objective function, supplied in the subroutine FCAP, is expressed as

$$\sum_i \left[\frac{P_i - P_{c,i}}{P_i} \right]^2 \tag{8.13}$$

where P_i and $P_{c,i}$ are the experimentally observed and the corresponding calculated pressures, respectively. The denominator is the weighting factor and reflects the expectation that the pressure measuring device will have constant relative error. The minimization of the objective function with simple bounds on its parameters is accomplished by the IMSL routine DBCONF using a quasi-Newton method and a finite-difference gradient [38]. The calculated pressure $P_{c,i}$ is determined by substituting Equation 8.8 into Equation 8.4 to get

$$P_{c,i} = \beta_1 \dot{\gamma}_{a,u}^{\beta_2} + \frac{2L}{R} \sigma_w(\dot{\gamma}_a) \tag{8.14}$$

where the wall shear stress $\sigma_w(\dot{\gamma}_a)$ is calculated from the chosen flow function (Equation 8.10 or 8.11) in the function subprogram STR. The apparent shear rate is derived by substituting the slip-velocity function (Equation 8.9) into the Mooney slip equation (Equation 8.5):

$$\dot{\gamma}_a = \dot{\gamma}_{a,u} - \frac{4}{R} \beta_3 [\sigma_w(\dot{\gamma}_a)]^m \tag{8.15}$$

The solution for $\dot{\gamma}_a$ is obtained by using the Newton-Raphson iterative method [37] in the subroutine EQSOLN with the apparent, uncorrected shear rate as the initial estimate. Once again, information from the chosen flow function is relayed into the preceding relation.

For this type of optimization to succeed, good starting values for the parameters are extremely important. From a survey of available data [24-27,39,40], the entrance pressure function parameters expressed in Equation 8.8 were found to scale with the relation

$$P \sim \beta_1^o (\tau\dot{\gamma}_{a,u})^{\beta_2^o} \tag{8.16}$$

where P is the total pressure drop, β_i^o are the initial estimates for the search parameters, β_i, and τ is a characteristic time constant. If the Cross model is chosen as the suitable flow function, τ is set to the estimate of the time constant parameter β_5 in Equation 8.11; otherwise, τ is set to the reciprocal of the lowest shear rate from the data set. From a logarithmic plot of Equation 8.16, β_i^o can be found from linear regression.

Dealy and Wissbrun [8] and Kalika and Denn [33] discuss a method for estimating the slip velocity on the basis of power-law flow behavior. The method was used as a basis in developing a correlation, found from available data sets [27-34], to obtain a parameter estimate for the slip-velocity coefficient β_3. With respect to the power-law flow function parameters defined in Equation 8.10, the estimation is expressed as

$$\log \beta_3^o = 1.2 \log\left(\beta_4^{-1/\beta_5} \frac{\beta_5}{3\beta_5 + 1} \right) \tag{8.17}$$

where β_3^o is the initial estimate for the slip-velocity coefficient with slip velocity defined in millimeters per second and β_4 and β_5 are the power-law parameters in the apparent flow function ($RP/2L$-versus-$4Q/\pi R^3$) above 100 kPa. The slip-velocity exponent, m, also estimated from this flow region, is just $1/\beta_5$.

The slip-velocity parameter, β_3, turned out to be the most critical of all the parameters in the minimization procedure. A range of 21 decades for β_3 was found by examining plots from the literature; this contrasts to the three decades for the entrance pressure parameter β_1. Fixing the slip-velocity function exponent, m, to a constant value during the minimization procedure turns out to be a small compromise in limiting the number of function parameters and it has the additional effect of reducing the range of β_3. Attempts to float m or defining it in terms of the power-law index from the flow function do not provide better results from either simulated or real data – a consequence of parameter interaction.

The determination of the flow function better suited to the data is handled in the same fashion as that outlined in Section 8.2.1, and provides the initial estimates for the parameters in the flow function used in the minimization process.

Once the flow characterization parameters are known, the corresponding errors are approximated by the method in Section 2.2 via the subroutine ERROR. The error in each parameter is estimated by the expression

$$\pm\ t_{95,\,\nu}\ s_{\varepsilon}\sqrt{C_{ii}} \tag{2.9}$$

where $t_{95,\,\nu}$ is the t-statistic as a function of the degrees of freedom, ν; s_{ε} is the standard error of estimate, and C_{ii} is the diagonal element for the estimated covariance matrix of the standard errors in the fitted parameters [37].

To obtain the viscosity at the wall, $\eta(\dot{\gamma}_w)$, the velocity-profile correction is accomplished with the derivative-finding routine SLOPE introduced in the discussion of CAPSEQ. For each value of $\dot{\gamma}_w$, the viscosity is also calculated by using the alternative velocity-profile correction scheme (Equation 8.7). Recall that Equation 8.7 uses the derivative of the inverse flow function, $g'(\sigma_w)$, to perform the correction. The equation-solving routine EQSOLN is used to solve for $\dot{\gamma}_a$ given the $\dot{\gamma}_w$ information found from SLOPE. The expressions for Equation 8.7, based on either the power-law or Cross flow function, are contained in the function subprogram FVPC. Once the solution to $\dot{\gamma}_a$ is found, the wall shear stress is extracted from the flow function such that the viscosity may be calculated.

8.3 Execution of the Programs

The execution of both programs require the external subroutines ALINR, EQSOLN, GAUSSN, SLOPE, and SORT2 found in Appendix B. Refer to Appendix A for the program dependencies. For example, the routine SLOPE calls POLYN to fit a polynomial function. In turn, POLYN uses the *LU* matrix decomposition routines available from *Numerical Recipes* [37]. In addition, the subroutines CAPCON (Section 8.5.3) and FXCR (imbedded in VISFIT, Section 9.5) are used in both programs while the optimization and error analysis procedures in CAPSIM require the IMSL routines DBCONF and DFDHES [38] and the subroutines ERROR and SPECC from Appendix B.

The data file format requirements are identical and must be set up by using an editor or text processor that outputs ASCII code. The first record is the title of the file that will be copied into the output file; insert anything up to 79 characters containing identification information. The second record requires an integer entry denoting the data's dimensional units. The integer is chosen from Table 8.1, which features four different combinations of dimensional units corresponding to length, force, and time. The subroutine CAPCON proceeds to convert the data into the SI

Table 8.1 The Group Number, to Be Entered into the Second Record of the Data File, Corresponding to a Combination of Dimensional Units

Group No.	Length	Force	Time
1	in	lb_f	min
2	cm	kg_f	min
3	mm	kg_f	s
4	mm	kN	min

system. The third record is for the diameter of the reservoir or barrel. The succeeding records are for data sets, in no particular order, from each capillary. For each set, the top record includes the radius and length of the capillary plus an integer corresponding to the number of extrusion velocities. The succeeding records are for pairs of extrusion velocity and force data. A sample listing of a data file is provided in Section 8.6.1.

8.3.1 CAPSEQ

The execution of CAPSEQ is noninteractive; only the names for the input/output files must be provided. In instances where the data are found to be insufficient to proceed properly with sequential correction, CAPSEQ will stop with a message to that effect. The contents of the output file include a tabulation, sorted in ascending order of stress, of the apparent shear rate, the shear rate and shear stress at the wall, and the viscosity. A sample listing of an output file is provided in Section 8.6.3.

8.3.2 CAPSIM

The execution of CAPSIM requires three additional steps of user interaction. A sample listing of the terminal dialog involved in running CAPSIM is provided in Section 8.6.2. The first interaction in CAPSIM requires the input of the convergence criterion for the GAUSSN subroutine used in the fitting of the Cross flow function. The default value is 10^{-4}. However, when searching for the proper flow function, convergence for the Cross model may be difficult with data sets exhibiting strictly power-law flow behavior and may result in computational errors such as DOMAIN (-SQRT) or OVERFLOW. A larger value for the convergence criterion, but less than 1, should then be used. The second interaction is required if correction for slip velocity appears feasible. Unlike the restrictions imposed on the Mooney slip equation by CAPSEQ, CAPSIM makes an arbitrary decision based on the number of capillaries, the range of radius sizes, and the number of data points above a critical apparent shear stress of 100 kPa. At this point, the option to suppress or continue with the correction becomes available. Suppression may be needed for comparison with literature results where no slip correction had been introduced.

If one wishes to continue with the slip correction, CAPSIM provides an estimate and the 95% confidence interval of the slip-velocity exponent, m, based on the power-law index of the $RP/2L$-versus-$4Q/\pi R^3$ flow curve above 100 kPa. In terms of the true flow curve, expect the estimate of m to be on the low side. The upper bound of the confidence interval may be a better alternative, or, if the information is available beforehand, a new value for m may be entered. For a variety of melts, Shaw [9] observed an average value of 3, whereas Hatzikiriakos and Dealy [14] list a range of 1.92 - 6.00 for several polyethylenes.

The output file provides a tabulation of data similar to that produced in CAPSEQ. In addition, the tabulation includes the calculated values for viscosity, based on the flow function $\sigma_w(\dot{\gamma}_a)$ chosen, and the corresponding residuals. CAPSIM provides results for the flow-characterization parameters β_i, the slip-velocity exponent, the corresponding 95% confidence intervals, and the mean relative deviation from Equation 8.13. A sample listing of an output file is provided in Section 8.6.4.

8.4 Application of the Programs

The first sets of data were numerically simulated to test the ability of CAPSIM to recover the simulation flow parameters and to ensure that no systematic problems were present in the structure of either CAPSEQ or CAPSIM. For the latter, it provided a test to ensure that global convergence was being attained. A full complement of six flow-characterization parameters, listed in Table 8.2, was applied to generate 90 data points over 4.5 decades of shear rate from nine different capillaries. The simulation values were recovered with narrow confidence intervals. In running CAPSIM, the data set required a computation time of about 135 s on an 80386, 25-MHz microcomputer and 5.6 s on an IBM ES/9000 mainframe.

Additional simulations were produced by adding random deviates to the pressure data generated from the assumed functions. The results for the simulated data with absolute deviations of 1% and 5% are summarized in Table 8.2. With the simulated viscosity curve as a reference, Figure 8.4 compares the uncorrected and corrected data using CAPSIM on the set with 5% deviation. The Figure 8.4 inset shows the CAPSEQ results using the original, noise-free data set. In both plots, the agreement is excellent. The final flow-characterization parameter values are within error of the simulation values. Based on a flow curve above 100 kPa, the calculated slip-velocity exponent was around 2.4 ± 0.2 for both sets. This is 20% off the simulation value, but was not unexpected since the program, prior to the minimization procedure, only has *RP/2L*, not σ_w, information. Noteworthy is that, regardless of the slip-velocity exponent value, the errors for the slip-velocity coefficient β_3 were large when error was introduced into the data sets. Nevertheless, the overall performance in minimizing the objective function (Equation 8.13) and recovering the simulated parameters was satisfactory. The important parameters, β_4 to β_6, are recovered with reasonable accuracy, even at the 5% error level.

Table 8.2 Application of `CAPSIM` to Numerically Simulated Data (90 Points over 4.5 Decades of Shear Rate) Using a Flow Function Based on the Cross Viscosity Model [35,36]

		Deviation, %[a]		
Function parameter	Simulation value	1[b]	1	5
Entrance pressure				
β_1, kPa•s$^{\beta_2}$	150	156. ± 7.5	156. ± 7.5	186. ± 41
β_2	0.600	0.596 ± 0.006	0.596 ± 0.006	0.578 ± 0.032
Slip velocity				
β_3, kPa^{-m} nm s^{-1}	1.0	9.50 ± 12.1	0.84 ± 1.20	0.56 ± 6.58
m	3.00	2.41 ± 0.24	3.00	3.00
Flow				
β_4, kPa•s	15.0	14.8 ± 0.34	14.9 ± 0.03	14.1 ± 1.69
β_5, s	1.0	0.977 ± 0.058	0.990 ± 0.053	0.937 ± 0.264
β_6	0.667	0.669 ± 0.004	0.668 ± 0.003	0.671 ± 0.018
Sum of relative errors, Equation 8.13	0.0101 0.2520	0.0095 –	0.0096 –	– 0.2451

[a] The pressure data, generated from the assumed flow-characteristic functions, were adjusted by the addition of random deviates to obtain the corresponding deviations. Parameter errors correspond to the 95% confidence intervals.

[b] The slip-velocity exponent, m, calculated from the apparent (Newtonian) flow curve was used instead of the simulation value.

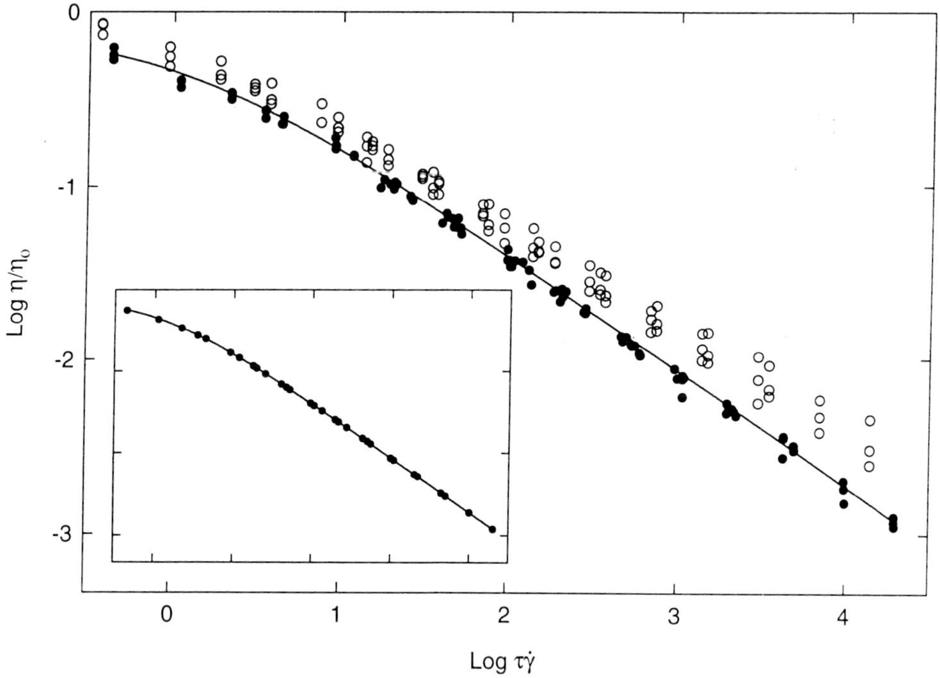

Figure 8.4 Results of a numerical simulation exercise (–) showing uncorrected data (○), generated by assuming realistic flow characterization functions and adding 5% random deviations, and the subsequent analysis by CAPSIM (●). Inset is a comparison of the results from CAPSIM (–) and CAPSEQ (●) using the noise-free data set.

The second example for illustrating the capabilities of both algorithms was supplied by Hatzikiriakos and Dealy [27] for a 0.35-MI HDPE resin (Dupont Canada Sclair 56B/3830, M_w = 1.78 × 10^5) at 200 °C. The data were collected from an Instron rheometer by using three constant-radius capillary sets with an assortment of aspect ratios: R = 0.026 in (L/R = 40, 80, 120), 0.015 in (L/R = 40, 80, 160), and 0.005 in (L/R = 20, 80, 120). The listings of the input/output files are provided in Section 8.6.

Figure 8.5 compares the results from CAPSEQ with that provided by Hatzikiriakos and Dealy using sequential data reduction, and the agreement is excellent. In correcting for slip, CAPSEQ chose the Cross model for the two larger capillaries due to curvature in the data at low shear rates. Not unexpectedly, CAPSIM also chose the Cross model as the flow function. In Figure 8.6, a comparison is made between results from CAPSEQ and CAPSIM. This comparison highlights the fact that, whereas data reduction takes place in CAPSEQ when

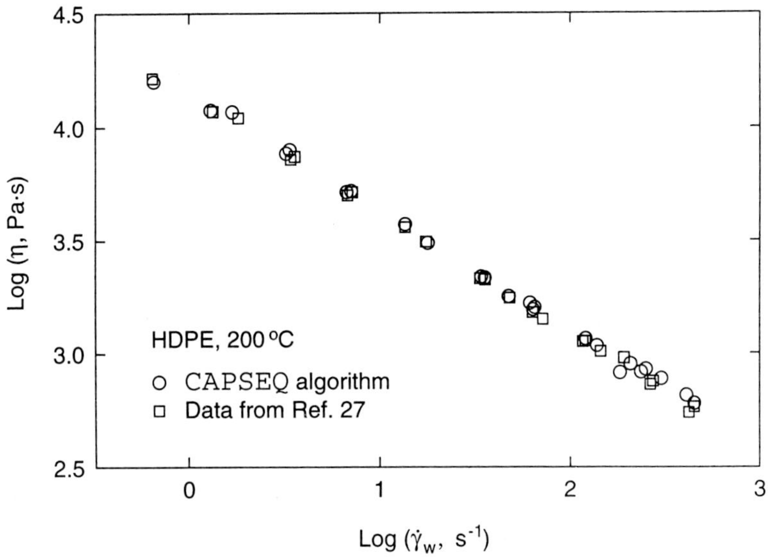

Figure 8.5 Comparison of capillary data analyses based on sequential corrections for a 0.35-MI HDPE at 200 °C [27]. Reprinted by permission.

extracting σ_w information from the slopes of the Bagley plot, CAPSIM performs the individual corrections on all the experimental points collected. In the latter, the mean relative deviation of the pressure data, based on Equation 8.13, was 0.5%. The slip-function exponent, *m*, calculated by CAPSIM was 2.6 ± 0.2 as compared to the reported value of 2.95 obtained by using *L*/*R* ratios greater than 120 [14, Table III]. As is evident from the Figure 8.6 inset, there is no significant difference in the viscosity curve if the reported value (2.95) were used, even though it appears that CAPSIM is undercorrecting for slip at high shear rates. The entrance pressure coefficient, β_1 = 79.7 ± 75.6 kPa·s$^{\beta_2}$, suggests that the entrance pressure losses are quite low.

The third example uses data for a 0.20-MI LDPE resin (Union Carbide DFD-6201) at 193 °C. The data were collected from an Instron rheometer by using three constant-radius capillary sets; *R* = 0.05 and 0.025 in (*L*/*R* = 40, 80, 120), and *R* = 0.015 in (*L*/*R* = 40, 80, 160). The flow function chosen by the CAPSIM algorithm was the power-law model because the curvature in the data was insufficient to justify the Cross model. The CAPSEQ output indicated that the slip correction was skipped for all points; that is, the tabulated values for the apparent shear rate correspond to $4Q/\pi R^3$. As shown in Figure 8.7, the algorithms agree when correction for slip was suppressed in CAPSIM. The mean relative deviation of the pressure data from CAPSIM, without slip correction, was 0.8%. Unlike the previous example, the

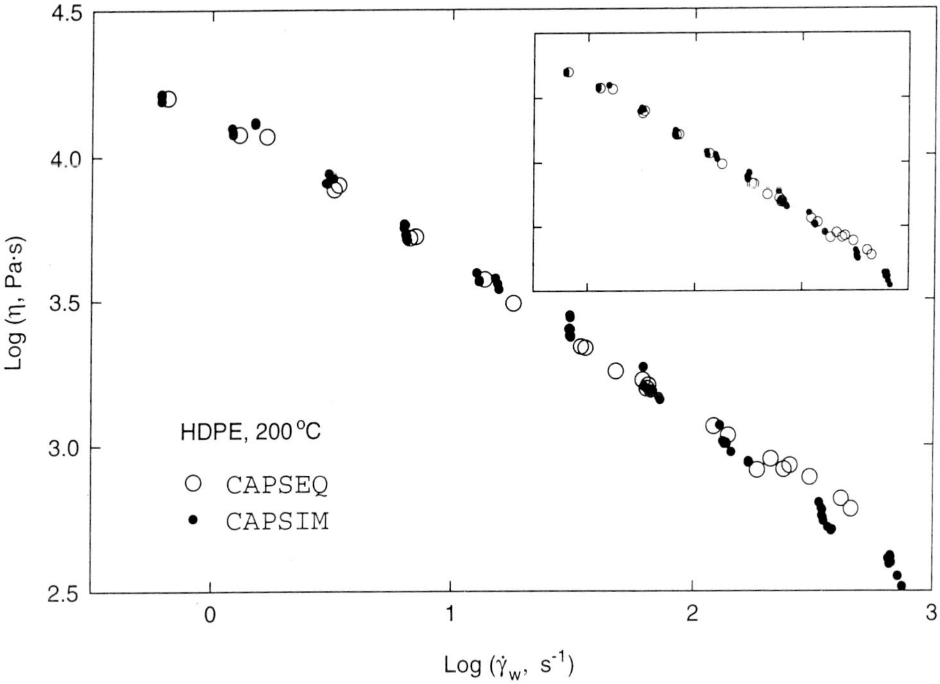

Figure 8.6 Comparison of capillary data analyses from CAPSEQ and CAPSIM for a 0.35-MI HDPE at 200 °C [27]. The flow function in CAPSIM is based on the Cross model [35,36] and the slip-function exponent, *m*, found was 2.60 ± 0.20. Inset is a comparison between CAPSEQ and CAPSIM using the reported value for *m* of 2.95 [14].

entrance pressure losses were significant as reflected in value obtained for β_1, 154 ± 28 kPa•s$^{\beta_2}$.

The last data set, polypropylene (M_v = 3.2 × 10^5) at 232 °C, is a textbook example designed to illustrate end corrections [40,41]. The raw data [40], along with the assortment of seven capillaries used, is shown in Figure 8.8 as our rendition of Figure 12-15(a) from Reference 41. Figure 8.9 illustrates CAPSIM's ability to analyze data without reducing the shear-rate range, a result of interpolation at arbitrary $\dot{\gamma}_{a,u}$ to obtain the constant-shear-rate data necessary for performing the end correction (refer to Figure 12-15(b) from Reference 41). The flow function chosen by CAPSIM was the Cross model, and the mean relative deviation was 0.4%.

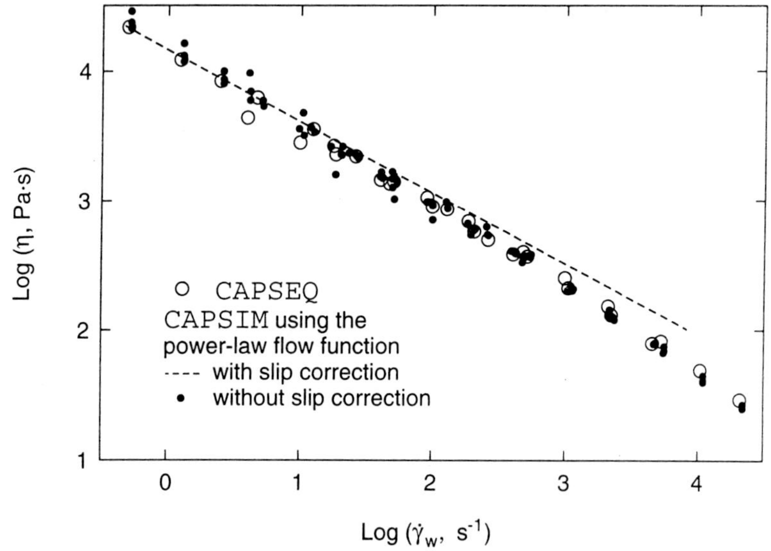

Figure 8.7 Comparison of capillary data analyses from CAPSEQ and CAPSIM for a 0.2-MI LDPE at 193 °C. The flow function in CAPSIM is based on the power-law model.

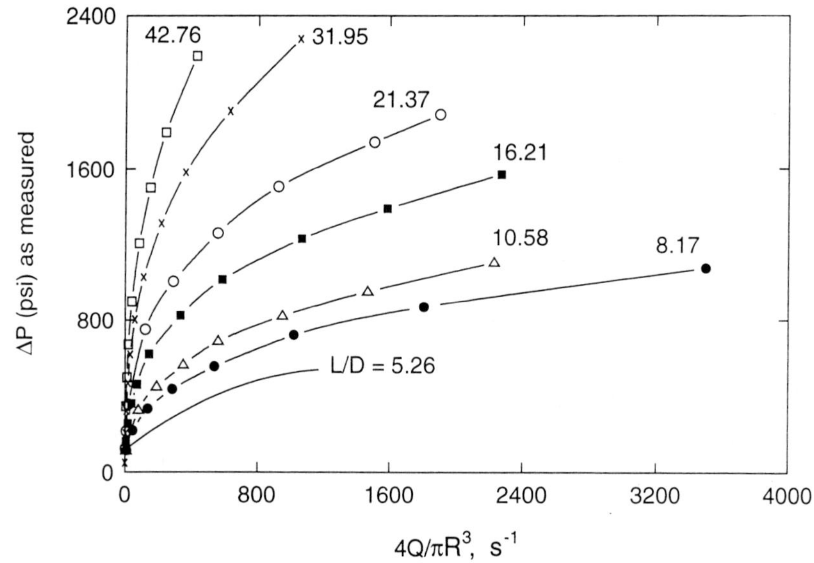

Figure 8.8 Polypropylene (M_v = 320,000) data at 232 °C [40]. Adapted from Figure 12-15(a) of Ref. 41, © 1971 by permission of Prentice-Hall, Inc., Englewood Cliffs, New Jersey.

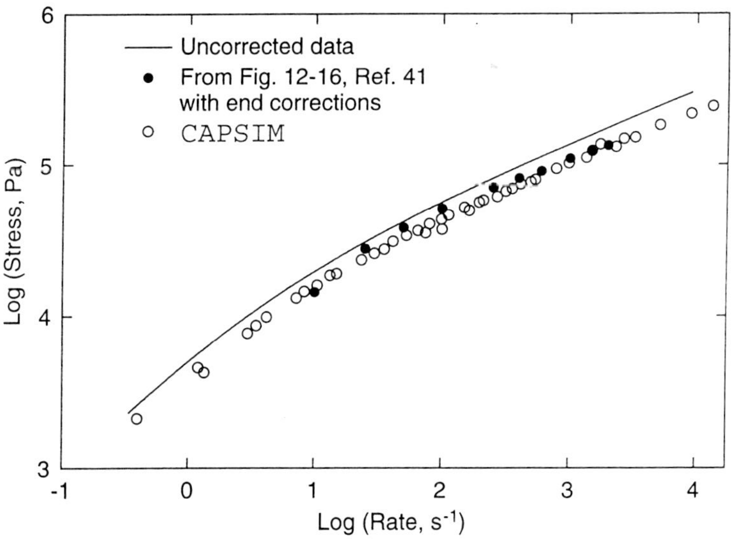

Figure 8.9 CAPSIM's analysis (O) of the polypropylene data [40] shown in Figure 8.8. The uncorrected data (–) is based on Newtonian calculations. CAPSIM covers a shear-rate range that is two decades wider than that given in the literature (●) [41]. Adapted from Ref. 41, © 1971 by permission of Prentice-Hall, Inc., Englewood Cliffs, New Jersey.

8.5 Program Listings

8.5.1 CAPSEQ

Imbedded into the code is the function FSV inherent to the execution of CAPSEQ.

```
      PROGRAM CAPSEQ
C Performs sequential calculation of end, slip and velocity profile
C  corrections for capillary rheometer data.
C Requires the subroutines ALINR, CAPCON, EQSOLN, FXCR, GAUSSN, SLOPE
C  and SORT2.
C The subroutine CAPCON uses the COMMON block BLOC1.
C Data file format:
C   1st record -  Title, up to 79 characters.
C   2nd record -  the data file's dimensional units group number.
C                    Group No.   Length    Force    Time
C                        1          in       lb      min
C                        2          cm       kg      min
C                        3          mm       kg       s
C                        4          mm       kN      min
C   3rd record -  reservoir (barrel) diameter.
C   4th record -  capillary radius, length, and no. of plunger velocities.
C   5th record -  plunger velocity, force.
C Repeat for each plunger velocity, then each capillary.
```

```
      IMPLICIT DOUBLE PRECISION (A-H,O-Z)
      PARAMETER (CC=1.D-4,NMX1=13,NMX2=NMX1*NMX1,NPAR=3,NPD=10)
      INTEGER IR(NMX1),KR(NMX1),KV(NMX1,NMX1),MODEL(NMX1),NCV(NMX1)
      DIMENSION A(NMX2,NPD),AT(NPD,NMX2),B(NMX1,NPAR),C(NPAR),CR(NMX1)
      DIMENSION CV(NMX1,NMX1),ERR(NPAR),RR(NMX2,1),RWM(NMX2)
      DIMENSION STR(NMX1,NMX1),SRW(NMX2),X(NMX2),Y(NMX2),Z(11)
      COMMON /BLOC1/ F(NMX1,NMX1),V(NMX1,NMX1),R(NMX1),XL(NMX1),NV(NMX1)
      EXTERNAL FSV,FXCR
C t-statistic for 80% confidence interval with N degrees of freedom.
      T80(N)=1.282+0.8309/N+0.6827/N**2+0.2823/N**4
10    CALL MSHEAD
      CALL UNITS(NTER,NINT,NOUT)
      WRITE(NOUT,20)
20    FORMAT(/14X,'Sequential Capillary Rheometry Analysis'/14X,39('-'))
      CALL CAPCON(NCAP)
      WRITE(NTER,'(/2X,A)')'Computing!'
C Sort data as a function of constant capillary radius.
      DO 30 I=1,NCAP
        KR(I)=1
      DO 30 J=1,NV(I)
        KV(I,J)=1
30    CONTINUE
      IDX=1
      DO 100 I=1,NCAP
C NCER is the number of capillaries of an equivalent radius size.
40      NCER=1
        IR(1)=IDX
        CR(I)=R(IDX)
C Search for capillaries with a radius equivalent to CR(I).
        ISET1=0
        DO 50 J=IDX+1,NCAP
          IF(KR(J).GT.0)THEN
            IF(ABS(R(J)-CR(I)).LE.1.D-6)THEN
              NCER=NCER+1
              IR(NCER)=J
              KR(J)=0
            ELSE IF(ISET1.EQ.0)THEN
              IDX=J
              ISET1=1
            ENDIF
          ENDIF
50      CONTINUE
C At least two capillaries of different aspect ratios but equivalent radius
C  are required for the end corrections.
        IF(NCER.LT.2)THEN
          IF(ISET1.GT.0)GOTO 40
          WRITE(NTER,*)' Data insufficient to proceed with CAPSEQ.'
          STOP
        ENDIF
C Sort out pressure and aspect ratio data as a function of the apparent,
C  uncorrected shear rate, CV(I,J).
        I1=IR(1)
        J1=1
        NSR=0
        DO 80 J=1,NMX1
60        CV(I,J)=V(I1,J1)
          ISET2=0
C NAR is the number of different aspect ratios at a constant CV(I,J).
          NAR=0
          DO 70 K=1,NCER
          DO 70 L=1,NV(IR(K))
            IF(KV(IR(K),L).GT.0)THEN
              IF(ABS(V(IR(K),L)-CV(I,J)).LE.1.D-4)THEN
                NAR=NAR+1
                X(NAR)=XL(IR(K))/CR(I)
                Y(NAR)=F(IR(K),L)
                KV(IR(K),L)=0
              ELSE IF(ISET2.EQ.0)THEN
                I1=IR(K)
```

```
                J1=L
                ISET2=1
              ENDIF
            ENDIF
70        CONTINUE
          IF(NAR.LT.2)THEN
            IF(ISET2.GT.0)GOTO 60
              WRITE(NTER,*)' Data insufficient to proceed with CAPSEQ.'
              STOP
            ELSE
C Proceed with the Bagley plot to obtain wall shear stress, STR(I,J).
              CALL ALINR(NAR,X,Y,Z)
              STR(I,J)=Z(2)/2.D0
              NSR=NSR+1
            ENDIF
            IF(ISET2.EQ.0)GOTO 90
80      CONTINUE
C NCV(I) is the number of shear rate data for a capillary radius size.
90      NCV(I)=NSR
        IF(ISET1.EQ.0)GOTO 110
100   CONTINUE
C NCR is the number of capillary radii.
110   NCR=I
C Check feasibility for correction of slip velocity (MODE=2).
      MODE=1
      DO 120 I=1,NCR-1
        IF(ABS(CR(I)/CR(I+1)-1.).GT..5.AND.NCR.GT.1)MODE=2
120   CONTINUE
      IF(MODE.EQ.2)THEN
C Set up stress-rate data as a function of capillary radius size.
        DO 150 I=1,NCR
          DO 130 J=1,NCV(I)
            X(J)=LOG10(CV(I,J))
            Y(J)=LOG10(STR(I,J))
130       CONTINUE
          CALL SORT2(NCV(I),X,Y,X,Y)
C Obtain estimates for two-parameter power-law model using ALINR.
          CALL ALINR(NCV(I),X,Y,Z)
C Obtain estimates for three-parameter Cross model using GAUSSN.
          C(1)=Y(1)-X(1)
          C(3)=1.-Z(2)
          C(2)=(C(1)-Z(1))/C(3)
          CALL GAUSSN(FXCR,CC,NPAR,NCV(I),X,Y,C,A,AT,RR,ERR,SEE)
C Determine flow function better suited to constant-radius data.
          IF(SEE.GT.Z(8))THEN
            MODEL(I)=NPAR-1
            B(I,1)=1.D1**Z(1)
            B(I,2)=Z(2)
          ELSE
            MODEL(I)=NPAR
            DO 140 J=1,NPAR
              B(I,J)=C(J)
140         CONTINUE
          ENDIF
150     CONTINUE
C Obtain apparent, uncorrected shear rate at constant wall shear stress.
        DO 180 I=1,NCR
        DO 180 J=1,NCV(I)
          DO 170 K=1,NCR
            X(K)=1.D0/CR(K)
            IF(K.EQ.I)THEN
              Y(K)=CV(I,J)
            ELSE IF(MODEL(K).EQ.2)THEN
              Y(K)=(STR(I,J)/B(K,1))**(1.D0/B(K,2))
            ELSE IF(MODEL(K).EQ.3)THEN
            DO 160 L=1,NPAR
              C(L)=B(K,L)
160         CONTINUE
              YI=STR(I,J)*(1.D0+(1.D1**C(2)*CV(I,J))**C(3))/1.D1**C(1)
```

```
              CALL EQSOLN(FSV,NPAR,C,STR(I,J),YI,CC,Y(K))
            ENDIF
170       CONTINUE
C Proceed with Mooney slip correction.
          CALL ALINR(NCR,X,Y,Z)
          IF(Z(1).GT.0.AND.Z(2).GT.0.)THEN
            IF(NCR.EQ.2)THEN
              CV(I,J)=Z(1)
            ELSE IF(ABS(Z(2)/Z(10)).GT.T80(NCR-2))THEN
              CV(I,J)=Z(1)
            ENDIF
          ENDIF
180     CONTINUE
      ENDIF
C Proceed with the velocity-profile correction.
      NP=0
      DO 190 I=1,NCR
      DO 190 J=1,NCV(I)
        NP=NP+1
        X(NP)=LOG10(CV(I,J))
        Y(NP)=LOG10(STR(I,J))
190   CONTINUE
      CALL SORT2(NP,Y,X,Y,X)
      NO=NCAP
      IF(NO.LT.9)NO=9
      IF(NO.GE.NP)NO=NP/2
      CALL SLOPE(NP,Y,X,NO,2,RWM,SRW)
      WRITE(NOUT,200)
200   FORMAT(/11X,'Apparent',4X,'Corrected'/12X,'Shear',2(8X,'Shear'),
     &        7X,'Viscosity,'/6X,2(4X,'Rate, 1/s'),4X,'Stress,kPa',
     &        6X,'kPa-s'/6X,2(4X,9('-')),2(4X,10('-'))/)
      DO 220 I=1,NP
        IF(RWM(I).GT.0.)THEN
          SRW(I)=1.D1**X(I)*(.75D0+.25D0*RWM(I))
        ELSE
          SRW(I)=1.D1**X(I)
        ENDIF
        WRITE(NOUT,210)1.D1**X(I),SRW(I),1.D1**Y(I),1.D1**Y(I)/SRW(I)
210     FORMAT(6X,1P,3(4X,E9.2),5X,E9.2)
220   CONTINUE
      CALL MSTAIL(*10)
      STOP
      END
C
      DOUBLE PRECISION FUNCTION FSV(N,B,SV)
C Evaluates shear rate (SV) given FSV for a Cross flow function.
      DOUBLE PRECISION B(N),SV
      FSV=1.D1**B(1)*SV/(1.D0+(1.D1**B(2)*SV)**B(3))
      RETURN
      END
```

8.5.2 CAPSIM

Imbedded into the code are the subprograms FASR, FCAP, FVPC, and STR inherent to the execution of CAPSIM.

```
      PROGRAM CAPSIM
C Simultaneous calculation of end, slip and velocity profile corrections
C  for capillary rheometer data.
C Requires the subroutines ALINR, CAPCON, EQSOLN, ERROR, FXCR, GAUSSN,
C  SLOPE, SORT2, SPECC, and the IMSL routine DBCONF.
C The subroutine CAPCON uses the COMMON block BLOC1.
C Data file format:
```

```
C   1st record -  Title, up to 79 characters.
C   2nd record -  the data file's dimensional units group number.
C                    Group No.    Length    Force    Time
C                        1          in        lb      min
C                        2          cm        kg      min
C                        3          mm        kg       s
C                        4          mm        kN      min
C   3rd record -  reservoir (barrel) diameter.
C   4th record -  capillary radius, length and no. of plunger velocities.
C   5th record -  plunger velocity, force.
C Repeat for each plunger velocity, then each capillary.
      IMPLICIT DOUBLE PRECISION (A-H,O-Z)
      PARAMETER (NMX1=13,NMX2=NMX1*NMX1,NPD=10,NPR1=6,NPR2=3)
      CHARACTER ANS*1
      DIMENSION A(NMX2,NPD),AT(NPD,NMX2),BGUESS(NPR1),BSCALE(NPR1)
      DIMENSION BLB(NPR1),BUB(NPR1),B(NPR1),CI(NPR1),COV(NPR1),CR(NPR2)
      DIMENSION ERR(NPR2),H(NPR1,NPR1),HI(NPR1,NPR1),IPARAM(7),RPARAM(7)
      DIMENSION PL(11),PL2(11),RR(NMX2,1),RWM(NMX2)
      DIMENSION FL(NMX2),VL(NMX2),FL2(NMX2),VL2(NMX2)
      COMMON /BLOC1/ F(NMX1,NMX1),V(NMX1,NMX1),R(NMX1),XL(NMX1),NV(NMX1)
      COMMON /BLOC2/ I,MODE,MODEL,NCAP,SVE
      EXTERNAL FASR,FXCR,FCAP,FVPC
C t-statistic for 95% confidence interval with n degrees of freedom.
      T95(N)=1.96+2.3539/N+3.1902/N**2+0.7004/N**3+4.5015/N**4
10    CALL MSHEAD
      CALL UNITS(NTER,NINT,NOUT)
      WRITE(NOUT,20)
20    FORMAT(/,21X,'Capillary Rheometry Analysis'/,21X,28('-'))
      CALL CAPCON(NCAP)
      NP=0
      NP2=0
      DO 30 I=1,NCAP
        FAC=R(I)/XL(I)/2.D0
      DO 30 J=1,NV(I)
        RP2L=F(I,J)*FAC
        IF(RP2L.GT.1.0D2)THEN
C Set up data for apparent stress > 100 kPa in case of slip correction.
          NP2=NP2+1
          FL2(NP2)=LOG10(RP2L)
          VL2(NP2)=LOG10(V(I,J))
        ENDIF
        NP=NP+1
        FL(NP)=LOG10(RP2L)
        VL(NP)=LOG10(V(I,J))
30    CONTINUE
C Obtain estimates for two-parameter power law model using ALINR.
      CALL ALINR(NP,VL,FL,PL)
C Obtain estimates for three-parameter Cross model using GAUSSN.
      CALL SORT2(NP,VL,FL,VL,FL)
      CR(1)=FL(1)-VL(1)
      CR(3)=1.-PL(2)
      CR(2)=(CR(1)-PL(1))/CR(3)
      CALL SPECC(NTER,CC)
      CALL GAUSSN(FXCR,CC,NPR2,NP,VL,FL,CR,A,AT,RR,ERR,SEE)
C Determine flow function better suited to data.
      MODEL=2
      IF(PL(8).GT.SEE)MODEL=3
      IF(MODEL.EQ.2)THEN
        TC=10.**(-VL(1))
      ELSE
        TC=10.**(CR(2))
      ENDIF
      NP=0
      DO 40 I=1,NCAP
      DO 40 J=1,NV(I)
        NP=NP+1
        FL(NP)=LOG10(F(I,J))
        VL(NP)=LOG10(V(I,J)*TC)
40    CONTINUE
```

```
      CALL ALINR(NP,VL,FL,PL2)
C Set up call to DBCONF for optimization of objective function FCAP.
      BGUESS(1)=0.8*PL2(1)
        BLB(1)=BGUESS(1)-1.0
        BUB(1)=PL2(1)+0.5
      BGUESS(2)=PL2(2)
        BLB(2)=0.10
        BUB(2)=0.99
C Check feasibility for correction of slip velocity (MODE=3).
      MODE=2
      DO 50 I=1,NCAP-1
        IF(ABS(R(I)/R(I+1)-1.).GT..5)MODE=3
50    CONTINUE
      IF(NP2.LT.NP/4)MODE=2
      IF(MODE.EQ.3)THEN
        WRITE(NTER,*)' Correction for slip is feasible, proceed? (Y/N)'
        READ(NTER,'(A1)')ANS
        IF(ANS.EQ.'Y'.OR.ANS.EQ.'y')THEN
          CALL ALINR(NP2,VL2,FL2,PL2)
          SVE=1.D0/PL2(2)
          WRITE(NTER,60)SVE,T95(NP2-N)*PL2(10)*SVE**2
60        FORMAT(2X,'Use calculated slip-velocity exponent, m =',F5.2,
     &               ' +/-',F5.2,' ? (Y/N)')
          READ(NTER,'(A1)')ANS
          IF(ANS.EQ.'N'.OR.ANS.EQ.'n')THEN
            WRITE(NTER,*)' Enter value for slip-velocity exponent.'
            READ(NTER,*)SVE
          ENDIF
          PL2(1)=1.D1**PL2(1)
          BGUESS(3)=1.2*LOG10(PL2(1)**(-SVE)*(PL2(2)/(3.*PL2(2)+1.)))
            BLB(3)=BGUESS(3)-2.
            BUB(3)=BGUESS(3)+2.
        ELSE
          MODE=2
        ENDIF
      ENDIF
      IF(MODEL.EQ.2)THEN
        BGUESS(MODE+1)=PL(1)
        BGUESS(MODE+2)=PL(2)
          BLB(MODE+2)=0.10
          BUB(MODE+2)=0.60
      ELSE
       BGUESS(MODE+1)=CR(1)
       BGUESS(MODE+2)=CR(2)
          BLB(MODE+2)=BGUESS(MODE+2)-1.
          BUB(MODE+2)=BGUESS(MODE+2)+1.
       BGUESS(MODE+3)=CR(3)
          BLB(MODE+3)=0.40
          BUB(MODE+3)=1.00
      ENDIF
          BLB(MODE+1)=BGUESS(MODE+1)-1.
          BUB(MODE+1)=BGUESS(MODE+1)+1.
      N=MODE+MODEL
      IBTYPE=0
      DO 70 I=1,N
        BSCALE(I)=1.D0
70    CONTINUE
      FSCALE=1.D0
      IPARAM(1)=0
      WRITE(NTER,'(/2X,A)')'Computing!'
      CALL DBCONF(FCAP,N,BGUESS,IBTYPE,BLB,BUB,BSCALE,FSCALE,IPARAM,
     &            RPARAM,B,FX)
      SEE=SQRT(FX/FLOAT(NP-N))
C Calculate the errors in the parameter estimates.
      CALL ERROR(FCAP,N,B,BSCALE,FX,H,HI,COV)
      DO 90 I=1,N
        BGUESS(I)=B(I)
        CI(I)=T95(NP-N)*SEE*SQRT(COV(I))
```

```
        IF(I.EQ.1)THEN
          WRITE(NOUT,*)'Parameters for the entrance pressure function'
          BGUESS(I)=1.D1**BGUESS(I)
          CI(I)=BGUESS(I)*LOG(1.D1)*CI(I)
        ELSE IF(I.EQ.3.AND.MODE.EQ.3)THEN
          WRITE(NOUT,'(/A)')' Parameters for the slip velocity function'
          BGUESS(I)=1.D1**BGUESS(I)
          CI(I)=BGUESS(I)*LOG(1.D1)*CI(I)
          WRITE(NOUT,'(3X,A,F10.2)')'Exponent =',SVE
        ELSE IF(I.EQ.MODE+1)THEN
          BGUESS(I)=1.D1**BGUESS(I)
          CI(I)=BGUESS(I)*LOG(1.D1)*CI(I)
          IF(MODEL.EQ.2)THEN
            WRITE(NOUT,'(/A)')
     &                        ' Parameters for the power-law flow function'
          ELSE
            WRITE(NOUT,'(/A)')' Parameters for the Cross flow function'
          ENDIF
        ELSE IF(I.EQ.MODE+2.AND.MODEL.EQ.3)THEN
          BGUESS(I)=1.D1**BGUESS(I)
          CI(I)=BGUESS(I)*LOG(1.D1)*CI(I)
        ENDIF
      WRITE(NOUT,80)I,BGUESS(I),CI(I)
80    FORMAT(1P,7X,'B(',I1,') = ',E9.2,' +/-',E9.2)
90    CONTINUE
      WRITE(NOUT,100)SQRT(FX)/NP*1.D2
100   FORMAT(/1X,'Average of pressure deviation =',F6.1,' %'/)
      NP=0
      DO 110 I=1,NCAP
        FAC=R(I)/XL(I)/2.D0
C Perform corrections on actual pressure and apparent shear rate.
      DO 110 J=1,NV(I)
C Calculate shear stress.
        F(I,J)=(F(I,J)-BGUESS(1)*V(I,J)**BGUESS(2))*FAC
C Calculate slip-free, apparent shear rate.
        IF(MODE.EQ.3)V(I,J)=V(I,J)-4.D0*BGUESS(3)*F(I,J)**SVE/R(I)
C Calculate velocity profile correction.
          IF(F(I,J).GT.0..AND.V(I,J).GT.0.)THEN
            NP=NP+1
            VL(NP)=LOG10(V(I,J))
            FL(NP)=LOG10(F(I,J))
          ENDIF
110   CONTINUE
      CALL SORT2(NP,FL,VL,FL,VL)
      NO=NCAP*2
      IF(NO.LT.15)NO=15
      IF(NO.GE.NP)NO=NP/2
      CALL SLOPE(NP,FL,VL,NO,2,RWM,VL2)
      WRITE(NOUT,120)
120   FORMAT(/3X,'Apparent',2X,'Corrected',26X,'Calculated'/
     & 4X,'Shear',2(6X,'Shear'),5X,'Viscosity,',2X,'Viscosity,'/
     & 2(2X,'Rate, 1/s'),2X,'Stress,kPa',4X,'kPa-s',7X,'kPa-s',
     & 6X,'Residual'/2(2X,9('-')),4(2X,10('-'))/)
      DO 140 I=1,NP
         IF(RWM(I).GT.0.)THEN
           SRW=1.D1**VL(I)*(.75D0+.25D0*RWM(I))
         ELSE
           SRW=1.D1**VL(I)
         ENDIF
         VIS=1.D1**FL(I)/SRW
C Calculate viscosity from the derivative of the flow function.
         CALL EQSOLN(FVPC,N,BGUESS,SRW,SRW,CC,ASR)
         VISC=STR(N,B,ASR)/SRW
         WRITE(NOUT,130)1.D1**VL(I),SRW,1.D1**FL(I),VIS,VISC,VIS-VISC
130      FORMAT(1P,2(2X,E9.2),4(2X,E10.2))
140   CONTINUE
      CALL MSTAIL(*10)
      STOP
      END
```

```
      DOUBLE PRECISION FUNCTION FASR(N,B,ASR)
C Evaluates the apparent shear rate corrected for slip (ASR) from
C  Equation 8.15.
      IMPLICIT DOUBLE PRECISION (A-H,O-Z)
      PARAMETER (NMX1=13)
      DIMENSION B(N)
      COMMON /BLOC1/ F(NMX1,NMX1),V(NMX1,NMX1),R(NMX1),XL(NMX1),NV(NMX1)
      COMMON /BLOC2/ I,MODE,MODEL,NCAP,SVE
      EXTERNAL STR
      FASR=4.D0*1.D1**B(3)/R(I)*STR(N,B,ASR)**SVE+ASR
      RETURN
      END
C
      SUBROUTINE FCAP(N,B,FX)
C Evaluates the function FX to be minimized (Equation 8.13).
C Requires the subroutine EQSOLN and the function subprograms FASR and STR.
      IMPLICIT DOUBLE PRECISION (A-H,O-Z)
      PARAMETER (NMX1=13)
      DIMENSION B(N)
      COMMON /BLOC1/ F(NMX1,NMX1),V(NMX1,NMX1),R(NMX1),XL(NMX1),NV(NMX1)
      COMMON /BLOC2/ I,MODE,MODEL,NCAP,SVE
      EXTERNAL FASR,STR
      FX=0.D0
      DO 20 I=1,NCAP
        DO 10 J=1,NV(I)
          IF(MODE.EQ.2)THEN
            ASR=V(I,J)
          ELSE IF(MODE.EQ.3)THEN
            ER=1.D-8
            CALL EQSOLN(FASR,N,B,V(I,J),V(I,J),ER,ASR)
          ENDIF
          PC=1.D1**B(1)*V(I,J)**B(2)+2.D0*XL(I)/R(I)*STR(N,B,ASR)
          FX=FX+((F(I,J)-PC)/F(I,J))**2
10      CONTINUE
20    CONTINUE
      RETURN
      END
C
      DOUBLE PRECISION FUNCTION FVPC(N,B,SR)
C Evaluates the apparent shear rate (SR) from the velocity profile
C  correction scheme expressed in Equation 8.7.
      IMPLICIT DOUBLE PRECISION (A-H,O-Z)
      PARAMETER (NMX1=13)
      DIMENSION B(N)
      COMMON /BLOC1/ F(NMX1,NMX1),V(NMX1,NMX1),R(NMX1),XL(NMX1),NV(NMX1)
      COMMON /BLOC2/ I,MODE,MODEL,NCAP,SVE
      IF(MODEL.EQ.2)THEN
        FVPC=.25D0*SR*(3.D0+(1.D0/B(MODE+2)))
      ELSE IF(MODEL.EQ.3)THEN
        FVPC=SR*(1.D0+(B(MODE+2)*SR**B(MODE+3)*(1.D0-.75D0*B(MODE+3))))/
     &         (1.D0+(B(MODE+2)*SR**B(MODE+3)*(1.D0-B(MODE+3))))
      ENDIF
      RETURN
      END
C
      DOUBLE PRECISION FUNCTION STR(N,B,ASR)
C Evaluates the wall shear stress (STR) based on the flow function.
      IMPLICIT DOUBLE PRECISION (A-H,O-Z)
      DIMENSION B(N)
      COMMON /BLOC2/ I,MODE,MODEL,NCAP,SVE
      IF(MODEL.EQ.2)THEN
        STR=1.D1**B(MODE+1)*ASR**B(MODE+2)
      ELSE IF(MODEL.EQ.3)THEN
        STR=1.D1**B(MODE+1)*ASR/(1.D0+(1.D1**B(MODE+2)*ASR)**B(MODE+3))
      ENDIF
      RETURN
      END
```

8.5.3 CAPCON

```
      SUBROUTINE CAPCON(NCAP)
C Calculates the apparent, uncorrected shear rate and pressure from
C  the plunger velocity and force data for use in CAPSEQ and CAPSIM.
C Converts the data file into SI units based on the group number NDUGN.
C         Group No.    Length   Force   Time
C            1           in       lb     min
C            2           cm       kg     min
C            3           mm       kg      s
C            4           mm       kN     min
C Requires the subroutines STOPP and UNITS.
C Carries the COMMON block BLOC1 from both CAPSEQ and CAPSIM.
      IMPLICIT DOUBLE PRECISION (A-H,O-Z)
      PARAMETER (NMX1=13)
      CHARACTER TITLE*80
      COMMON /BLOC1/ F(NMX1,NMX1),V(NMX1,NMX1),R(NMX1),XL(NMX1),NV(NMX1)
      CALL UNITS(NTER,NINT,NOUT)
      READ(NINT,'(A80)')TITLE
      WRITE(NOUT,'(/1X,A80/)')TITLE
      READ(NINT,*)NDUGN,RD
      NP=0
      DO 10 I=1,NMX1
        READ(NINT,*,END=20)R(I),XL(I),NV(I)
        IF(NV(I).GT.NMX1-1)CALL STOPP(NTER,NMX1-1)
        IF(I.EQ.NMX1)CALL STOPP(NTER,NMX1-1)
      DO 10 J=1,NV(I)
        NP=NP+1
        READ(NINT,*)V(I,J),F(I,J)
10    CONTINUE
20    NCAP=I-1
      WRITE(NOUT,30)NCAP
30    FORMAT(1X,'Number of capillaries =',I4)
      WRITE(NOUT,40)NP
40    FORMAT(1X,'Number of data points =',I4/)
C Define conversion factors.
      FFAC=1273239.545
      VFAC=6.D1
      CFAC=1.D0
      IF(NDUGN.EQ.1)THEN
        FFAC=8.778647
        CFAC=25.4
      ELSE IF(NDUGN.EQ.2)THEN
        FFAC=1.24862146D2
        CFAC=1.D1
      ELSE IF(NDUGN.EQ.3)THEN
        FFAC=1.24862146D4
        VFAC=1.D0
      ENDIF
      DO 60 I=1,NCAP
        DO 50 J=1,NV(I)
C Calculate apparent, uncorrected shear rate, 4Q/(πR³), in 1/s.
          V(I,J)=V(I,J)*RD**2/R(I)**3/VFAC
C Calculate pressure, in kPa.
          F(I,J)=F(I,J)*FFAC/RD**2
50      CONTINUE
C Convert capillary dimensions to millimeters.
        R(I)=R(I)*CFAC
        XL(I)=XL(I)*CFAC
60    CONTINUE
      RETURN
      END
```

8.6 Sample Listings of INPUT/OUTPUT Files

8.6.1 Data File

Shown here is a data file in the format required for running either CAPSEQ or CAPSIM. The set listed below was provided by Hatzikiriakos and Dealy [27]. and is reprinted by permission.

```
 HDPE at 473 K, Hatzikiriakos and Dealy, McGill Univ., 1990.
2
0.9525
        0.06604 2.6416 10
        10.       166.
         5.       123.
         2.        82.
         1.        60.
         0.5       44.
         0.2       29.
         0.1       20.5
         0.05      15.
         0.02       9.
         0.01       6.
        0.06604 5.2832 10
        10.       313.
         5.       243.
         2.       165.
         1.       118.
         0.5       86.
         0.2       56.5
         0.1       41.
         0.05      29.
         0.02      17.7
         0.01      12.
        0.06604 7.9248 10
        10.       482.
         5.       371.
         2.       245.
         1.       176.
         0.5      130.
         0.2       88.
         0.1       61.
         0.05      44.
         0.02      27.
         0.01      18.
        0.0381   1.524   9
         2.       155.
         1.       115.
         0.5       89.
         0.2       62.5
         0.1       46.5
         0.05      34.5
         0.02      22.5
         0.01      16.5
         0.005     12.
        0.0381   3.048   9
         2.       305.
         1.       229.
         0.5      176.
         0.2      122.
         0.1       90.
         0.05      64.
         0.02      42.5
         0.01      32.
         0.005     23.5
```

```
0.0381  6.096  9
 2.      620.
 1.      456.
 0.5     350.
 0.2     243.
 0.1     180.
 0.05    130.
 0.02     87.
 0.01     63.5
 0.005    46.5
0.0127  0.254  5
 0.1      80.
 0.05     61.
 0.02     43.
 0.01     36.
 0.005    27.
0.0127  1.016  5
 0.1     290.
 0.05    225.
 0.02    161.
 0.01    123.
 0.005   100.
0.0127  1.524  5
 0.1     420.
 0.05    333.
 0.02    265.
 0.01    185.
 0.005   150.
```

8.6.2 Terminal Dialog with CAPSIM

Reproduced here is the terminal dialog for running CAPSIM with the data file CAP.DAT listed in Section 8.6.1.

```
>CAPSIM
 Enter name of data file (N if not applicable)
CAP.DAT
 Results to be written to NEW file: CAP.RES
 OK (Y/N)?
Y
 Use the default convergence criterion, 1.E-4? (Y/N)
Y
 Correction for slip is feasible, proceed? (Y/N)
Y
 Use calculated slip-velocity exponent, m = 2.60 +/-  .20 ? (Y/N)
Y

 Computing!

 The results are in CAP.RES
 Run another set of data? (Y/N)
N
Stop - Program terminated.
```

8.6.3 Output File from CAPSEQ

Reproduced here is the output file from CAPSEQ using the data listed in Section 8.6.1. The results are plotted in Figures 8.5 and 8.6.

```
              Sequential Capillary Rheometry Analysis
              ---------------------------------------

HDPE at 473 K, Hatzikiriakos and Dealy, McGill Univ., 1990.

Number of capillaries =   9
Number of data points =  72

         Apparent     Corrected
          Shear         Shear          Shear        Viscosity,
         Rate, 1/s    Rate, 1/s     Stress,kPa        kPa-s
         ---------    ---------     ----------      ----------

         5.25E-01     6.51E-01      1.03E+01        1.59E+01
         1.05E+00     1.30E+00      1.55E+01        1.19E+01
         1.37E+00     1.69E+00      1.98E+01        1.17E+01
         2.62E+00     3.25E+00      2.49E+01        7.67E+00
         2.73E+00     3.39E+00      2.70E+01        7.96E+00
         5.25E+00     6.70E+00      3.48E+01        5.20E+00
         5.47E+00     7.08E+00      3.72E+01        5.25E+00
         1.05E+01     1.36E+01      5.07E+01        3.73E+00
         1.37E+01     1.79E+01      5.51E+01        3.08E+00
         2.62E+01     3.40E+01      7.40E+01        2.18E+00
         2.73E+01     3.55E+01      7.67E+01        2.16E+00
         3.69E+01     4.74E+01      8.46E+01        1.79E+00
         5.06E+01     6.39E+01      9.98E+01        1.56E+00
         4.74E+01     6.14E+01      1.02E+02        1.67E+00
         5.08E+01     6.49E+01      1.04E+02        1.60E+00
         9.51E+01     1.21E+02      1.40E+02        1.16E+00
         1.11E+02     1.39E+02      1.50E+02        1.08E+00
         1.48E+02     1.84E+02      1.51E+02        8.21E-01
         1.70E+02     2.09E+02      1.87E+02        8.96E-01
         1.96E+02     2.37E+02      1.96E+02        8.25E-01
         2.08E+02     2.51E+02      2.13E+02        8.51E-01
         2.51E+02     3.03E+02      2.35E+02        7.74E-01
         3.39E+02     4.10E+02      2.67E+02        6.53E-01
         3.74E+02     4.52E+02      2.72E+02        6.01E-01
```

8.6.4 Output File from CAPSIM

Reproduced below is the output file from CAPSIM using the data and terminal dialog listed in Sections 8.6.1 and 8.6.2, respectively. The results are plotted in Figure 8.6.

```
                    Capillary Rheometry Analysis
                    ----------------------------

HDPE at 473 K, Hatzikiriakos and Dealy, McGill Univ., 1990.

Number of capillaries =   9
Number of data points =  72
```

```
Parameters for the entrance pressure function
      B(1) =  7.97E+01 +/- 7.56E+01
      B(2) =  3.65E-01 +/- 1.94E-01

Parameters for the slip velocity function
  Exponent =        2.60
      B(3) =  4.18E-06 +/- 1.33E-06

Parameters for the Cross flow function
      B(4) =  4.53E+01 +/- 9.71E+00
      B(5) =  3.15E+00 +/- 1.52E+00
      B(6) =  6.02E-01 +/- 1.66E-02

Average of pressure deviation =     .5 %
```

Apparent Shear Rate, 1/s	Corrected Shear Rate, 1/s	Shear Stress,kPa	Viscosity, kPa-s	Calculated Viscosity, kPa-s	Residual
5.16E-01	6.14E-01	9.53E+00	1.55E+01	1.64E+01	-8.97E-01
5.15E-01	6.13E-01	9.93E+00	1.62E+01	1.64E+01	-2.37E-01
5.15E-01	6.13E-01	1.01E+01	1.64E+01	1.64E+01	-1.66E-02
1.02E+00	1.22E+00	1.45E+01	1.19E+01	1.25E+01	-5.88E-01
1.02E+00	1.22E+00	1.47E+01	1.21E+01	1.25E+01	-3.76E-01
1.02E+00	1.21E+00	1.51E+01	1.25E+01	1.25E+01	-1.04E-02
1.27E+00	1.51E+00	1.95E+01	1.29E+01	1.14E+01	1.59E+00
1.27E+00	1.51E+00	1.97E+01	1.30E+01	1.14E+01	1.68E+00
1.27E+00	1.51E+00	1.97E+01	1.31E+01	1.14E+01	1.73E+00
2.52E+00	3.00E+00	2.42E+01	8.07E+00	8.25E+00	-1.81E-01
2.52E+00	3.04E+00	2.44E+01	8.03E+00	8.20E+00	-1.76E-01
2.52E+00	3.06E+00	2.48E+01	8.10E+00	8.18E+00	-7.86E-02
2.51E+00	3.08E+00	2.68E+01	8.71E+00	8.15E+00	5.57E-01
2.50E+00	3.15E+00	2.69E+01	8.56E+00	8.06E+00	4.99E-01
2.50E+00	3.22E+00	2.70E+01	8.38E+00	7.98E+00	3.99E-01
5.02E+00	6.55E+00	3.34E+01	5.10E+00	5.57E+00	-4.65E-01
5.00E+00	6.50E+00	3.44E+01	5.29E+00	5.59E+00	-3.06E-01
5.00E+00	6.43E+00	3.44E+01	5.34E+00	5.62E+00	-2.79E-01
4.99E+00	6.32E+00	3.56E+01	5.64E+00	5.68E+00	-3.56E-02
4.95E+00	6.35E+00	3.69E+01	5.81E+00	5.66E+00	1.45E-01
4.95E+00	6.41E+00	3.70E+01	5.77E+00	5.63E+00	1.33E-01
9.92E+00	1.29E+01	4.74E+01	3.66E+00	3.87E+00	-2.08E-01
9.92E+00	1.29E+01	4.75E+01	3.70E+00	3.89E+00	-1.88E-01
9.85E+00	1.26E+01	4.97E+01	3.93E+00	3.92E+00	7.07E-03
1.23E+01	1.56E+01	5.38E+01	3.44E+00	3.49E+00	-5.02E-02
1.22E+01	1.54E+01	5.53E+01	3.60E+00	3.52E+00	7.18E-02
1.21E+01	1.51E+01	5.68E+01	3.76E+00	3.56E+00	2.01E-01
2.45E+01	3.04E+01	7.23E+01	2.38E+00	2.41E+00	-2.91E-02
2.45E+01	3.08E+01	7.24E+01	2.35E+00	2.39E+00	-3.87E-02
2.45E+01	3.08E+01	7.35E+01	2.38E+00	2.39E+00	-5.85E-03
2.40E+01	3.02E+01	7.58E+01	2.51E+00	2.42E+00	9.23E-02
2.39E+01	3.04E+01	7.66E+01	2.52E+00	2.41E+00	1.13E-01
2.39E+01	3.07E+01	7.67E+01	2.50E+00	2.39E+00	1.03E-01
2.36E+01	3.06E+01	8.42E+01	2.75E+00	2.40E+00	3.50E-01
2.34E+01	3.06E+01	8.48E+01	2.77E+00	2.40E+00	3.70E-01
2.31E+01	3.05E+01	8.55E+01	2.80E+00	2.40E+00	3.97E-01
4.86E+01	6.21E+01	9.90E+01	1.59E+00	1.60E+00	-5.38E-03
4.86E+01	6.48E+01	9.94E+01	1.53E+00	1.56E+00	-2.73E-02
4.86E+01	6.64E+01	9.95E+01	1.50E+00	1.54E+00	-4.01E-02
4.72E+01	6.62E+01	1.03E+02	1.55E+00	1.54E+00	1.09E-02
4.72E+01	6.65E+01	1.03E+02	1.55E+00	1.54E+00	1.51E-02
5.11E+01	7.25E+01	1.03E+02	1.43E+00	1.46E+00	-3.58E-02
4.71E+01	6.77E+01	1.03E+02	1.53E+00	1.52E+00	6.57E-03
5.05E+01	7.16E+01	1.04E+02	1.46E+00	1.47E+00	-1.35E-02
4.44E+01	6.17E+01	1.14E+02	1.85E+00	1.61E+00	2.48E-01
1.02E+02	1.34E+02	1.35E+02	1.01E+00	1.02E+00	-1.07E-02
9.62E+01	1.32E+02	1.36E+02	1.03E+00	1.03E+00	-3.57E-03
1.02E+02	1.43E+02	1.36E+02	9.47E-01	9.83E-01	-3.60E-02
9.56E+01	1.36E+02	1.39E+02	1.02E+00	1.01E+00	5.41E-03

9.56E+01	1.37E+02	1.39E+02	1.01E+00	1.01E+00	6.64E-03
1.18E+02	1.69E+02	1.47E+02	8.69E-01	8.91E-01	-2.18E-02
1.17E+02	1.70E+02	1.48E+02	8.72E-01	8.88E-01	-1.60E-02
1.17E+02	1.69E+02	1.49E+02	8.81E-01	8.92E-01	-1.01E-02
8.81E+01	1.28E+02	1.50E+02	1.17E+00	1.05E+00	1.23E-01
2.62E+02	3.61E+02	1.88E+02	5.21E-01	5.69E-01	-4.84E-02
2.60E+02	3.74E+02	1.89E+02	5.06E-01	5.58E-01	-5.16E-02
2.37E+02	3.47E+02	1.90E+02	5.48E-01	5.83E-01	-3.48E-02
2.55E+02	3.77E+02	1.93E+02	5.12E-01	5.55E-01	-4.37E-02
2.35E+02	3.44E+02	1.93E+02	5.61E-01	5.85E-01	-2.48E-02
2.35E+02	3.41E+02	1.94E+02	5.70E-01	5.89E-01	-1.95E-02
2.37E+02	3.42E+02	2.04E+02	5.97E-01	5.88E-01	8.76E-03
2.37E+02	3.39E+02	2.05E+02	6.05E-01	5.91E-01	1.41E-02
2.35E+02	3.32E+02	2.10E+02	6.33E-01	5.98E-01	3.47E-02
5.42E+02	7.66E+02	2.37E+02	3.10E-01	3.64E-01	-5.44E-02
5.27E+02	7.45E+02	2.44E+02	3.27E-01	3.70E-01	-4.27E-02
5.06E+02	7.15E+02	2.53E+02	3.54E-01	3.79E-01	-2.53E-02
4.66E+02	6.59E+02	2.57E+02	3.89E-01	3.98E-01	-8.63E-03
4.66E+02	6.59E+02	2.57E+02	3.91E-01	3.98E-01	-7.45E-03
4.60E+02	6.51E+02	2.64E+02	4.06E-01	4.01E-01	4.87E-03
4.75E+02	6.71E+02	2.64E+02	3.94E-01	3.94E-01	-3.04E-06
4.71E+02	6.65E+02	2.73E+02	4.11E-01	3.96E-01	1.47E-02
4.69E+02	6.63E+02	2.76E+02	4.16E-01	3.97E-01	1.93E-02

References

1. ASTM Standard Test Method D 3835-90, "Rheological Properties of Thermoplastics with a Capillary Rheometer," 1992 *Annual Book of ASTM Standards.* ASTM, Philadelphia, 1992, Vol. 08.03, pp. 201-206.
2. Brydson, J. A.: *Flow Properties of Polymer Melts.* Iliffe, London, 1970.
3. Walters, K.: *Rheometry.* Chapman and Hall, London, 1975.
4. Darby, R.: *Viscoelastic Fluids.* Marcel Dekker, New York, 1976.
5. Han, C. D.: *Rheology in Polymer Processing.* Academic Press, New York, 1976.
6. Cogswell, F. N.: *Polymer Melt Rheology.* Halsted Press, New York, 1981.
7. Dealy, J. M.: *Rheometers for Molten Plastics.* Van Nostrand Reinhold, New York, 1982.
8. Dealy, J. M.; Wissbrun, K. F.: *Melt Rheology and Its Role in Plastics Processing.* Van Nostrand Reinhold, New York, 1990.
9. Shaw, M. T.: "An Alternative Analysis of Capillary Rheometer Data," *SPE Tech. Papers* **32**, 707-710 (1986).
10. Bagley, E. B.: "End Corrections in the Capillary Flow of Polyethylene," *J. Appl. Phys.* **28**, 624-627 (1957).
11. Mooney, M.: "Explicit Formulas for Slip and Fluidity," *J. Rheol.* **2**, 210-222 (1931).
12. Weissenberg, K.; as cited by B. Rabinowitsch: "Ueber die Geschwindigkeitsfunktion der Viskosität disperser Systeme. I.," *Z. Phys. Chem.* A**145**, 1-26 (1929).

13. Hill, D. A.; Hasegawa, T.; Denn, M. M.: "On the Apparent Relation Between Adhesive Failure and Melt Fracture," *J. Rheol.* **34**, 891-918 (1990).
14. Hatzikiriakos, S. G.; Dealy, J. M.: "Wall Slip of Molten High Density Polyethylenes. II. Capillary Rheometer Studies," *J. Rheol.* **36**, 703-741 (1992).
15. Arai, T.; Aoyama, H.: "Die Wall Restriction on Elastic Shear Deformation in Viscoelastic Flow of Polymer Melt," *Trans. Soc. Rheol.* **7**, 333-355 (1963).
16. Nakajima, N.; Shida, M.: "Viscoelastic Behavior of Polyethylene in Capillary Flow Expressed with Three Material Functions," *Trans. Soc. Rheol.* **10**, 299-316 (1966).
17. Cogswell, F. N.; Lamb, P.: "The Mechanism of Melt Distortion," *Trans. J. Plastics Inst.* 809-813 December (1967).
18. Ballenger, T. F.; White, J. L.: "The Development of the Velocity Field in Polymer Melts in a Reservoir Approaching a Capillary Die," *J. Appl. Polym. Sci.* **15**, 1949-1962 (1971).
19. Han, C. D.; Kim, K. U.: "Influence of Reservoir Diameter on the Melt Elasticity in Capillary Flow," *Polym. Eng. Sci.* **11**, 395-400 (1971).
20. Han, C. D.: "Influence of the Die Entry Angle on the Entrance Pressure Drop, Recoverable Elastic Energy, and Onset of Flow Instability in Polymer Melt Flow," *J. Appl. Polym. Sci.* **17**, 1403-1413 (1973).
21. Leblanc, J. L.: "Factors Affecting the Extrudate Swell and Melt Fracture Phenomena of Rubber Compounds," *Rubber Chem. Technol.* **54**, 905-929 (1981).
22. Kannabiran, R.: "Correlation Between End Correction and Extrudate Swell for Some Raw Elastomers and Black-Filled Rubber Compounds," *Rubber Chem. Technol.* **57**, 1001-1012 (1984).
23. Utracki, L. A.; Schlund, B.: "Linear Low Density Polyethylenes and Their Blends: Part 2. Shear Flow of LLDPE's," *Polym. Eng. Sci.* **27**, 367-379 (1987).
24. Han, C. D.; Charles, M.: "Entrance- and Exit-Correction in Capillary Flow of Molten Polymers," *Trans. Soc. Rheol.* **15**, 371-384 (1971).
25. Hyun, K. S.: "End Corrections in the Capillary Flow of Polystyrene Melts," *Polym. Eng. Sci.* **14**, 666-673 (1974).
26. Shroff, R. N.; Shida, M.; Krigas, T. M.: "Die Swell, Recoverable Strain, and Entrance Pressure Drop in Polymer Extrusion," *SPE Tech. Papers* **22**, 411-413 (1976).
27. Hatzikiriakos, S. G.; Dealy, J. M.: McGill University, Montreal, personal communication, 1990.
28. Lupton, J. M.; Regester, J. W.: "Melt Flow of Polyethylene at High Rates," *Polym. Eng. Sci.* **5**, 235-245 (1965).
29. Blyler, L. L., Jr.; Hart, A. C., Jr.: "Capillary Flow Instability of Ethylene Polymer Melts," *Polym. Eng. Sci.* **10**, 193-203 (1970).
30. Lim, T. T. S.: "Capillary Extrusion of Composite Materials," *Polym. Eng. Sci.* **11**, 240-246 (1971).

31. Worth, R. A.; Parnaby, J.; Helmy, H. A. A.: "Wall Slip and Its Implications in the Design of Single Screw Melt-Fed Extruders," *Polym. Eng. Sci.* **17**, 257-265 (1977).
32. Ramamurthy, A. V.: "Wall Slip in Viscous Fluids and Influence of Materials of Construction," *J. Rheol.* **30**, 337-357 (1986).
33. Kalika, D. S.; Denn, M. M.: "Wall Slip and Extrudate Distortion in Linear Low-Density Polyethylene," *J. Rheol.* **31**, 815-834 (1987).
34. Hatzikiriakos, S. G.; Dealy, J. M.: "Wall Slip of Molten High Density Polyethylene. I. Sliding Plate Rheometer Studies," *J. Rheol.* **35**, 497-523 (1991).
35. Cross, M. M.: "Rheology of Non-Newtonian Fluids: A New Flow Equation for Pseudoplastic Systems," *J. Colloid Sci.* **20**, 417-437 (1965).
36. Cross, M. M.: "Analysis of Flow Data on Molten Polymers," *Eur. Polym. J.* **2**, 299-307 (1966).
37. Press, W. H.; Flannery, B. P.; Teukolsky, S. A.; Vetterling, W. T.: *Numerical Recipes in* FORTRAN: *The Art of Scientific Computing.* Cambridge University Press, Cambridge, 1986.
38. IMSL MATH/LIBRARY *Users' Manual*, IMSL, Houston, 1991; Vol. 3.
39. Anzini, D. J.: "The Effect of Polymer Rheology on the Residence Time Distribution of a Single Screw Plasticating Extruder," M.S. thesis, University of Connecticut, 1985.
40. Kowalski, R. C.: "Elastic Behavior of Molten Viscoelastic Polymeric Materials," Ph.D. thesis, Polytechnic Institute of Brooklyn, 1963.
41. Williams, D. J.: *Polymer Science and Engineering.* Prentice-Hall, Englewood Cliffs, 1971.

Chapter 9

Modeling the Viscosity Function

9.1 Introduction

The rewards for describing the shear-rate dependence of viscosity with a generalized Newtonian model include estimates of the zero-shear-rate viscosity, η_o, and the characteristic time constant τ, for the fluid. The model parameters may be useful for testing theories. The functionalization of shear stress or viscosity for interpolation or other manipulations is important for simulating fabrication processes such as extrusion.

Several empirical and semiempirical expressions are available for describing the generalized Newtonian viscosity of a polymeric melt or solution [1-18]. The applications and limitations of these empiricisms have been reviewed by Bird et al. [18]. Except for the widely used power-law model [19,20], all these expressions predict a Newtonian plateau at low shear rates. At a critical shear rate, the reciprocal of which is the characteristic time constant, the models describe the gradual decrease of viscosity or shear-thinning behavior of the fluid. Qualitatively, any deviation from Newtonian behavior results from deformation rates that remove the equilibrium entanglement structure of the polymer and align the molecular chains in the flow direction. At higher shear rates, a power law dependence of the viscosity on shear rate is often observed. Some viscosity functions incorporate power-law asymptotes at high shear rates, while others provide for a second Newtonian region.

Exact theory for the viscosity function is not easy to come by. One example derived from a molecular theory is the Eyring equation [6,7]

$$\eta_R = \frac{\sinh^{-1}(\tau\dot{\gamma})}{\tau\dot{\gamma}} \tag{9.1}$$

where $\eta_R = \eta/\eta_o$ is the reduced viscosity, and $\tau\dot{\gamma}$ is the reduced shear rate. The hyperbolic sine function is similarly seen in other transport equations, such as crazing and electrical conduction, involving an activated state. Sutterby increased the flexibility of the Eyring model by additing a third parameter to describe the breadth of the transition to non-Newtonian behavior [13,14]. Working with more chainlike molecules, Bird and co-workers derived asymptotic expressions for dilute suspensions of various model molecules [21]. Using less exact models, other investigators have derived numerous equations; these are often in nonexplicit or series forms. For example, the Bueche theory [1] is written

$$\eta_R = 1 - \frac{6}{\pi^2} \sum_{u=1}^{N} \frac{(\tau\dot{\gamma})^2}{u^2\left[u^4 + (\tau\dot{\gamma})^2\right]} \left[2 - \frac{(\tau\dot{\gamma})^2}{u^4 + (\tau\dot{\gamma})^2}\right] \tag{9.2}$$

The relaxation time τ is a function of the weight of the polymer molecule which acts as an imbedded spring. The index u refers to the vibration mode of the spring system in flow, and N is the number of segments in a single chain.

A summary of the more popular empirical models was compiled by Elbirli and Shaw [22] and is reproduced in Table 9.1. While not exhaustive, it provides forms that are general enough to fit most data for melts with unimodal molecular weight distributions. Chapter 4 describes additional functions for the treatment of multimodal distributions. As noted by the authors, the viscosity functions originally featuring a limiting viscosity η_∞ at high shear rates were specialized by setting this limit to zero, and several original notations were converted to a reduced form. The reduced forms for all the listed two-parameter models are shown in Figure 9.1. The ability of these models to fit viscosity data is determined by the breadth of the tran-

Table 9.1 Viscosity Functions in Terms of Reduced Viscosity $\eta_R = \eta/\eta_o$, Reduced Shear Rate $\tau\dot{\gamma}$, and Power-Law Index n [22]

Parameters	Code	Name	n	η_R	Ref.
2	Bu	Bueche-Harding	1/4	$[1 + (\tau\dot{\gamma})^{0.75}]^{-1}$	1,2
	Fy	Ferry	1/2	$[1 + \eta_R\tau\dot{\gamma}]^{-1}$	3
	Dh	DeHaven	1/3	$[1 + (\eta_R\tau\dot{\gamma})^2]^{-1}$	4
	Sp	Spencer-Dillon	0	$[\exp(\eta_R\tau\dot{\gamma})]^{-1}$	5
	Ey	Eyring	0	$\sinh^{-1}(\tau\dot{\gamma})/\tau\dot{\gamma}$	6,7
3	Ca	Carreau	n	$[1 + (\tau\dot{\gamma})^2]^{(n-1)/2}$	8
	Cr	Cross	n	$[1 + (\tau\dot{\gamma})^{1-n}]^{-1}$	9,10
	El	Ellis	n	$[1 + (\eta_R\tau\dot{\gamma})^{(1-n)/n}]^{-1}$	11
	Mi	Mieras	n	$[1 + (\eta_R\tau\dot{\gamma})^2]^{(n-1)/2n}$	12
	Su	Sutterby	n	$[\sinh^{-1}(\tau\dot{\gamma})/\tau\dot{\gamma}]^{1-n}$	13,14
	Qd	Quadratic	–	$\exp[-a(\ln \tau\dot{\gamma})^2]$	15
4	Sa	Sabia	n	$[1 + (\tau\dot{\gamma})^{(1-n)/a}]^{\eta_R - a}$	16
	Vn	Vinogradov	n	$[1 + a(\tau\dot{\gamma})^{(1-n)/2} + (\tau\dot{\gamma})^{1-n}]^{-1}$	17
	Gr	Generalized rate	$1-ab$	$[1 + (\tau\dot{\gamma})^a]^{-b}$	22
	Gs	Generalized stress	$\frac{1}{1+ab}$	$[1 + (\eta_R\tau\dot{\gamma})^a]^{-b}$	22

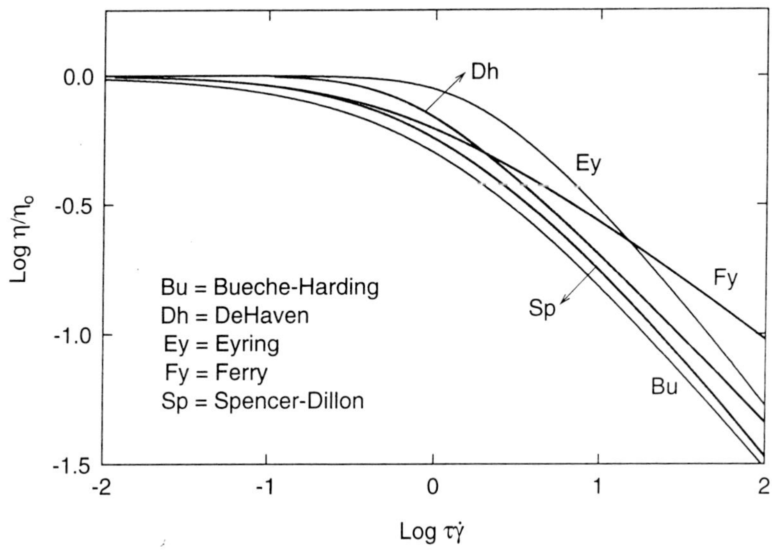

Figure 9.1 Forms for the two-parameter viscosity models in terms of reduced viscosity, η/η_o, and reduced shear rate, $\tau\dot{\gamma}$.

sition and the limiting slope in the power-law region, both of which are fixed. From Figure 9.1, the Eyring model provides the narrowest transition with the magnitude of the limiting slope approaching unity (or slope = -1), whereas the Ferry model provides the broadest transition with a limiting slope magnitude of 0.5.

A useful series of viscosity functions is

$$\eta_R = \frac{1}{1 + (\tau\dot{\gamma})^{0.75}} \tag{9.3}$$

$$\eta_R = \frac{1}{1 + (\tau\dot{\gamma})^{a}} \tag{9.4}$$

$$\eta_R = \frac{1}{\left[1 + (\tau\dot{\gamma})^{a}\right]^{b}} \tag{9.5}$$

which are the Bueche-Harding (BU), Cross (CR), and generalized-rate (GR) models, respectively. The parameter a in Equation 9.4 and the product ab in Equation 9.5 describe the shear-rate dependence of viscosity at high shear rates (power-law region). While all the model parameters ultimately depend on molecular weight, the nondimensional parameters a and b are expected to be functions of the molecular weight distribution. This series is handy because the first model, a two-parameter function, can give good starting values for the nonlinear solution of the next, more complex, model and so on.

9.2 Description of the Algorithm **VISFIT**

The design of program VISFIT is straightforward and is set up to be partially interactive, allowing the operator to control the modeling process. As written, it includes the Bueche-Harding, Eyring, Carreau, Cross, Sutterby, and the generalized-rate viscosity functions. However, this particular selection should not imply an endorsement. Based on 12 data sets, Elbirli and Shaw [22] reported that no one model listed in Table 9.1 was clearly superior. Other models can be easily added, as will be apparent from an examination of the source code listed in Section 9.5. Additional viscosity functions should be represented by a two-character code (see Table 9.1) while the corresponding subroutines to evaluate the objective function and its first partial derivatives with respect to each parameter must be declared EXTERNAL in the calling program VISFIT. As an illustrative example, specific areas in the code of the subroutine FXBU for the Bueche-Harding model have been boxed to aid custom modification. However, the superiority of any new model added to this selection must ultimately be judged by an analysis of residuals and errors relative to those from competing models.

For numerical reasons, the models explicit in viscosity are the easiest to handle. It is not usually critical whether viscosity $\eta(\dot{\gamma})$ or shear stress $\sigma_{21}(\dot{\gamma})$ is considered the dependent variable. However, since the relative errors in $\sigma_{21}(\dot{\gamma})$ are normally of constant magnitude, as opposed to the absolute errors, the regression should be done on $\log \eta(\dot{\gamma})$ or $\log \sigma_{21}(\dot{\gamma})$. To avoid distortion, a logarithmic $\dot{\gamma}$ scale is appropriate for the independent variable. Using the generalized-rate viscosity function as a representative case, we can rewrite Equation 9.5 as

$$\log \sigma_{21} = \log \dot{\gamma} + \beta_1 - \beta_4 \log\left[1 + 10^{\beta_3(\beta_2 + \log \dot{\gamma})}\right] \tag{9.6}$$

where $\beta_1 = \log \eta_o$, $\beta_2 = \log \tau$, $\beta_3 = a$, and $\beta_4 = b$. Note that the definitions of the program's search parameters β_3 and β_4 are model dependent. Refer to Table 9.2 for the correspondence between the dimensionless parameters in the viscosity models available from VISFIT and the search parameters.

The nonlinear regression analysis is based on the Gauss-Newton [23] method, introduced in Chapter 1. It is carried out via the subroutine GAUSSN. To use GAUSSN, we need the analytical forms of the partial first derivatives of the model with respect to each parameter and the initial estimates of the parameters. The derivatives are entered in the same subroutine containing the transformed function for the particular model. For an example, refer to the subroutine FXBU imbedded in

Table 9.2 Correspondence Between the Dimensionless Parameters for the Viscosity Models Available From VISFIT and the Program Parameters β_3 and β_4

Parameters	Code	Name	β_3	β_4
2	Bu	Bueche-Harding	–	–
	Ey	Eyring	–	–
3	Ca	Carreau	$\frac{1-n}{2}$	–
	Cr	Cross	$1-n$	–
	Su	Sutterby	$1-n$	–
4	Gr	Generalized rate	a	b

the code for VISFIT.

To start the search in the right direction, VISFIT will suggest initial parameter estimates. However, unlike the programs in Chapters 1 and 3, which also use GAUSSN, no elaborate linearization of the model is used. For the zero-shear-rate viscosity β_1, VISFIT simply uses the viscosity at the lowest shear rate after sorting the data in order of increasing shear rate via SORT2. The initial estimate for the second parameter, β_2, is taken as the reciprocal of the arithmetic average of the lowest and highest shear rates. For the more complex models, the estimates for β_3 and β_4 are initialized to 0.75 and 1.0, respectively. These initial estimates reduce Equations 9.4 and 9.5 to Equation 9.3 and should be adequate as starting values. These estimates are provided for the first call to GAUSSN and are updated to the converged parameter values regardless of the next model chosen to be fit to the same set of data.

Expressed in terms of 95% confidence intervals, the errors in the fitted parameters are estimated from the diagonal components of the covariance matrix by using the expression

$$\pm\, t_{95,\nu}\; s_{\varepsilon} \sqrt{C_{ii}} \tag{1.16}$$

where $t_{95,\nu}$ is the t-statistic, as a function of the appropriate degrees of freedom ν, s_ε is the standard error of estimate for the data set, and C_{ii} is the corresponding diagonal element for the estimated covariance matrix of the standard error [23,24]. In turn, the 95% confidence intervals associated with the zero-shear-rate viscosity, η_o, the time constant, τ, and the power-law index, n, are obtained from error propagation analysis.

9.3 Execution of VISFIT

The program VISFIT requires the subroutines GAUSSN, SORT2 and SPECC found in Appendix B. Refer to Appendix A for the program dependencies. For example, MATINV, called by GAUSSN, requires the *LU* matrix decomposition routines available from *Numerical Recipes* [24].

Before executing the program VISFIT, a data file must be set up by using an editor or text processor that outputs ASCII code. The format of the file is identical to that used by MWD in Chapter 4 to obtain a molecular weight distribution from a viscosity function. The first record is set aside for the title of the file; include anything up to 79 characters containing identification information. The second record specifies the variable representing the ordinate. Enter V or S for viscosity or shear stress, respectively. Data may be entered as dimensional numbers or in the corresponding logarithmic values. The particular format should be specified in the third record by entering N for dimensional numbers or L for logarithmic values. The succeeding records are either pairs of shear-rate and viscosity data, or shear-rate and shear-stress data. The data pairs may be entered in any sequence; they are subsequently sorted by shear rate by the subroutine SORT2. The tabulation in the output file is also in order of increasing shear rate to facilitate an examination of the residuals for nonrandomness. A sample listing of a data file is provided in Section 9.6.1.

The first interaction required is the specification of the relative convergence criterion. As defined in Equation 1.15, the iterations are said to have converged if, for all parameters, the criterion is greater than the absolute value of $\delta\beta_j / \beta_j$. While the criterion should be a small positive number, it should be no smaller than the square root of the machine's floating-point precision [24]. The default value is 10^{-4}.

The second interaction is to specify the viscosity model by using the codes in Table 9.1. In the execution of VISFIT, the model is the most critical choice. For the available two-parameter models, the Bueche-Harding model (Equation 9.3) has been found to be best suited for polymers with most-probable distributions ($M_w / M_n \rightarrow 2$) [2,25,26]. The Bueche-Harding model also agrees moderately well with Graessley's entanglement-theory predictions for most-probable distributions [26]. The Eyring model has the narrowest transition to power-law behavior, and was reported to have a slight edge among the two-parameter models [22]. However, starting with the Bueche-Harding model is usually the safest step regardless of the molecular weight distribution.

Figure 9.2 shows the reduced forms for the three-parameter models available in VISFIT. All three models were plotted with a limiting power-law slope of -9/11. The Cross model has the broadest response and will work best for typical commercial polymers. In comparing the Carreau and Cross models on a variety of polymer melts, Hieber and Chiang [27] found that the Cross model provided a better

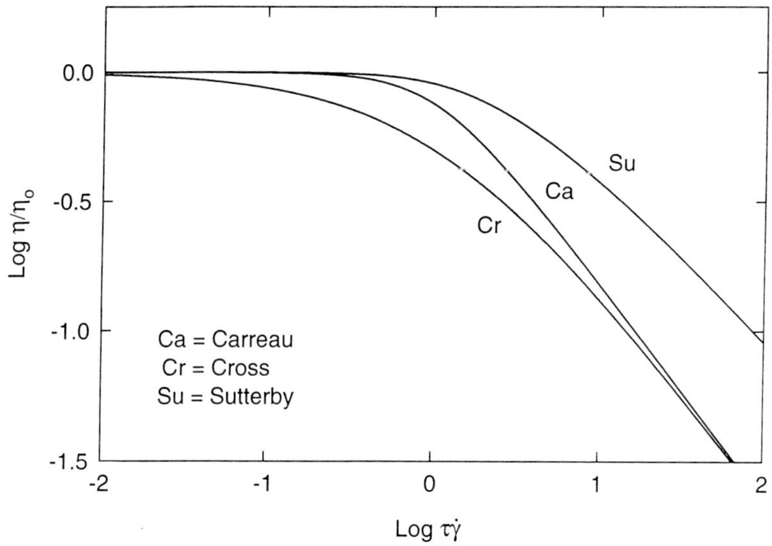

Figure 9.2 Forms for the available three-parameter viscosity models in VISFIT expressed in terms of reduced viscosity, η/η_o, reduced shear rate, $\tau\dot{\gamma}$, and a limiting slope of -9/11 in the power-law region. Note that the Sutterby model has the characteristic of having a gradual approach to the limiting slope value.

overall fit. On the other hand, the Carreau and Sutterby models are better suited to describe the more distinct transitions typical of narrowly distributed polymers. For the Carreau model, there is a sharp transition to the power-law region, whereas the approach of the Sutterby model to the limiting slope is more gradual.

After the appropriate model is chosen and entered, the third interaction is a prompt for initial estimates to the parameters defined, for example, by Equation 9.6 for the generalized-rate model. For the first iteration, the values suggested by VISFIT refer to the logarithmic value of the highest viscosity datum for β_1 ($\log \eta_o$) and the negative logarithm of an intermediate shear rate for β_2 ($\log \tau$). In the event a more complex model is chosen, the parameters β_3 and β_4 are initialized to 0.75 and 1.0, respectively,

The convergence to a global minimum is usually rapid if the initial estimates are reasonably close to the true solution. While checking with different starting values would be wise, Elbirli and Shaw [22] observed the insensitivity of the final values to the initial estimates for many data sets. If all goes well, the program will display the parameters and respective errors, and the standard error of estimate of $\log \sigma_{21}$. This information is also included in the output file along with the zero-shear-rate viscosity and time constant. The power-law index is supplied for the three- and four-parameter models. The remainder of the output file contains a summary table consisting of the logarithmic values of shear rate, observed shear stress and viscosity, calculated shear stress and viscosity, and the residual of shear stress. The units of the variables are consistent with those supplied in the data. Sample listings of a typical terminal dialog and an output file are provided in Sections 9.6.2 and 9.6.3, respectively.

If the iterative process diverges, an exponential out-of-bounds error will occur, or occasionally a divide check in the matrix inversion subroutine. In both cases, try different starting values for the parameter estimates. Clearly, if the data are strictly power-law (that is, no curvature), it will be difficult for the Bueche-Harding model, or any nonlinear function, to fit.

If the Bueche-Harding model can be successfully used, check the errors in the parameters and in the model (s_ε) before moving to more complex functions. The program first provides the option to try new parameter estimates before inquiring if another model is desired; the suggested values are based on the converged values of the preceding run. If the errors remain large, an examination of the residuals in the output file is the next recourse. For example, the presence of just three strings of signs is an indication of the inadequacy of the model; therefore, a different choice of model is recommended. However, if the residuals are randomly positive and negative, it is usually inappropriate to try a more complex model and such is unlikely to be successful. More extensive and precise data are probably called for. With respect to the precision of η_o, aside from the magnitude of the 95% confidence interval, a glance at the residuals at low shear rates may be instructive; the residuals should be small and random.

To start up the optimization for the three-parameter Carreau or Cross model, use the final values of β_1 and β_2 found with the Bueche-Harding model as initial estimates. For the parameter β_3, an initial estimate between 0.67 and 0.82 should suffice. The suggested value for β_3, as initially provided by `VISFIT`, is 0.75 and corresponds to the magnitude of the limiting slope of the Bueche-Harding model. Table 9.2 lists the correspondence between the model parameters and the program parameters β_3 and β_4. For example, β_3 for the Carreau model corresponds to $(1 - n)/2$, and has a range between 0 and 0.5. The other available option for data showing a distinct transition from Newtonian to power-law behavior is the three-parameter Sutterby model; here, start by fitting the data to the Eyring model.

If the data are extremely good and extend over a wide range of shear rate, moving to the generalized-rate model may be justified using 1.0 as an initial estimate for β_4. However, such sets of data are rare and the optimization is prone to parameter interaction; divergence and parameters with large errors are not uncommon. The effects of the parameters a and b are shown in Figure 9.3; increasing the ratio b/a at constant ab (which determines the power-law slope at high shear rates) broadens the distribution.

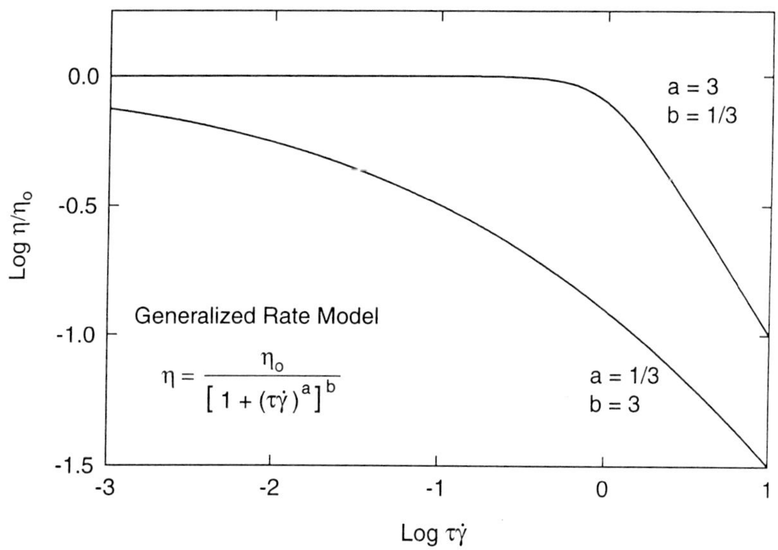

Figure 9.3 Illustration of the influence of the nondimensional parameters *a* and *b* on the response of the generalized-rate (GR) viscosity model.

9.4 Application of **VISFIT**

The first example to illustrate the capabilities of VISFIT uses data for a silicone gum (General Electric SE-30, $M_w = 5 \times 10^5$) at 30 °C [28]. Measurements were taken with a mechanical spectrometer, using a cone-and-plate geometry (0.06 to 2.5 s^{-1}), and a capillary rheometer (0.39 to 27.5 s^{-1}). This is a narrowly distributed material, and the transition from Newtonian to power-law behavior is quite distinct. From the available models, the Sutterby and the Carreau models provided the best fits based on the standard error of estimate and analyses of the residuals – not unexpected based on the discussion in Section 9.3.

The results for the Eyring and Sutterby models are shown in Figure 9.4 while the input/output files using the Sutterby model are given in Section 9.6. The quality of the fit for these models is illustrated through a plot of the residuals in Figure 9.5. There are only three strings of signs for the Eyring model (s_ε = 0.038 kPa), while the Sutterby model (s_ε = 0.024 kPa) produces nine sign changes, a good indicator of more randomness in the residuals. (The large residuals at high shear rates are an indication that the cone-and-plate data point at 2.5 s^{-1} could be discarded. In doing so, the standard error of estimate from the Sutterby model is reduced by 25%.) The values for the zero-shear-rate viscosity are equivalent for each model, but there are significant differences in the quality of estimates for the time constant: 1.48 ± 0.27 and 2.47 ± 0.75 s for the Eyring and Sutterby models, respectively. Even with the

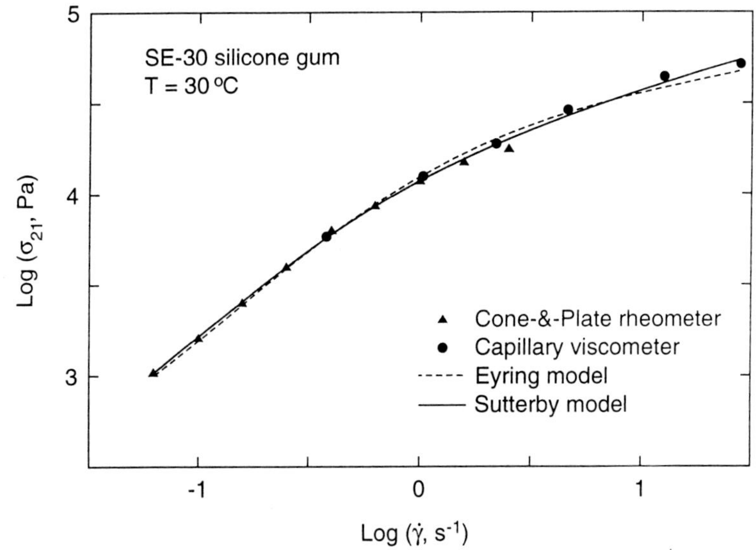

Figure 9.4 The flow curve for a silicone gum (General Electric SE-30, $M_w = 5 \times 10^5$) at 30 °C [28]. Using the Sutterby viscosity function: $\eta_o = 16.6 \pm 1.0$ kPa•s; $\tau = 2.47 \pm 0.75$ s; $n = 0.19 \pm 0.08$.

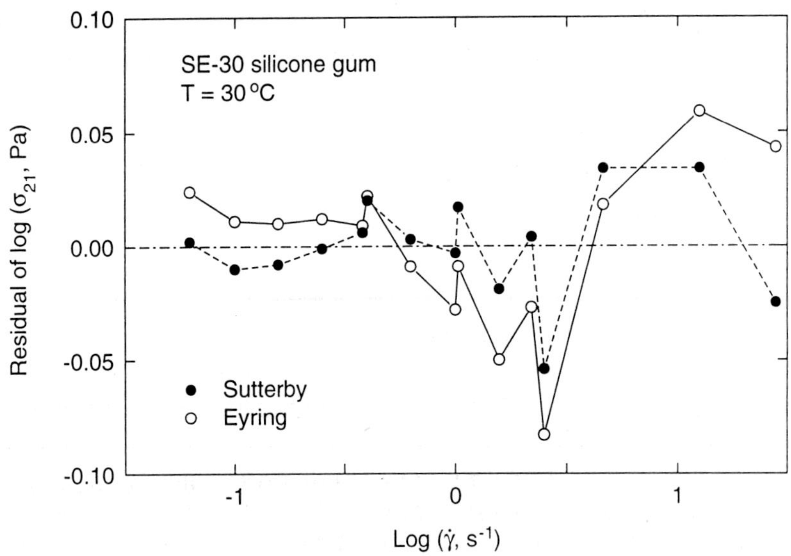

Figure 9.5 The residuals from Figure 9.4 illustrating the quality of fit using the Eyring (s_ε = 0.038 kPa) and Sutterby (s_ε = 0.024 kPa) models.

better fit, a larger error in τ from the three-parameter model confirms the observation by Elbirli and Shaw [22] that the more complex models do not always provide time constants with as low an expected error as the better two-parameter models. Note that the magnitude of the time constant, as opposed to the zero-shear-rate viscosity, is intrinsically model dependent.

The second illustration, shown in Figure 9.6, is based on published data sets derived using three different polymer melts [29-31]. For the Carreau model, the results from VISFIT are compared with those reported by Abdel-Khalik et al. [32] in Table 9.3. Aside from using nonlinear least-squares fitting, the authors report an alternative method for calculating the model parameters based on master plots of reduced viscosity versus reduced shear rate. Unfortunately, the actual procedure used was not specified. Nevertheless, the agreement is satisfactory. The discrepancies are also undoubtedly attributable to systematic errors incurred during the digitization of the data.

Most commercial polymers will have a broad transition, the shear-rate zone between the Newtonian and power-law regions of the viscosity curve, that is a characteristic of a broad molecular weight distribution (see Chapter 4). The three to four decades of shear rate obtained with a typical capillary rheometer usually produces data exclusively in or near the power-law region. The program VISFIT and other optimization algorithms can converge on such sets as long as there is the slightest curvature on a logarithmic scale of the viscosity curve. However, the model parameters will have very high uncertainties. The best results usually require a combination of data sets from two rheological instruments or the superposition of sets taken at different temperatures (see Chapter 5). The data should emphasize low shear rates to minimize the error in estimating the zero-shear-rate viscosity, and low and intermediate shear rates for the best estimates of the time constant [22].

Table 9.3 Comparison of the Carreau Viscosity Model Parameters and Corresponding 95% Confidence Intervals Obtained via VISFIT with Literature Values [32][a] for Three Polymer Melts

			η_o, kPa•s		τ, s		$(1 - n)/2$	
Material	Ref.	T, °C	VISFIT	Lit.	VISFIT	Lit.	VISFIT	Lit.
Polystyrene	29	180	15.2 ± 2.5	14.8	1.01 ± 0.5	1.04	0.302 ± .02	0.301
Phenoxy-A	30	212	12.2 ± 0.3	12.4	6.78 ± 1.7	7.44	0.134 ± .01	0.136
LDPE	31	180	12.8 ± 0.3	12.1	9.55 ± 1.4	5.23	0.173 ± .01	0.217

[a] The method used to obtain the model parameters was not specified. The 95% confidence intervals for the literature values were not available.

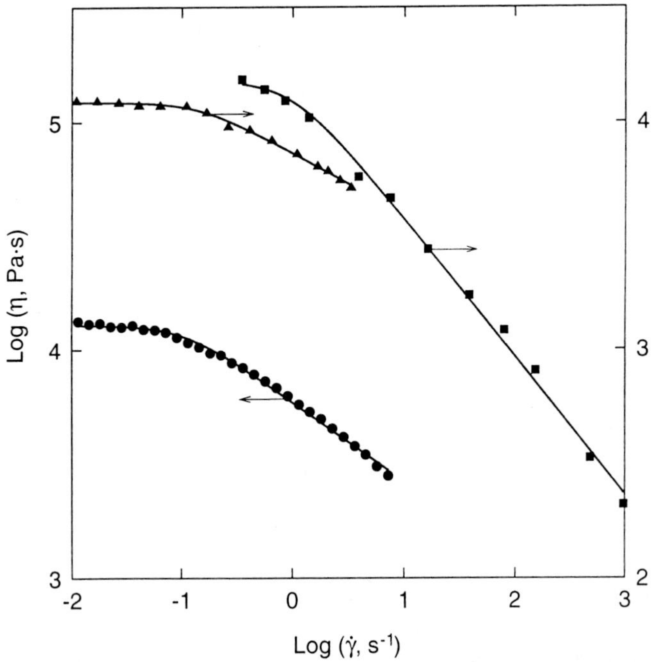

Figure 9.6 VISFIT results for three polymer melts using the Carreau viscosity function [8]. ■ Polystyrene: η_o = 15.2 ± 2.5 kPa•s, τ = 1.01 ± 0.47 s, n = 0.40 ± 0.04 (Adapted from Ref. 29. Copyright © 1971 by The Society of Rheology, Inc. Reprinted by permission of the American Institute of Physics.). ▲ Phenoxy-A: η_o = 12.2 ± 0.3 kPa•s, τ = 6.8 ± 1.7 s, n = 0.73 ± 0.02 (Adapted and reprinted with permission from *Chem. Eng. Sci.,* **23**, P. J. Carreau, I. F. MacDonald and R. B. Bird, "A Nonlinear Viscoelastic Model for Polymer Solutions and Melts II" Copyright © 1968, Pergamon Press Ltd.) ● Low-density polyethylene: η_o = 12.8 ± 0.3 kPa•s, τ = 9.6 ± 1.4 s, n = 0.65 ± 0.01. (Adapted from Ref. 31. Copyright © 1972 by The Society of Rheology, Inc. Reprinted by permission of the American Institute of Physics.)

9.5 Program Listing for VISFIT

Imbedded into the code are the subprograms FXBU, FXEY, FXSU, FXCA, FXCR, FXGR, and ASINH. Specific areas of the subroutine FXBU have been boxed to aid custom modification.

```
      PROGRAM VISFIT
C Fits viscosity functions to data. Program will ask for model key:
C    BU = Bueche-Harding model                CA = Carreau model
C    EY = Eyring model                        CR = Cross model
C    SU = Sutterby model                      GR = Generalized rate model
C Requires the subroutines GAUSSN, SORT2, SPECC, and STOPP.
C Data file format:
C         1st record - Title, up to 79 characters.
C         2nd record - V(iscosity) or S(tress).
C         3rd record - L(ogarithmic) or N(ormal) values.
C succeeding records - shear rate (1/s), viscosity.
C                 or,   shear rate (1/s), shear stress.
      IMPLICIT DOUBLE PRECISION (A-H,O-Z)
      PARAMETER (NMAX=201,NPD=10,NPR=4)
      CHARACTER ANS*1,FORM*80,FUN*2,TITLE*80,TYPE*80
      DIMENSION A(NMAX,NPD),AT(NPD,NMAX),B(NPR),ERR(NPR),R(NMAX,1)
      DIMENSION X(NMAX),Y(NMAX)
      EXTERNAL FXBU,FXCA,FXCR,FXEY,FXGR,FXSU
C t-statistic for 95% confidence interval with N degrees of freedom.
      T95(N)=1.96+2.3539/N+3.1902/N**2+0.7004/N**3+4.5015/N**4
10    CALL MSHEAD
      CALL UNITS(NTER,NINT,NOUT)
      WRITE(NOUT,'(/23X,A)')'Modeling the Viscosity Function'
      WRITE(NOUT,'(23X,31(''-'')/)')
      READ(NINT,'(A80)')TITLE
      WRITE(NOUT,'(/A80/)')TITLE
      READ(NINT,'(A80)')TYPE
      READ(NINT,'(A80)')FORM
      ITYPE=1
      IFORM=1
      DO 20 I=1,80
        IF(TYPE(I:I).EQ.'S'.OR.TYPE(I:I).EQ.'s')ITYPE=2
        IF(FORM(I:I).EQ.'N'.OR.FORM(I:I).EQ.'n')IFORM=2
20    CONTINUE
C Read in viscosity data.
      DO 30 I=1,NMAX
        READ(NINT,*,END=40)X(I),Y(I)
        IF(I.EQ.NMAX)CALL STOPP(NTER,NMAX-1)
C Transform viscosity data into stress data.
        IF(IFORM.EQ.2)THEN
          X(I)=LOG10(X(I))
          IF(ITYPE.EQ.1)THEN
            Y(I)=X(I)+LOG10(Y(I))
          ELSE
            Y(I)=LOG10(Y(I))
          ENDIF
        ELSE IF(ITYPE.EQ.1)THEN
          Y(I)=X(I)+Y(I)
        ENDIF
30    CONTINUE
40    NP=I-1
      CALL SORT2(NP,X,Y,X,Y)
C Initialize starting values.
      B(1)=Y(1)-X(1)
      B(2)=-(X(1)+X(NP))/2.
      B(3)=0.75
      B(4)=1.0
C Specify convergence criterion.
```

```
      CALL SPECC(NTER,CC)
50    WRITE(NTER,*)' Enter model ID'
      READ(NTER,'(A2)')FUN
C Fit viscosity model chosen.
60    IF(FUN.EQ.'BU'.OR.FUN.EQ.'bu'.OR.FUN.EQ.'Bu')THEN
        N=2
        WRITE(NTER,70)N,(B(I),I=1,N)
        READ(NTER,*)(B(I),I=1,N)
        CALL GAUSSN(FXBU,CC,N,NP,X,Y,B,A,AT,R,ERR,SEE)
        WRITE(NOUT,'(/24X,A/)')'Bueche-Harding Model Results'
      ELSE IF (FUN.EQ.'EY'.OR.FUN.EQ.'ey'.OR.FUN.EQ.'Ey') THEN
        N=2
        WRITE(NTER,70)N,(B(I),I=1,N)
        READ(NTER,*)(B(I),I=1,N)
        CALL GAUSSN(FXEY,CC,N,NP,X,Y,B,A,AT,R,ERR,SEE)
        WRITE(NOUT,'(/28X,A/)')'Eyring Model Results'
      ELSE IF(FUN.EQ.'SU'.OR.FUN.EQ.'su'.OR.FUN.EQ.'Su')THEN
        N=3
        WRITE(NTER,70)N,(B(I),I=1,N)
        READ(NTER,*)(B(I),I=1,N)
        CALL GAUSSN(FXSU,CC,N,NP,X,Y,B,A,AT,R,ERR,SEE)
        PLI=1.-B(3)
        PLIE=ERR(3)
        WRITE(NOUT,'(/27X,A/)')'Sutterby Model Results'
      ELSE IF(FUN.EQ.'CA'.OR.FUN.EQ.'ca'.OR.FUN.EQ.'Ca')THEN
        N=3
        WRITE(NTER,70)N,(B(I),I=1,N)
        READ(NTER,*)(B(I),I=1,N)
        CALL GAUSSN(FXCA,CC,N,NP,X,Y,B,A,AT,R,ERR,SEE)
        PLI=1.-2.*B(3)
        PLIE=2.*ERR(3)
        WRITE(NOUT,'(/28X,A/)')'Carreau Model Results'
      ELSE IF(FUN.EQ.'CR'.OR.FUN.EQ.'cr'.OR.FUN.EQ.'Cr')THEN
        N=3
        WRITE(NTER,70)N,(B(I),I=1,N)
        READ(NTER,*)(B(I),I=1,N)
        CALL GAUSSN(FXCR,CC,N,NP,X,Y,B,A,AT,R,ERR,SEE)
        PLI=1.-B(3)
        PLIE=ERR(3)
        WRITE(NOUT,'(/29X,A/)')'Cross Model Results'
      ELSE IF(FUN.EQ.'GR'.OR.FUN.EQ.'gr'.OR.FUN.EQ.'Gr')THEN
        N=4
        WRITE(NTER,70)N,(B(I),I=1,N)
        READ(NTER,*)(B(I),I=1,N)
        CALL GAUSSN(FXGR,CC,N,NP,X,Y,B,A,AT,R,ERR,SEE)
        PLI=1.-B(3)*B(4)
        PLIE=B(3)*ERR(4)+B(4)*ERR(3)
        WRITE(NOUT,'(/23X,A/)')'Generalized Rate Model Results'
      ELSE
       GOTO 50
      ENDIF
70    FORMAT(2X,'Input initial guesses for the ',I1,' parameters.'/
     &       2X,'Suggested values are:',4F8.2/)
      WRITE(NTER,80)(I,B(I),T95(NP-N)*SEE*ERR(I),I=1,N)
      WRITE(NOUT,80)(I,B(I),T95(NP-N)*SEE*ERR(I),I=1,N)
80    FORMAT(23X,'B(',I1,') =',F10.3,' +/-',F10.3)
      WRITE(NOUT,90)
     &    (10.**B(I),10.**B(I)*LOG(10.)*ERR(I)*T95(NP-N)*SEE,I=1,2)
90    FORMAT(/2X,'Zero-shear-rate viscosity =',1P,E10.3,' +/-',E10.3
     &    /14X,'Time constant =',1P,E10.3,' +/-',E10.3)
      IF(N.GE.3)THEN
        WRITE(NOUT,100)PLI,T95(NP-N)*SEE*PLIE
100     FORMAT(12X,'Power-law index =',F10.3,' +/-',F10.3)
      ENDIF
      WRITE(NOUT,110)
110   FORMAT(/19X,'Observed',14X,'Calculated'/12X,21('-'),2X,21('-'),
     &       2X,'Residual of'/5X,'Log',2(8X,'Log'),1X,2(8X,'Log'),
     &       9X,'Log'/4X,'Rate   ',2(4X,'Stress    Viscosity'),4X,
     &       'Stress'/2X,8('-'),2(2X,(10('-'),2X,9('-'))),2X,11('-')/)
```

```
      DO 130 I=1,NP
        WRITE(NOUT,120)X(I),Y(I),Y(I)-X(I),Y(I)-R(I,1),Y(I)-R(I,1)-X(I),
     &                 R(I,1)
120     FORMAT(2X,F7.3,2(4X,F7.3),1X,2(4X,F7.3),5X,F7.3)
130   CONTINUE
      WRITE(NTER,140)SEE
      WRITE(NOUT,140)SEE
140   FORMAT(/2X,'Standard error of estimate (log stress) = ',G9.3/)
      WRITE(NTER,*)' Another initial guess? (Y/N)'
      READ(NTER,'(A1)')ANS
      IF(ANS.EQ.'Y'.OR.ANS.EQ.'y')GOTO 60
      WRITE(NTER,*)' Try another model? (Y/N)'
      READ(NTER,'(A1)')ANS
      IF(ANS.EQ.'Y'.OR.ANS.EQ.'y')GOTO 50
      CALL MSTAIL(*10)
      STOP
      END
C
      SUBROUTINE FXBU(N,X,B,F,DER)
C Bueche-Harding viscosity function.
      IMPLICIT DOUBLE PRECISION (A-H,O-Z)
      DIMENSION B(N),DER(N)
C Determine the function value F(X).
      F=X+B(1)-LOG10(1.D0+1.D1**(.75D0*(B(2)+X)))
C Determine the partial derivatives of the Bueche-Harding model.
C DER(1) is the derivative of F with respect to the parameter B(1).
      DER(1)=1.D0
C DER(2) is the derivative of F with respect to the parameter B(2).
      DER(2)=-.75D0/(1.D0+1.D1**(-.75D0*(B(2)+X)))
      RETURN
      END
C
      SUBROUTINE FXEY(N,X,B,F,DER)
C Eyring viscosity function.
      IMPLICIT DOUBLE PRECISION (A-H,O-Z)
      DIMENSION B(N),DER(N)
      EY=ASINH(1.D1**(B(2)+X))
C Determine the function value F(X).
      F=B(1)+LOG10(EY)-B(2)
C Determine the partial derivatives of the Eyring model.
C DER(1) is the derivative of F with respect to the parameter B(1).
      DER(1)=1.D0
C DER(2) is the derivative of F with respect to the parameter B(2).
      DER(2)=1./EY/SQRT(1.D0+1.D1**(-2.D0*(B(2)+X)))-1.D0
      RETURN
      END
C
      SUBROUTINE FXSU(N,X,B,F,DER)
      IMPLICIT DOUBLE PRECISION (A-H,O-Z)
C Sutterby viscosity function.
      DIMENSION B(N),DER(N)
      SU=ASINH(1.D1**(B(2)+X))
C Determine the function value F(X).
      F=B(1)+X+B(3)*(LOG10(SU)-B(2)-X)
C Determine the partial derivatives of the Sutterby model.
C DER(1) is the derivative of F with respect to the parameter B(1).
      DER(1)=1.D0
C DER(2) is the derivative of F with respect to the parameter B(2).
      DER(2)=-B(3)+B(3)/SU/SQRT(1.D0+1.D1**(-2.D0*(B(2)+X)))
C DER(3) is the derivative of F with respect to the parameter B(3).
      DER(3)=LOG10(SU)-B(2)-X
      RETURN
      END
```

```
      SUBROUTINE FXCA(N,X,B,F,DER)
C Carreau viscosity function.
      IMPLICIT DOUBLE PRECISION (A-H,O-Z)
      DIMENSION B(N),DER(N)
      CA=LOG10(1.D0+1.D1**(2.D0*(X+B(2))))
C Determine the function value F(X).
      F=X+B(1)-B(3)*CA
C Determine the partial derivatives of the Carreau model.
C DER(1) is the derivative of F with respect to the parameter B(1).
      DER(1)=1.D0
C DER(2) is the derivative of F with respect to the parameter B(2).
      DER(2)=-2.D0*B(3)/(1.D0+1.D1**(-2.D0*(B(2)+X)))
C DER(3) is the derivative of F with respect to the parameter B(3).
      DER(3)=-CA
      RETURN
      END
C
      SUBROUTINE FXCR(N,X,B,F,DER)
C Cross viscosity function.
      IMPLICIT DOUBLE PRECISION (A-H,O-Z)
      DIMENSION B(N),DER(N)
      CR=B(3)*(B(2)+X)
C Determine the function value F(X).
      F=X+B(1)-LOG10(1.D0+1.D1**CR)
C Determine the partial derivatives of the Cross model.
C DER(1) is the derivative of F with respect to the parameter B(1).
      DER(1)=1.D0
C DER(2) is the derivative of F with respect to the parameter B(2).
      DER(2)=-B(3)/(1.D0+1.D1**(-CR))
C DER(3) is the derivative of F with respect to the parameter B(3).
      DER(3)=-(X+B(2))/(1.D0+1.D1**(-CR))
      RETURN
      END
C
      SUBROUTINE FXGR(N,X,B,F,DER)
C Generalized rate viscosity function.
      IMPLICIT DOUBLE PRECISION (A-H,O-Z)
      DIMENSION B(N),DER(N)
      GR=B(3)*(B(2)+X)
C Determine the function value F(X).
      F=X+B(1)-B(4)*LOG10(1.D0+1.D1**GR)
C Determine the partial derivatives of the generalized rate model.
C DER(1) is the derivative of F with respect to the parameter B(1).
      DER(1)=1.D0
C DER(2) is the derivative of F with respect to the parameter B(2).
      DER(2)=-B(4)*B(3)/(1.D0+1.D1**(-GR))
C DER(3) is the derivative of F with respect to the parameter B(3).
      DER(3)=-B(4)*(X+B(2))/(1.D0+1.D1**(-GR))
C DER(4) is the derivative of F with respect to the parameter B(4).
      DER(4)=-LOG10(1.D0+1.D1**GR)
      RETURN
      END
C
      DOUBLE PRECISION FUNCTION ASINH(X)
C Returns the inverse hyperbolic sine of variable X.
      DOUBLE PRECISION X
      ASINH=LOG(X+SQRT(1.D0+X**2))
      RETURN
      END
```

9.6 Sample Listings of INPUT/OUTPUT Files for VISFIT

9.6.1 Data File for VISFIT

Shown below is a data file in the format required for running VISFIT.

```
GE SE-30 PDMS (Mw=500,000) viscosity (kPa-s) data at 30 C.
V
N
  .0625    16.575
  .1000    16.045
  .1575    15.935
  .2500    15.788
  .4000    15.672
  .6250    13.755
  1.000    11.775
  1.575     9.503
  2.500     7.086
  0.38     15.3
  1.03     12.2
  2.2       8.6
  4.6       6.3
 12.6       3.5
 28.        1.85
```

9.6.2 Terminal Dialog with VISFIT

Reproduced below is the terminal dialog for running VISFIT with the data file VISFIT.DAT listed in Section 9.6.1.

```
>VISFIT
 Enter name of data file (N, if not applicable).
VISFIT.DAT
 Results to be written to NEW file: VISFIT.RES
 OK (Y/N)?
Y
 Use default convergence criterion, 1.E-4? (Y/N)
Y
 Enter model ID
SU
 Input initial guesses for the 3 parameters.
 Suggested values are:     1.22     -.12      .75
1.22 -0.12 0.75
                        B(1) =      1.219 +/-      .026
                        B(2) =       .393 +/-      .131
                        B(3) =       .808 +/-      .085

 Standard error of estimate (log stress) = .243E-01

 Another initial guess? (Y/N)
N
 Try another model?
N

 The results are in VISFIT.RES
 Run another set of data? (Y/N)
N
Stop - Program terminated.
```

9.6.3 Output File from **VISFIT**

Reproduced below is the output file from VISFIT using the Sutterby viscosity model for the data and terminal dialog listed in Sections 9.6.1 and 9.6.2, respectively.

```
                       Modeling the Viscosity Function
                       -------------------------------

GE SE-30 PDMS (Mw=500,000) viscosity (kPa-s) data at 30 C.

                           Sutterby Model Results

                       B(1) =      1.219 +/-      .026
                       B(2) =       .393 +/-      .131
                       B(3) =       .808 +/-      .085

    Zero-shear-rate viscosity = 1.655E+01 +/- 1.003E+00
                Time constant = 2.473E+00 +/- 7.479E-01
              Power-law index =       .192 +/-      .085

                   Observed                   Calculated
             --------------------    --------------------    Residual of
    Log         Log        Log           Log        Log          Log
    Rate       Stress    Viscosity      Stress    Viscosity     Stress
  --------   ---------   ---------   ---------   ---------   -----------

   -1.204       .015      1.219         .013      1.217          .002
   -1.000       .205      1.205         .215      1.215        -.010
    -.803       .400      1.202         .408      1.210        -.008
    -.602       .596      1.198         .597      1.199        -.001
    -.420       .764      1.185         .758      1.179          .006
    -.398       .797      1.195         .777      1.175          .020
    -.204       .934      1.138         .932      1.136          .003
     .000      1.071      1.071        1.074      1.074        -.003
     .013      1.099      1.086        1.082      1.070          .017
     .197      1.175       .978        1.194       .997        -.019
     .342      1.277       .934        1.273       .931          .004
     .398      1.248       .850        1.302       .904        -.054
     .663      1.462       .799        1.428       .766          .034
    1.100      1.644       .544        1.610       .510          .034
    1.447      1.714       .267        1.739       .292        -.025

  Standard error of estimate (log stress) =  .243E-01
```

References

1. Bueche, F.: "Influence of Rate of Shear on the Apparent Viscosity of A-Dilute Polymer Solutions, and B-Bulk Polymers," *J. Chem. Phys.* **22**, 1570-1576 (1954).
2. Bueche, F.; Harding, S. W.: "A New Absolute Molecular Weight Method for Linear Polymers," *J. Polym. Sci.* **32**, 177-186 (1958).

3. Ferry, J. D.: "Mechanical Properties of Substances of High Molecular Weight III. Viscosities of the System Polystyrene-Xylene," *J. Amer. Chem. Soc.* **64**, 1330-1336 (1942).
4. DeHaven, E. S.: "Extruder Design for a Pseudoplastic Fluid," *Ind. Eng. Chem.* **51**, 813-816 (1959).
5. Spencer, R. S.; Dillon, R. E.: "The Viscous Flow of Molten Polystyrene," *J. Colloid Sci.* **3**, 163-180 (1948).
6. Eyring, H.: "Viscosity, Plasticity, and Diffusion as Examples of Absolute Reaction Rates," *J. Chem. Phys.* **4**, 283-291 (1936).
7. Ree, F. H.; Ree, T.; Eyring, H.: "Relaxation Theory of Transport Problems in Condensed Systems," *Ind. Eng. Chem.* **50**, 1036-1040 (1958).
8. Carreau, P. J.: "Rheological Equations from Molecular Network Theories," Ph.D. thesis, University of Wisconsin, 1968.
9. Cross, M. M.: "Rheology of Non-Newtonian Fluids: A New Flow Equation for Pseudoplastic Systems," *J. Colloid Sci.* **20**, 417-437 (1965).
10. Cross, M. M.: "Analysis of Flow Data on Molten Polymers," *Eur. Polym. J.* **2**, 299-307 (1966).
11. Ellis, S. B.: (see Reiner, M.: *Deformation, Strain, and Flow.* Interscience, New York, 1960.)
12. Mieras, H. J. M. A.: "Solution and Bulk Flow Properties of Linear and Branched Styrene-Butadiene Rubbers," *J. Polym. Sci. Symp.* **42**, 987-1000 (1973).
13. Sutterby, J. L.: "Laminar Converging Flow of Dilute Polymer Solutions in Conical Sections. II," *Trans. Soc. Rheol.* **9**, 227-241 (1965).
14. Sutterby, J. L.: "Laminar Converging Flow of Dilute Polymer Solutions in Conical Sections: Part 1. Viscosity Data, New Viscosity Model, Tube Flow Solution," *AIChE J.* **12**, 63-68 (1966).
15. Klein, I.; Marshall, D. I.; Friehe, C. A.: "Metering Screw Performance with Temperature Gradients," *SPE J.* **21**, 1299-1304 (1965).
16. Sabia, R.: "On the Characterization of Non-Newtonian Flow," *J. Appl. Polym. Sci.* **7**, 347-355 (1963).
17. Vinogradov, G. V.; Malkin, A. Y.; Prozovovskaya, N. V.; Kargin, V. A.: "The Rheology of Polymers. The Universal Nature of the Temperature Invariant Characteristic of the Viscosity of Polymers in the Condensed State," *Dokl. Akad. Nauk SSSR.* **154**, 890-893 (1964).
18. Bird, R. B.; Armstrong, R. C.; Hassager, O.: *Dynamics of Polymeric Liquids.* Wiley, New York, 1987.
19. Ostwald, W.: "Viscosität und Elastizität von Solen," *Kolloid-Z.* **36**, 99-117 (1925).
20. de Waele, A.: "Viscometry and Plastometry," *Oil Color Chem. Assoc. J.* **6**, 33-88 (1923).

21. Bird, R. B.; Armstrong, R. C.; Hassager, O.: *Dynamics of Polymeric Liquids.* Wiley-Interscience, New York, 1977, Vol. 1, pp. 80-81.
22. Elbirli, B.; Shaw, M. T.: "Time Constants from Shear Viscosity Data," *J. Rheol.* **22**, 561-570 (1978).
23. Himmelblau, D. M.: *Process Analysis by Statistical Methods.* Wiley, New York, 1970.
24. Press, W. H.; Flannery, B. P.; Teukolsky, S. A.; Vetterling, W. T.: *Numerical Recipes in* FORTRAN: *The Art of Scientific Computing.* Cambridge University Press, Cambridge, 1986.
25. Wagner, H. L.; Wissbrun, K. F.: "Molecular Weight and Rheology of Acetal Copolymers," *Makromol. Chemie* **81**, 14-31 (1965).
26. Graessley, W. W.: "Viscosity of Entangling Polydisperse Polymers," *J. Chem. Phys.* **47**, 1942-1953 (1967).
27. Hieber, C. A.; Chiang, H. H.: "Shear-Rate-Dependence Modeling of Polymer Melt Viscosity," *Polym. Eng. Sci.* **32**, 931-938 (1992).
28. Shaw, M. T.; Giovanetti, R. C.: "Analysis of an Ellis Fluid by Squeezing Flow," *SPE Tech. Papers* **27**, 190-192 (1981).
29. Ballenger, T. F.; Chen, I-J.; Crowder, J. W.; Hagler, G. E.; Bogue, D. C.; White, J. L.: "Polymer Melt Flow Instabilities in Extrusion: Investigation of the Mechanism and Material and Geometric Variables," *Trans. Soc. Rheol.* **15**, 195-215 (1971).
30. Marsh, B.D.: (1967) (as cited by Carreau, P. J.; MacDonald, I. F.; Bird, R. B.: "A Nonlinear Viscoelastic Model for Polymer Solutions and Melts II," *Chem. Eng. Sci.* **23**, 901-911 (1968).)
31. Chen, I-J.; Bogue, D. C.: "Time-Dependent Stress in Polymer Melts and Review of Viscoelastic Theory," *Trans. Soc. Rheol.,* **16**, 59-78 (1972).
32. Abdel-Khalik, S. I.; Hassager, O.; Bird, R. B.: "Prediction of Melt Elasticity From Viscosity Data," *Polym. Eng. Sci.* **14**, 859-867 (1974).

Chapter 10

Predicting Stresses with Nonlinear Integral Models

10.1 Introduction

While it would be most useful to predict the spatial variation of stress in a nonuniform steady flow process or the temporal evolution of stress at a fixed point in an unsteady, nonuniform flow, these are difficult calculations if the fluid is viscoelastic. However, useful qualitative or rank-order information can be obtained if the more complicated process is simulated with a single or series of steps involving homogeneous flows. For example, the stresses leading to the neck-in of a web being coated on a substrate during extrusion coating can be examined by looking at the similar process of planar extension [1]. In this way it might be possible to predict the rank order of neck-in of resins during extrusion coating if the nonlinear viscoelastic characteristics of each resin are known and a constitutive equation for each has been determined.

With this in mind we provide a program, `STRESS`, which will predict the stresses in many important rheometric flows for a category of deformation histories. Examples of the types of information that `STRESS` can provide are given in Figures 10.1-10.3. Figure 10.1 illustrates what happens when a Lodge rubberlike liquid is stretched in a tensile testing machine with constant crosshead velocity. If the same total stretch were applied in the same time but with constant velocity gradient, all the curves would show continually increasing stress. The second example (Figure 10.2) shows the start-up in simple shear of a fluid with an n-power (Seth) strain function [2]. Both normal stress differences are plotted along with the shear stress. This model (Model 2) predicts a negative second normal stress difference. In Figure 10.3 are plotted the steady shear viscosities calculated by shearing three "fluids" to very high strains. The -2 slope of Model 5 in the power-law region is noteworthy.

Table 10.1 lists the six nonlinear constitutive models that are programmed, while others can be added by substituting modified statements for those found in the sample subprograms `STRSS`n. Two relaxation functions are provided: the Kohlrausch [3] and the Segalman [4, p. 285]. Both of these can be set at a single exponential or to extremely broad distributions by varying a single parameter in the input file. For the models incorporating the Kohlrausch function, the Segalman function can be substituted in with very few changes in the code.

There is a more fundamental restriction to single-integral models and a further restriction to those using the Finger (inverse Cauchy) strain tensor, although experienced modelers will see how the strain subroutine can be used to calculate components of the Cauchy tensor as well. The reader is referred to the comprehensive table in Bird et al. (Table C.1, pp. 622-624 of Reference 3) describing the

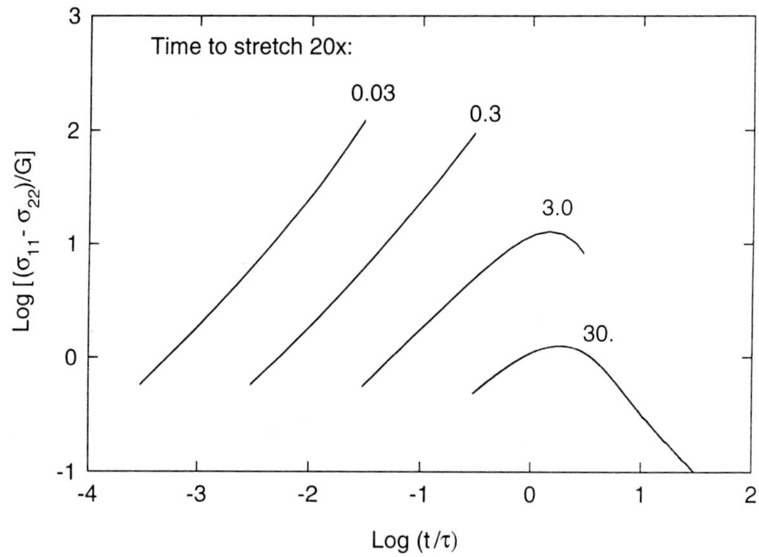

Figure 10.1 What happens when a Lodge rubberlike liquid is stretched in a tensile testing machine with constant crosshead velocity, according to program STRESS. G and τ are model parameters.

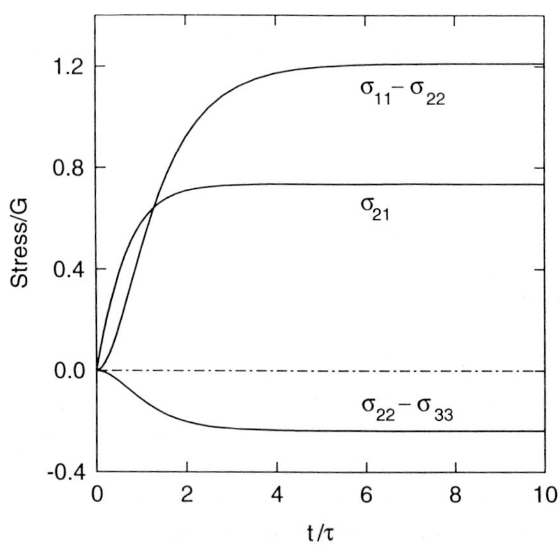

Figure 10.2 Predictions of STRESS for the growth of shear stress and the first and second normal stress differences for a model based on the Seth strain function (Model 2).

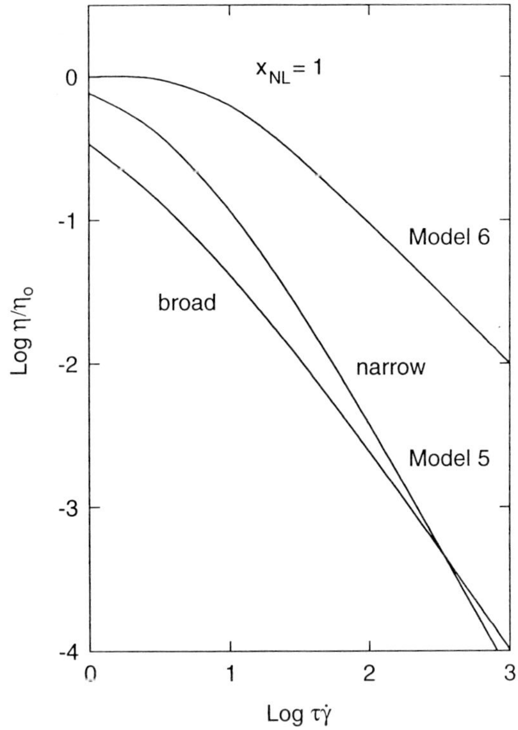

Figure 10.3 STRESS can predict steady flow viscosities; examples are shown here for the modified Tanner-Simmons model (Model 6) and the Papanastasiou-Scriven-Macosko model (Model 5). With Model 5, narrow and broader relaxation functions have been employed for illustration.

components of these tensors in rheometric flows. In this table, **B** is the Finger tensor and $\mathbf{B}^{-1}$ is the Cauchy strain tensor, while much of the older literature refers to these as $\mathbf{C}^{-1}$ and **C**, respectively.

In principle, the program STRESS could be used to match experimental data; that is, the parameters could be varied until a best-fit match between the predicted and observed stresses is obtained. It does not matter whether or not the deformation rates are constant. While this sounds like a straightforward task, it turns out that a complete and precise set of data is needed to determine the parameters accurately. In fact, it does not take much discrepancy between what the constitutive model can predict and what the material actually does before the entire scheme fails. For the adventuresome, we have included the program FITSTR that will attempt the task, but the reader should understand that this program can be expected to create a certain amount of frustration. Required for the execution of FITSTR is a table of strain and stress versus time, a great deal of patience, and a fast computer!

Table 10.1 Available Nonlinear Single-Integral Models in STRESS

ID	Description
1	Viscoelastic solid with Kohlrausch relaxation
2	Rubberlike liquid with Kohlrausch relaxation
3	Rubberlike liquid with Segalman relaxation
4	Tanner-Simmons model with Kohlrausch relaxation
5	Papanastasiou-Scriven-Macosko model with Kohlrausch relaxation
6	Modified Tanner-Simmons model with Kohlrausch relaxation

An important outcome of this approach is that with FITSTR and STRESS in combination, the rheologist can subject a material to arbitrary strain histories and transform the resulting data to a normal constant-rate experiment. This does require, however, some faith in the accuracy of the model.

10.2 Theory

Most of us find the single-integral approach to the computation of stress appealing because the stress normally appears explicitly and the equations are usually easy to understand and use. A place to start is with the Boltzmann expression for the response of a linear viscoelastic fluid:

$$\sigma_{ij}(t) = \int_{-\infty}^{t} G(t-s)\,\frac{d\gamma_{ij}}{ds}\,ds - p\delta_{ij} \tag{10.1}$$

where $\sigma_{ij}(t)$ is the total stress tensor, G is the relaxation modulus, γ_{ij} is the infinitesimal strain tensor, and t is the time at which the stress is needed. The second term, where δ_{ij} is the Kronecker delta, describes the isotropic-pressure p contribution to the normal stresses. In principle, all nonlinear equations should reduce to this expression as the magnitude of the deformation, and in some cases its derivatives, become small. Thus, the modeling of nonlinear responses should, at least, aid in the extrapolation of the response to the linear regime.

The extension to larger deformations requires the introduction of finite strain measures. The *base case* is the Lodge rubberlike liquid which uses the Finger finite-strain tensor **B**,

$$\sigma_{ij}(t) = -\int_{-\infty}^{t} G(t-s)\,\frac{dB_{ij}(t,s)}{ds}\,ds - p\delta_{ij} \tag{10.2}$$

or its equivalent form upon integrating by parts:

$$\sigma_{ij}(t) = \int_{-\infty}^{t} m(t-s)\, B_{ij}(t,s)\, ds - p\delta_{ij} \tag{10.3}$$

where $m(t-s)$ is the memory function, $\partial G(t-s)/\partial s$. The arguments for **B** shown in Equations 10.2-10.3 are very important; the first is the reference time and the second is the time at which the strain is to be evaluated. The uv term from the integration-by-parts procedure is missing from Equation 10.3. This is a result of making three assumptions: the strain at $-\infty$ is finite, the reference for strain is the present time t, and the modulus goes to zero at infinite time. When strain is referenced to the present time, the contribution of the part of the integral extending from $-\infty$ to 0 is very important even though the sample is not touched during that time. However, the $-\infty$ limit is awkward for numerical integration, so the contribution from $-\infty$ to 0 is normally split off and integrated to give the modulus multiplied by the strain at zero time:

$$\sigma_{ij}(t) = G(t)\, B_{ij}(t,0) + \int_{0}^{t} m(t-s)\, B_{ij}(t,s)\, ds - p\delta_{ij} \tag{10.4}$$

This all assumes, of course, that zero time has the usual meaning of being the start of the deformation.

Alternatively, Equation 10.2 can be split into summations of $-\infty$ to 0 and 0 to t; the first of which can be dropped because the fluid is not being deformed and changes in strain are zero. This eliminates the problem of the ∞ lower limit, but leaves the numerically undesirable (at least for general flows) aspect of taking derivatives of **B**. To vanquish this problem, one again integrates by parts, but this time part of the uv term remains. As one might expect, this turns out to be the same as the first term in Equation 10.4.

In addition to the introduction of a finite-strain measure, other changes are needed to give realistic results. Two popular options are the modification of the linear memory function and the introduction of a multiplicative strain-invariant-dependent function (damping function). A great deal of effort and ingenuity has gone into creating expressions that will apply equally well to all types of deformations. It has been especially difficult to reconcile the differences between shearing and extensional deformations while maintaining reasonable simplicity. This subject has been given extensive treatment by several rheologists, including Bird et al. [4], Larson [5], and Dealy and Wissbrun [6].

Bird and co-workers have emphasized the desirability of using forms suggested by molecular theory because the parameters of these equations can be regarded as quite fundamental material properties. Otherwise the parameters are just that and must be used only to compare materials. For example, the time constant from one treatment will not be quantitatively the same as that from another equation, although the ratio of time constants for two different materials is likely to be relatively independent of the equation used to fit the data.

While the equations used for the modeling in the programs in this chapter are perhaps less fashionable than others, they do enjoy the property of describing most nonlinear effects with only one nonlinear parameter. By "nonlinear parameter" we mean a material constant that is needed only as the deformations become high. The linear response of the material requires at least three parameters; one to scale stress, another to scale time, and a measure of the breadth of the time response of the material. We use the popular Kohlrausch [3] equation (or stretched exponential, or Williams-Watts [7], or KWW) for the linear response:

$$G(t) = G \exp\left[-\left(\frac{t}{\tau}\right)^{1/2B}\right] \tag{10.5}$$

where G scales the stress, τ scales the time, and B is the breadth parameter. Equation 10.5 is written such that a value of $B = 1$ gives the observed (at least with polystyrene [8]) $t^{-0.5}$ response for very narrow ($M_w/M_n \approx 1.0$) fractions. However, it is not our intention to suggest that the value of B will depend linearly on M_w/M_n. A value of $B = 0.5$ will give an exponential response; this is useful for testing the equations because the single exponential is the easiest to integrate analytically.

We have also included the Segalman [4, p. 285] relaxation function

$$G(t) = \frac{G}{\Gamma(1-\nu)}\left(\frac{t}{\tau}\right)^{\nu} e^{-t/\tau} \tag{10.6}$$

which has the characteristic of giving the following expression for the complex dynamic viscosity η^*:

$$\eta^* = \frac{G\tau}{(1 + i\tau\omega)^{1-\nu}} \tag{10.7}$$

where ω is the radial frequency. This allows the parameters for the relaxation function to be determined directly from linear dynamic data without the need for integral transformations. The Segalman relaxation function does have the somewhat awkward feature of an unbounded modulus at zero time if $\nu > 0$. This singularity shows up in the memory function, too, but it does not cause too much difficulty in the calculation of stress because it is multiplied by a zero strain. The Kohlrausch memory function also has this feature if $B > 0.5$. For the Segalman function, we have made the substitution $\nu = 0.5B^{-1} - 1$ to give it a similar dependence on the breadth parameter B as with the Kohlrausch function shown in Equation 10.5. Except at $B = 0.5$, the Kohlrausch and Segalman functions will give different values of the relaxation function.

10.2.1 Seth Strain-Energy Function (Models 1 - 3)

Used extensively by Tschoegl and co-workers [9-12] for describing the nonlinear properties of relaxing elastomers, the Seth strain-energy function has been applied to melts [13]. The form of the function is

$$\sigma_\alpha(t) = -\frac{2}{n}\int_{-\infty}^{t} G(t-s)\,\frac{d\lambda_\alpha^n(t,s)}{ds}\,ds - p\delta_{ij} \qquad \alpha = 1,2,3 \tag{10.8}$$

where n is the nonlinear parameter and λ is the principal elongation ratio ($\lambda_\alpha^n(t,s) \equiv [L(t)/L(s)]^n$, where L is the sample length in the α direction). A useful feature of this model is that if $n = 2$, the Lodge rubberlike liquid is recovered. Again, this is helpful for testing. The Lodge rubberlike liquid is an example of a highly strain-hardening material and most melts will be much "softer," corresponding to values of n varying toward zero. As n is decreased, the material becomes increasingly shear thinning and in extension does not develop the infinite stresses characteristic of the Lodge rubberlike liquid. We have provided two versions of this model corresponding to the solid (Model 1) and liquid (Models 2 and 3) version described by Chang, Bloch, and Tschoegl [10-12]. Models 2 and 3 simply incorporate different relaxation functions for the purpose of illustration.

10.2.2 Network-Rupture Model (Models 4 and 6)

An alternative model is based on that proposed by Tanner and Simmons in 1967 [14]. They argued that the nonlinear response of a fluid could be regarded as resulting from complete destruction of the entanglement network which had been strained for longer than a characteristic rupture time, t_r, or more than a critical strain, γ_c, where $t_r = \gamma_c/\dot{\gamma}$. In its most simplified form, this model uses the Lodge rubberlike liquid and simply replaces the lower integration limit with $t - t_r$. The resulting equation is

$$\sigma_{ij}(t) = \int_{t-t_r}^{t} m(t-s)\,B_{ij}(t,s)\,ds - p\delta_{ij} \tag{10.9}$$

It is logical that t_r should become shorter as the stresses grow, but to keep the equation explicit in stress it is convenient to have t_r depend on an invariant of the strain tensor. In keeping with the suggestions of Tanner and Simmons, we have, in Model 4, used the trace of **B** (its first invariant I_1) at time t referenced to time zero; consequently, the rupture time is

$$t_r = \frac{kt}{\sqrt{I_1 - 3}} \tag{10.10}$$

where k is the sole nonlinear parameter. In steady shear, Equation 10.10 reduces to $k/\dot{\gamma}$.

In Model 6 we have taken the liberty of introducing two slight changes, but one that has some appealing features. The first change arises because, if the lower limit of integration is no longer $-\infty$, the uv term resulting from the integration by parts of the rate form of the integral (Equation 10.2) is no longer zero; it features $G(k/\dot{\gamma})$ instead of $G(\infty)$. Our modification results in an important difference in the steady shear viscosity in the power-law region; the slope approaches -1 instead of -2. Another peculiarity of the network-rupture concept is that when the stress approaches its steady-state value, it does so with a bang because the integral from $t - t_r$ to t becomes independent of time when $t - t_r$ exceeds zero. When this happens, the stressing viscosity becomes

$$\int_0^{t_r} u\, m(u)\, du \tag{10.11}$$

for the original Tanner-Simmons form or

$$\int_0^{t_r} G(u)\, du \tag{10.12}$$

in the form coded here as Model 6. In Model 6, we have smoothed out the stress growth by replacing the discontinuous function for the bottom limit of the integral with the continuous function

$$t - t_r = t \exp\left[-\frac{k}{\sqrt{I_1 - 3}} \right] \tag{10.13}$$

The influence of these changes is illustrated in Figure 10.4.

10.2.3 Damping Function Model (Model 5)

We have included one example from a very large set of models where the nonlinearity is introduced via a strain-dependent damping function. The example is based on the findings of Papanastasiou et al. [15]. To allow the class of models using damping functions to be used, we again need the invariants of the Finger strain tensor. The definitions used for the invariants are given in Table 10.2. With this

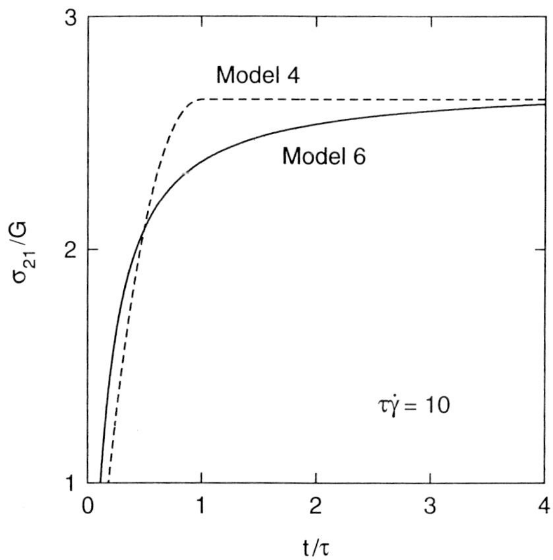

Figure 10.4 Shear stress growth for the Tanner-Simmons model in the original form (Model 4) and with some modifications (Model 6).

example in hand the user should be able to modify the code with reasonable ease to incorporate other variations of the damping function [5].

Damping-function models assume that the nonlinearity in the fluid can be described by a separable factor that is independent of time and that time-dependent effects are not altered by the level of strain. Most simply, in a stress-relaxation experiment, the stress will decay at the same rate after application of any strain, but will be multiplied by a factor that depends on the level of strain and not on time. This factor, referred to as the damping function, is formulated to decrease from an initial value of 1.0 as the strain increases.

Table 10.2 Definitions Used for the Invariants of the Finger Strain Tensor

Invariant	Shear	Extension
I_1	$\gamma^2 + 3$	$\sum_{i=1}^{3} \lambda_i^2$
I_2	$\gamma^2 + 3$	$\sum_{i=1}^{3} \lambda_i^{-2}$
I_3	1	1

The damping function, h, in the Papanastasiou-Scriven-Macosko (PSM) model is given by

$$h(I) = \frac{1}{1 + \alpha(I - 3)} \tag{10.14}$$

where I refers to the Wagner general invariant [16]:

$$I = \beta I_1 + (1 - \beta) I_2 \tag{10.15}$$

Thus, there are two parameters here: α and β. To maintain our restriction to one nonlinear parameter, we have incorporated the observation [6] that β often hovers around 0.02. Thus, α becomes the sole parameter, with values typically around 0.1. At a shear strain of 1, the damping function will reduce the stress by about 10% if α is 0.1.

10.3 Description of the Algorithms

10.3.1 STRESS

The general layout of the program is shown in Figure 10.5. The main module STRESS does very little but read in the data, which describes the kinematics, and write out the calculated stresses. Various deformations may be run in a batch mode by just stacking the data files together with an editor. The interactive aspects are limited to the choice of the model and variation of the material parameters; the kinematics are fixed by the data file. The reason for the batch mode is that some of the integrations can be very long, whereupon there is a strong motivation to walk away and let the computation run by itself.

The module STRESS calls a model-selection subroutine, STRMOD, which must be modified if new models are to be added. (STRESS, which is more complicated, does not have to be touched.) STRMOD is short and easy to understand. Section 10.6.3 contains documentation which shows exactly where the changes must be put. In turn, STRMOD calls the appropriate model function, STRSS*n*, which states how the stresses are to be calculated. It is here that the important aspects of the model such as reference state and damping function are described.

As an example, we will examine the module STRSS2 which incorporates the ideas of the Lodge rubberlike liquid with a Kohlrausch relaxation function. The expression for the Lodge rubberlike liquid from Equation 10.2 can be integrated by parts to give Equation 10.4. This assumes that the material is a fluid (it relaxes to zero at infinite time) and that all deformations start at time 0. In STRSS2 the corresponding statement

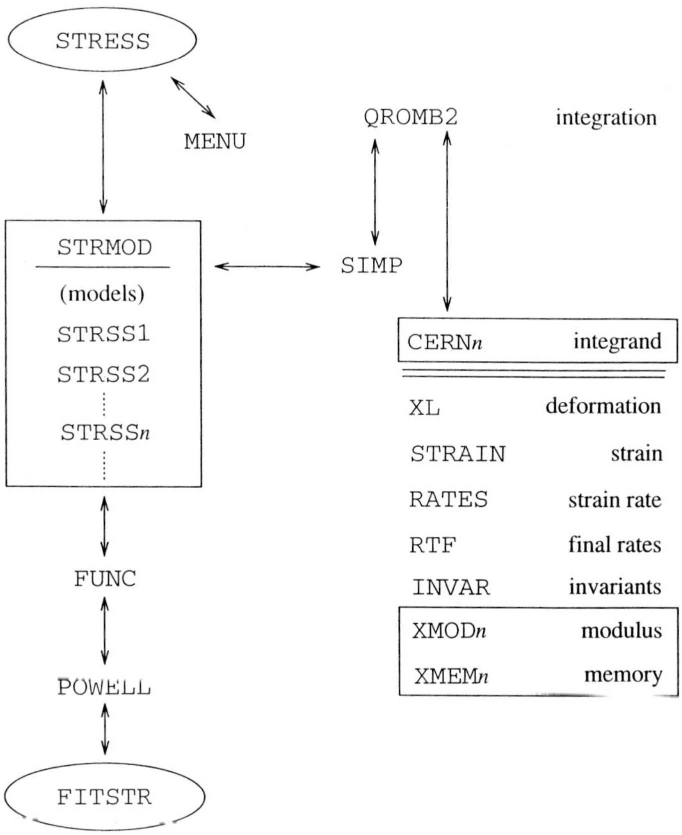

Figure 10.5 Layout of STRESS and FITSTR. STRESS takes model parameters and predicts stresses, while FITSTR inverts this process.

```
STRSS2 =  STRAIN(XLR,XLZ,XN,C,D,IJ)*XMOD1(TF,G,B,TAU,GF)
          +SIMP(TZ,TF,CERN2)
```

is found, where the first term is the $G(t)\ B_{ij}(t,0)$ term of Equation 10.4 and the second term is the integral. SIMP is a function which controls the integration between the zero time TZ and the final time TF. CERN2 is the integrand of the integral and is described in the function attached to STRSS2. The all-important strain function (STRAIN) is not supplied with times but with displacement gradients. This simplifies the structure of the program and makes writing of new models easier. Seen in the program statement are the reference displacement XLR and the zero-time displacement XLZ. STRAIN always calculates strain differences for the normal stresses; thus, the pressure term in Equation 10.4 disappears.

The function subprogram `XL` calculates the laboratory displacement gradients: shear strain in the case of shear and stretch ratio in the case of extension. The displacement is then passed with the reference displacement to the function subprogram `STRAIN`, which does the real work. For fluids, the reference displacement is that at the final time `TF`, the time at which the stress is to be calculated. For solids, the reference state can be the initial undeformed state. `STRSS1` is set up to simulate solids.

To add new models, one needs to copy one of the `STRSS`*n* functions, including its `CERN`*n* attachment, and change or add the necessary statements. This process can be seen by referring to the program unit `STRSS4`, where the Tanner-Simmons model has been written. Here, the lower limit of integration is influenced by the invariant of the deformation tensor divided by the final time t (`TF` in the code). Specifically, we start with Equation 10.3, but substitute $t - t_r$ for $-\infty$ to give Equation 10.9. If $t - t_r < 0$, the preceding equation can be broken into two integrals (with limits of $t - t_r$ to 0 and 0 to t) giving

$$\sigma_{ij}(t) = B_{ij}(t,0)\,[G(t) - G(t_r)] + \int_0^t m(t-s)\,B_{ij}(t,s)\,ds - p\delta_{ij} \qquad (10.16)$$

As noted, the stress calculated from Equation 10.9 has the peculiar feature of being independent of time if the strain rate is constant; the fluid reaches its breaking strain as it is sheared and then the stress instantly plateaus. The constancy of the integral can be seen by substituting $u = t - s$, whereupon

$$\sigma_{21} = \dot{\gamma}\int_0^{t_r} u\,m(u)\,du \qquad (10.17)$$

For those who wish to add their own favorite models, we have appended expanded listings (Section 10.6.3) of the two modules that need to be worked on. In these listings, we have drawn a box around each part of the code that needs to be changed in a routine fashion so these new models can be called properly.

The integration is done with the Romberg extrapolation technique, which is described thoroughly in *Numerical Recipes* [17]. With a few minor modifications to the *Numerical Recipes* routine `QROMB`, the Romberg technique, implemented in here as `QROMB2`, does a series of trapezoidal integrations of smaller and smaller step size and extrapolates the result to zero step size. For a given accuracy for smooth integrands, this algorithm proves to be very fast. However, if the deformation path is changed abruptly, the extrapolation would be very uncertain. For such integrals an adaptive step-size method (e.g., Runge-Kutta) would be needed. We have found that by integrating over each block in a multiple-block deformation history, the Romberg method outperforms the Runge-Kutta method by a large margin. `SIMP` handles the blocks for such histories. Of course, one needs to know beforehand where the blocks begin and end, and we do, so it would be foolhardy not take advantage of this information.

Integrations of high accuracy for sharply curved integrands can be very slow. For this reason we have tried to shorten the steps needed to calculate the integrand. The statements in the unit CERN*n* are the ones that get called repeatedly as the Romberg integration attempts to reduce the integration error down to 1 part in 10^6. Unavoidably, this consists of one call to the displacement gradient function XL to get the value at s, the dummy integration variable; one call to the strain function STRAIN; and one call to the memory function XMEM*n* just to get one value of the function being integrated. Thus, XL, STRAIN, and XMEM*n* are filled with approximations to avoid calls to math functions whenever possible; this makes their structure quite messy. However, they never need be changed when adding new models.

In addition, the integration accuracy is varied for each block such that the blocks contributing the most to the integral are evaluated the most accurately. The effort required to reach a certain relative error can be just as great for a puny little integral as for a massive one, yet the contribution of the small integral to the overall error could be insignificant. This is equivalent to saying that the absolute error from each block should be the same.

We have experimented quite a bit with methods of integration and have found ways to hasten the integration considerably, but always with a sacrifice in accuracy. If one is interested in the behavior of models, then it would be very unfortunate to have the conclusions tainted by inaccurate results, however consistent they may be.

10.3.2 FITSTR

FITSTR, which stands for "fit stress," appears at the bottom of Figure 10.5 and calls all the subprograms used by STRESS, plus a few others. Again, the purpose of FITSTR is to take a list of time-strain-stress data and work out the best fit to these data by using the chosen model. Inputs made by the user start with the name of a file of data describing how the data were taken and the data itself. The user then gets to choose the model, except that the program has been written only for fluids. (Those who care to examine the code will easily see how this can be circumvented, but we feel that the result will be very unsatisfactory; four parameters will provide enough excitement!)

After choosing the model, the user must then supply bracketing parameter values for the optimization process. Next, Powell's method (POWELL [17, pp. 294-301]) attempts the minimization of sum of squares of the differences between the observed stresses and the stresses calculated by the model (objective function). Powell's method was found by tedious labor to be a good one for this type of problem [18]. However, it is not perfect and can get one into a lot of trouble. In particular, to work it *must* find a minimum along the designated search direction at each iteration.

However, with constraints it is possible to avoid early disasters and get some useful information for a second attempt. We will provide some hints on the "art" of looking at the results and deciding what to do next in the section on using the programs.

The objective function to be minimized is written in subprogram FUNC. It is the sum of squares of differences (residuals) between the observed stresses and the calculated stresses. This difference is normalized for the observed stress; if the calculated stress is used for the normalization, the program becomes unstable. Stress is calculated at each observation; to give reasonable running time we have limited the calculation steps to 200. Remember that each stress calculation involves a separate integration over the entire time. Cannot information from the previous integrations be used? Well, the process can be speeded up if one is fortunate enough to have data that are equally spaced and is content to use trapezoidal integration. But equally spaced data leaves out information at short times or, alternatively, loads up the longer times with useless points. The method chosen is thus a compromise, but, we think, a good one.

Powell's method can be described as an intelligent modification of the ad hoc optimization method of varying one parameter at a time until an optimum is obtained, and then moving on to the next parameter, etc. This ad hoc method works poorly because it zigzags down the slope toward the optimum. Powell's method keeps changing the "direction" in which the parameters are modified and does a single-variable optimization along this new direction. Of course, this single variable causes the values of all the parameters to change. While not as fast near the optimum as gradient methods, it has good stability. As with all optimization methods, Powell's method is not immune from the possibility of converging to a local minimum, that is, reaching an apparent solution which does not provide the best possible description to the data.

Because gradients are not calculated in Powell's method, it is not straightforward to estimate the error in the parameters. The Monte Carlo method, introduced in Chapter 5, could be used, but computational time may be prohibitive. To get at the errors, one needs to establish the boundaries of the χ^2 region around the optimum corresponding to the probability of interest [17]. The merit function χ^2 is relatable to the objective function described earlier. The χ^2's inside this region of parameter space are no larger than what one might expect by chance, given the precision and amount of the data. While this is described at some length by Press et al. [17], we illustrate the concept in two dimensions in Figure 10.6.

Figure 10.6 shows imaginary contours of χ^2 given values of G and τ, the zero-time modulus and relaxation time, respectively. We expect χ^2 to change rapidly in a direction where both G and τ increase, and not so rapidly in a direction where one increases and the other decreases, because their effects tend to cancel. The confidence interval for the parameters is the projection of the appropriate $\Delta\chi^2$ contour onto the axis. We have labeled this contour χ_c^2 in Figure 10.6. Note that

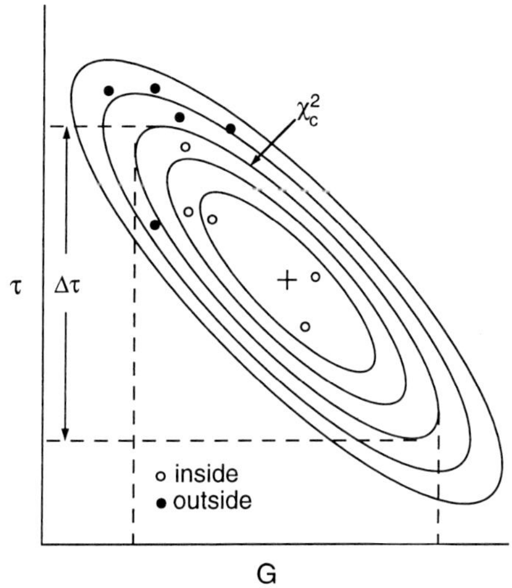

Figure 10.6 Schematic of the process used by FITSTR for determining the confidence interval for the parameters. The cross in the center of the contours represents the best values of G and τ. Trial values are represented by ○ and ● symbols; the latter have resulted in χ^2's greater than the critical value for the given degrees of freedom and probability. The range of the ○ symbols gives an estimate of the confidence interval for the parameters.

the confidence interval is not the length of the cut through this contour at the optimum.

How can this contour be established? One could go back and calculate the double and mixed partial derivatives of χ^2 with respect to the parameters and invert this curvature matrix. However, without derivative information, this must be done numerically. It also assumes that the bottom of the well is parabolic in nature, which is probably not the case.

An alternative is to use the statistical information already available during the optimization; that is, the values of χ^2 and the parameters. A huge number of these will be produced, and it makes sense to use them. Thus, FUNC records every evaluation of the objective function in the arrays SSI and PI before the subroutine CCEROR searches through these to find the ones that are within the critical contour. This step takes almost no time at all. The penalty? The maximum and minimum values of the parameters used in these included calls always fall within the true confidence interval; thus, the interval reported is not conservative. In fact, the search may never reach the critical χ^2 contours on the far side of the optimum from where the search was started. Again, this is illustrated schematically in Figure 10.6. Counteracting this somewhat is the assumption that the inherent variance in the data points is revealed by the final values of the residuals. The latter are likely to be high, which will push the critical χ^2 contour out, yielding a more conservative estimate of the confidence interval for the parameters.

If one has several sets of data (using the same material, of course), FITSTR should be applied to each of them separately. If no trends of the parameters are observed with the independent variables (e.g., stretch rate or deformation mode), this can be taken as evidence that the model is a good one. Furthermore, the distribution of the parameters so obtained will be an excellent indication of their accuracy because each set can be regarded as an independent observation drawn from the population of all possible tests within the ranges of the independent variables.

10.4 Execution of the Programs

Inherent to the execution of STRESS and FITSTR are the subroutines RATES and STRMOD and the function subprograms INVAR, SIMP, STRAIN, XL, XMEM1, XMEM2, XMOD1, and XMOD2. The listings of the source codes are provided in Sections 10.6.1 to 10.6.10. In addition, FITSTR requires the subroutine POWELL available from *Numerical Recipes* [17]. Refer to Appendix A for the program dependencies. For example, the *Numerical Recipes* routines may use additional routines which, in turn, call other routines. The subroutine LINMIN, used by POWELL to do line minimization, calls the routines MNBRAK and BRENT and instructs them to minimize the function F1DIM which is the value of the user-supplied function FUNC (Section 10.6.10).

10.4.1 STRESS

Following the examples in Sections 10.7.1 and 10.7.2, use a text processor or editor that outputs ASCII code to create a data file. Note that a number should be entered even if it does not apply for the particular deformation; simply use any value. The first record is set aside for the title of the file; insert anything up to 79 characters containing identification information.

The next record contains starting values for the material properties. In order, these properties are the characteristic modulus G, the breadth parameter B, the relaxation time τ, the infinite-time modulus G_∞ (for cross-linked materials), and the nonlinear parameter x_{NL}. In Models 1-3, x_{NL} is assigned to exponent n in the Seth strain function (Equation 10.8). In Models 4 and 6, x_{NL} is the log of the constant k in Equations 10.10 and 10.13, respectively. For Model 5, x_{NL} is related to α in Equation 10.14 by

$$\alpha = e^{-x_{NL}/3} \tag{10.18}$$

Given these transforms, x_{NL} can take on any value but will usually fall between 0 and 2.

In the third record are placed the parameters that define the deformation. Refer to the listing of the `STRESS` code in Section 10.6.1 for a breakdown of assignments given to each parameter. The first parameter has the following values: 0. for simple shear, 0.5 for simple extension, 1.0 for planar extension, and 2.0 for biaxial extension. The reason for these particular values is that they are used directly in the computation of the strain in the module `STRAIN`. The next two parameters are switches and should be integers (although the program will tolerate an errant decimal point). The first switch, if 1, gives the primary stress difference in planar extension or, if 2, gives the stress required to hold the sides of the stretching sheet from moving inward. (The free surface of the sheet is considered to have zero stress.) Earlier texts often listed these as $\sigma_{11} - \sigma_{22}$ and $\sigma_{33} - \sigma_{22}$, respectively, while more recent books have settled on $\sigma_{33} - \sigma_{11}$ and $\sigma_{22} - \sigma_{11}$, respectively. Use whichever switch is preferable; it does not matter as long as you remember that switch 1 will give the stress required to stretch out the sheet and switch 2 will give the stress required to hold the edges fixed. If switch 3 is input, the output will be the difference between the first two switches; this combination can show up in some problems.

The last parameter has a similar function, but works for simple shear. A value of 0 will give the shear stress, and a value of 1 will give the primary normal stress difference. A value of 2 will give the second normal stress difference; however, only Models 1-3 will give this normal stress, and only if the nonlinear parameter is not 2. None of the other models provided predict anything but a zero. Does 3 do anything? Yes, it will give the difference $N_1 - N_2$, which shows up in torsional flow. Refer to Chapter 7 for details.

The next records of the file give the strain history. Always required are the initial time, zero strain (or stretch ratio of 1 for extensional flows), and the path-shape parameter for each block. The path-shape parameter can be anything, but there are values that are useful to remember. For example, if the flow is simple shear, a value of 1 will give a constant shear rate. If 0, the history will be "exponential shear" given by the equation

$$\gamma(t) = e^{\dot{\gamma}_0 t} - 1 \tag{10.19}$$

Anything between 0 and 1 will give paths between these two extremes. For extension, 0 will give constant rate (or exponential increases in the length of a clamped sample), while 1 will give a constant-clamp-velocity experiment as if one were to stretch the sample in a tensile testing machine. Higher values of the path-shape parameter will give more deformation near the beginning, less at the end, while lower values will give the opposite effect, as illustrated in Figure 10.7.

The equations we use to calculate the deformations between the starting point and the next point are compiled in Appendix 10.1. A summary of these is:

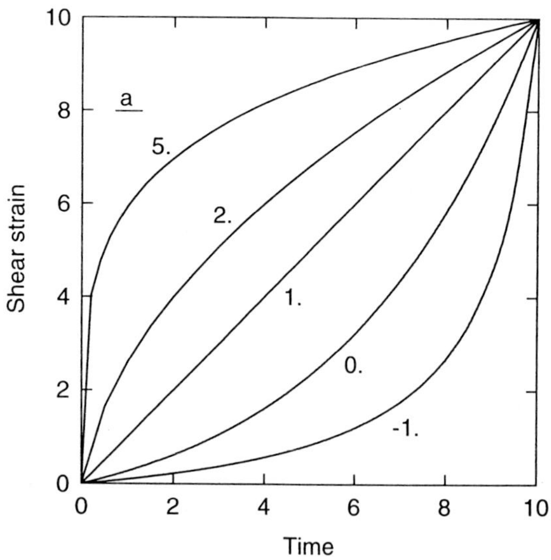

Figure 10.7 Illustration of the influence of the path-shape parameter a on the way the strain changes with time. A value of 1.0 gives constant shear rate, while a value of 0. gives exponential shear.

Shear

$$\gamma = \mathrm{SGN}(\dot{\gamma}_0)\left[(1 + a\dot{\gamma}_0 t)^{1/a} - 1\right] \tag{10.20}$$

Extension

$$\lambda = (1 + a\dot{\varepsilon}_0 t)^{1/a} \tag{10.21}$$

where a is the path-shape parameter and the subscript zero signifies the appropriate velocity gradient at zero time. In the case of shear, the reason for the absolute value is to provide symmetry with respect to shearing direction; that is, one should expect that the effects will be the same whether the sample is sheared to the right or to the left. In the case of extension, one is either stretching or squashing the sample, and one expects perfectly smooth behavior going from one of these to the other.

The next and subsequent records are similar: time, deformation, and path shape for the next section of the deformation. The material can be sheared one way and then the other by just putting in the appropriate strains. For example, the entries

```
0.      0.      1.
1.      1.      1.
2.      0.   9999.
```

will shear the material 1 unit of strain in 1 unit of time and return the material to zero strain in the same amount of time. An experiment to allow some relaxation before returning the material to its original configuration would look like

```
0.    0.       1.
1.    1.       1.
2.    1.       1.
3.    0.    9999.
```

The `9999` for the path shape appearing in the last record is a flag for the end of the deformation history. We chose not to use any end-of-file markers because it can be very useful to stack many different files and run them in a batch mode (say overnight!). If the batch mode is desired, a `B` answer should be given to the batch-mode question when starting the program.

There are two final records in the input file. The next-to-last record requires two numbers; the first number is the time after the start of the block where the first value of stress is to be calculated. The units will be the same as the time values in the preceding records. This time is used to calculate the stress after each break in the history; that is, after each time in the preceding records. The choice of this time is important for revealing the desired details of the stress history. For example, if the chosen time is much longer than the relaxation time, most of the transient behavior will be lost. The second number is a multiplier for the first time and must be greater than 1.0. To illustrate, if the first time is 0.1 s and the multiplier is 2, the corresponding stresses will be calculated at 0.1, 0.2, 0.4, 0.8, ... s. To space points evenly on a common logarithmic scale, set the first time to be a power of 10 and the multiplier to be some fractional power of 10. A few of these multipliers are listed in Table 10.3. The final record is the model number. The available models, served up on the display terminal by the subroutine `MENU`, are listed in Table 10.1.

Table 10.3 Multipliers for the Common Logarithmic Scale

Points per Decade	Multiplier
1	10.0
2	3.1623
5	1.5849
10	1.2589
20	1.12202

Upon executing the program (type in STRESS at the prompt), the first questions are the filenames for the data (discussed above) and the output file. After these are supplied, the program inquires about the mode of operation, batch or interactive. If batch mode is selected, sit back and wait for the results. If the interactive mode is selected, you will be given the opportunity to change everything except the deformation paths. Of most interest will be the influence of model and material properties on the stress growth, viscosity function, etc. In the interactive mode, the program displays a menu: a property can be selected, and a new value entered. Models can be changed by so indicating. When a new model is selected, the parameters are all reset to the base case, that is, those in the original file. The reason for this is that it allows stacked input files to be rerun with a new model in the interactive mode.

Appended to this chapter are several sample data files for various deformations and materials. One example, listed in Section 10.7.2, illustrates the stacking of files, but it has another purpose: it produces steady shear stresses by imposing large shear strains. The net result is a viscosity function for the material over the range of $0.01/\tau$ to $100/\tau$ units of shear rate.

10.4.2 FITSTR

Running FITSTR is similar to running STRESS except that the data file is substantially different. With FITSTR, the input is a file of time, strain, and stress data. All three values are required, and there must be at least six entries (although many more would be a very good idea).

To run FITSTR use an editor or word processor to create an ASCII file with the following structure: The first record identifies the file; use any character string up to 79 characters long. The second record describes the mode of deformation and is identical to the third record in the input file for STRESS. The first entry defines the type of deformation: 0. for shear, 0.5 for simple extension, 1.0 for planar extension, and 2.0 for biaxial extension. The second entry refers to planar extension; it should be 1 if the stresses are those required to pull the film, 2 if they are those required to hold the edges of the film, and 3 if the difference between these has been measured. The third entry should be 0 for shear stresses and 1 for normal stresses if the deformation is simple shear. (We cannot envision second-normal-stress data being used, but if so, enter 2.) A value of 3 is useful if normal-force data from torsional flow are available because it results in the calculation of $N_1 - N_2$. Note, these numbers should never be left out; if they do not apply, just put in anything. If there is no entry, an error like NAN (not a number), or something equally catastrophic, will result.

The third and subsequent records are time, strain and stress data sets in that order. The first of these records should be 0. 0. 0. if the deformation is simple shear, 0. 1.

0. if extension. Here, the term "strain" refers to laboratory displacement gradients: elongation ratio (starting at 1) for extension and shear strain for shear. Some of the stresses might be zero if the experiment involves a recovery step.

The program execution starts with the usual queries concerning the input and output files. Estimates are made of the shape parameters for the strain history and are subsequently displayed on the terminal. If the estimates are close to either 0. or 1.0, it would be highly beneficial, with reference to computational speed, to change them all to exactly this value. The change should have a very slight effect on the results.

The hard question is thrown out at the next interactive step; what are the bracketing values of G, B, τ, and x_{NL}? Good guesses versus almost good guesses are a matter of success versus failure. Use information from previous runs or through examination of linear data on the same material. Other than this, we have little advice to offer – good advice, that is. Note that the starting values of the parameters are the means of the bracketing values (arithmetic for nondimensional parameters, geometric for dimensional ones). Thus, if you think the modulus scale parameter will be about 1 MPa, enter lower and upper limits around 0.1 and 10 MPa, respectively. Watch the display terminal as the sum of squares of the relative residuals is calculated for the starting point; if it is large (e.g., 10 - 100), you know that things are bad. Also, scrutinize the row of plus or minus signs under this report (see Section 10.7.4 for an illustrative example), these are the signs of the residuals. If they are all positive, try a higher modulus; if they start out negative and go positive, try a higher value of B. Do not work on this too hard; this is what `FITSTR` is supposed to do. The program will ask if you wish to see a similar report at each function call. While the report produces a screen full of numbers, it will often provide an advance warning of problems.

As the program progresses, examine carefully the residuals displayed on the terminal. If the number of sign changes are few, the data may not be converging on the global minimum; try different starting values or a different model. If the parameters seem to be sliding off in the direction of the ridiculous, the model is probably inappropriate or the data are insufficient. Try another model, or get more data.

The relative change of the objective function that is tolerated by `POWELL` before exiting is controlled by `FITSTR`, but you will have the opportunity to change this. It starts out at 10%, which will give a rough estimate, but takes very little computational time. By asking for more tolerance (see sample outputs in Section 10.7), the precision is increased by two orders of magnitude a leap until it gets to one part in 10^{11}. This is effected interactively by answering `Y` to the query concerning increases in precision. A precision of 10^{-11} is almost guaranteed to be beyond the precision justified by any set of real experimental data.

Patience is needed; sometimes Powell's method will look like it is slowing down (observe the listed sum of squares of the residuals in the stresses at each iteration), but it will speed up as it gets the proper search direction sorted out. This can happen several times. Occasionally, a call for higher precision will return a very quick result; this means that there was an unlucky change in parameters that did not produce much change in the objective function. Keep going by asking for more precision; it does not mean convergence was attained. With the chosen model, there may be no minimum and `POWELL` will simply keep sliding forever downhill. Break the calculation by pressing the `Ctrl` and `C` keys simultaneously to stop at the next iteration. However, write down the values on the display terminal, because a `Ctrl C` break stops data from being written to the results file.

A somewhat philosophical question concerns the value, if any, of a model which is not the optimum. It can be argued that if it still describes the data well and has reasonable parameters, it may be adequate for mild extrapolation. While you would not enjoy the same snug feeling that comes with having convergence, remember that your *optimum* model may only be a local one. Therefore, use everything that comes out with due caution.

10.5 Application of the Programs

10.5.1 STRESS

Several sample data files have been appended to this chapter; some of them have been used to produce the results discussed here.

As always, one should worry about the accuracy of any numerical process, and this must be checked with examples having known results. Analytical solutions are available for several simple cases; one of these is the stress-growth function for extension of a Lodge rubberlike liquid for which the analytical solution is

$$\sigma_T = G\dot{\varepsilon}\tau\left[\frac{3}{(1-2\dot{\varepsilon}\tau)(1+\dot{\varepsilon}\tau)} - \frac{2e^{-(1-2\dot{\varepsilon}\tau)t/\tau}}{1-2\dot{\varepsilon}\tau} - \frac{e^{-(1+\dot{\varepsilon}\tau)t/\tau}}{1+\dot{\varepsilon}\tau}\right] \tag{10.22}$$

In Table 10.4 are the results for $\dot{\varepsilon}\tau = 1$; these were produced by running the input file shown in Section 10.7.1. Although this is probably not a worst case, it is one in which the strain and memory parts of the integrand are changing rapidly, a bad situation for numerical integration. Thus, we have some confidence that the integration errors are being kept low, at least at reasonable stresses. If more accuracy is needed, one can change the error parameter in the module `SIMP` (Section 10.6.9) and recompile it.

The second example demonstrates the calculation of the viscosity function, that is, the dependence of the steady-state shear viscosity on shear rate. This calculation uses the stacked file shown in Section 10.7.2 and is run in a batch mode. The data in this file consist of strains from 10^2 to 10^6 that are achieved in times designed to

Table 10.4 STRESS Results at $\dot{\varepsilon}\tau = 1$ for Data Shown in Section 10.7.1

	σ_T/G		
t/τ	STRESS	Analytical	Error, %
0.01	0.030001	0.030000997	0.
0.1	0.30098	0.300976459	+0.001
1.0	3.8689	3.868896015	+0.0001
3.162	45.748	45.74778996	+0.0004
10.0	4.4051E4	4.40514315E4	−0.0009

give the desired shear rates. A strain of 100 is usually enough to give steady state, but one also has to be concerned about the Deborah number (De), the ratio of the relaxation time to the total time of the deformation. Therefore, we have made sure that De is less than 0.1 as well. Despite this, we should still worry about a systematic error which may change the shape of the viscosity function at high rates; this would be very misleading. How can we check this?

Again, checks on numerical accuracy rely on known results. One known result is the Lodge rubberlike liquid model, which must give no shear-rate dependence of the viscosity. This corresponds to Model 2 with a nonlinear parameter of exactly 2. Using the data file in Section 10.7.2 and following the precautions mentioned, the results shown in Table 10.5 are obtained. These numbers are really $\tau\dot{\gamma}$ and $\eta(\dot{\gamma})/G\tau$. There is no deviation from the exact value of 1.000000 as the shear rate increases. There will be some deviation with single precision calculations; however, the deviations do not get progressively worse with increasing rates.

A more difficult test involves a broadening of the relaxation spectrum. While computation takes longer, the results are about the same. A value of the breadth parameter of 1.0 gives an $\exp(-t^{0.5})$ relaxation which is considerably broader. This can be analytically integrated to give a viscosity of $2G\tau$. (In general, the viscosity is $(2B)!G\tau$ or the appropriate gamma function if $2B$ is nonintegral.) The result of the numerical integration was $1.9985G\tau$ when the integration was carried to 1000τ. The longer time is necessary because of the much broadened distribution. (Clearly, there is a message here for impatient experimentalists!)

For shear-thinning fluids, we need something with an analytical solution. Our somewhat modified Tanner-Simmons model (Model 6) fills the bill; its steady shear viscosity function is given by

$$\eta(\dot{\gamma}) = G\tau\left(1 - e^{-t_r/\tau}\right) \tag{10.23}$$

Table 10.5 Results to Check Against Systematic Error Using Model 2 with the Nonlinear Parameter = 2 Using the File Shown in Section 10.7.2

Shear Rate	Viscosity from STRESS
0.1	1.000000
1.0	1.000000
10.0	1.000000
100.0	1.000000
1000.0	1.000000

for a single-exponential relaxation function. Here, the rupture time t_r is

$$t_r = t\left(1 - e^{-k/\dot{\gamma}t}\right) \tag{10.24}$$

At long times relative to the reciprocal of the shear rate, the rupture time is simply $k/\dot{\gamma}$, and the viscosity, at low shear rates, is just $G\tau$. At high shear rates the exponential in Equation 10.23 can be expanded to give

$$\eta(\dot{\gamma}) = \frac{G\tau k}{\tau\dot{\gamma}} = \frac{Gk}{\dot{\gamma}} \tag{10.25}$$

Because of our custom modifications, these equations differ from the expressions in Bird et al. [4]; the power-law limit for the version cited has a -2 slope instead of a -1. (The difference can be seen by referring to Equations 10.11 and 10.12; they give identical results when t_r is large but not when t_r is small, that is, at high shear rates.) The results of the test for t/τ = 1000 are shown in Table 10.6; STRESS yields results that are accurate to about 1 part in 10^5.

The last example is for a cross-linked solid showing the influence of the reference state on results of a shear experiment. We will cross-link the material so that the modulus at infinite time is one-half of the modulus at short times. We can run such a material with Model 1 (reference is undeformed state) or Model 2 (present configuration is reference). In fact, such a material can be run with any of the models; they really are not very particular.

The results of such an experiment are shown in Figure 10.8. Both shear and normal stresses are plotted. The stresses were allowed to relax after a strain of 2 units had been imposed. At this low strain, the reference state makes little difference.

Table 10.6 Results Using the Modified Tanner-Simmons Model (Model 6) for $t/\tau = 1000$

	$\eta(\dot{\gamma})$	
$\dot{\gamma}$	STRESS	Analytical
0.1	1.00000	1.000000
1.	1.00000	1.000000
10.	0.99995	0.999954
100.	0.63212	0.632120
1000.	0.095163	0.0951626

10.5.2 FITSTR

With `FITSTR`, we expect that if we use data generated by `STRESS`, we should be able to recover the model parameters used to generate this data. If the data are precise, the parameters recovered by `FITSTR` should be just those used to generate the stresses. If the data are imprecise, the outcome may be large parameter errors or several unique sets of parameters that essentially give equal fits to the data.

Our first example illustrates this operation. With `STRESS`, a set of stresses and strains was generated with Model 2 and the model parameters

$$\begin{aligned} G &= 1.0 \\ B &= 0.5 \\ \tau &= 1.0 \\ G_\infty &= 0.0 \\ x_{\mathrm{NL}} &= 2.0 \end{aligned}$$

This is the Lodge rubberlike liquid model with a single-exponential relaxation function. Eight stresses were generated starting at $t = 0.1$ and extending to $t = 10$.

The data file and the terminal dialog for `FITSTR` are shown in Sections 10.7.3 and 10.7.4, respectively. The truncation of the stress values to four significant figures introduced error with a rectangular distribution. Because `FITSTR` uses a least-squares minimization, this error distribution is not optimal. Nevertheless, `FITSTR` does not have too much trouble converging back to reasonable estimates for the parameters starting from several search directions; although one of the initial estimates took nearly 100 iterations and 13,200 function calls (about 68.7 s on an 80386, 25-MHz microcomputer, or 3.5 s on an IBM ES/9000 mainframe). The

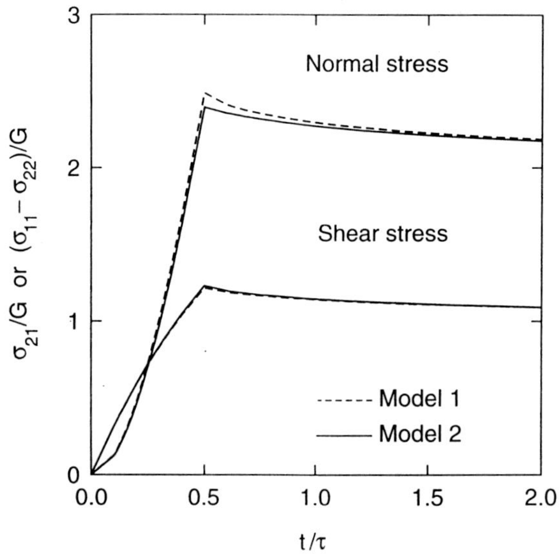

Figure 10.8 Illustration of what happens when an equilibrium modulus (due to cross-linking, for example) is introduced.

display terminal looked like that shown in Section 10.7.5 when a more favorable starting point was chosen.

The estimated confidence intervals place roughly 10% errors on each of the four parameters. Remember, the safest procedure is to take the confidence limit that is furthest away from the converged value and double its distance to give the confidence interval. Thus, if the minimum is symmetrical, the nonlinear parameter x_{NL} might be as low as 1.9 and as high as 2.1 given this set of data.

We can ask the very legitimate question concerning the amount of random noise the data can have before the method will fail. (If the errors are correlated, the method will, of course, give the wrong values of the parameters.) In Section 10.7.6, the same set of data is shown with 1% random noise. The results are different, as expected, but certainly in the same ball park. With 10% noise, things do not go well with so few points. In real transient experiments, the random error of the data is normally extremely low if the actuator and electronics are working properly. However, systematic error is another question.

As a check against the work of other rheologists, we considered the data of Chang, Bloch and Tschoegl [11] for an uncross-linked sample of styrene-butadiene rubber (SBR) stretched at 23 °C in a constant-velocity instrument (c = 0.5, a = 1.0). Sixteen points were available at an initial stretch rate of 0.04394 min^{-1} out to a final stretch ratio of just over 3.0.

The first convergence is shown in Section 10.7.7. Although this is a crude estimate, there are fully seven runs of signs in the residuals. Note, the listing on the

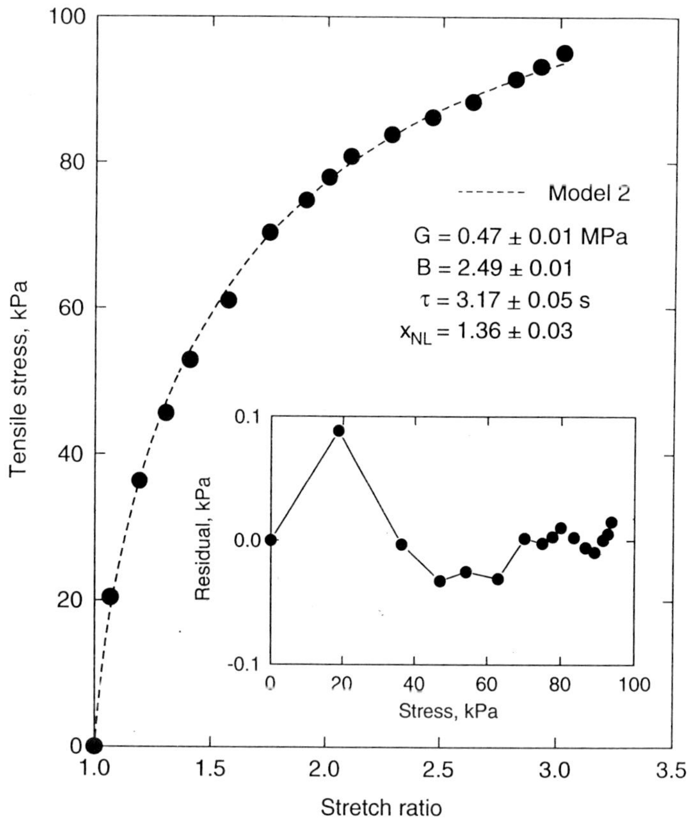

Figure 10.9 Fit found by `FITSTR` to the data of Chang et al. [11] for SBR gum stretched at 23 °C. Adapted from Ref. 11. Copyright © 1977 by John Wiley & Sons, Inc. Reprinted by permission.

display terminal provides only half the points (actually a maximum of 8). This limit keeps all the information on the screen to help in deciding if more iterations are needed. The results are plotted in Figure 10.9.

A far trickier example is from the 1978 manuscript written by the same authors [12]. They showed what happens when a sample of SBR is first stretched, then held fixed for 10 min, and finally returned to zero stress by reversing the crosshead of the testing machine. Section 10.7.8 shows this data ready for `FITSTR`, but getting things going is not easy. Choosing too low a relaxation time will cause negative stresses to be generated when the sample is pushed backwards; the result of this is that the best choice for the modulus scale parameter, which is optimized first, is zero! By watching the calculation, one can see this becomes evident quickly and can choose better starting values. The end result using Model 2 is shown in Figure

10.10, where one convergence of FITSTR has produced a fit to the data which compares very favorably to the original curve.

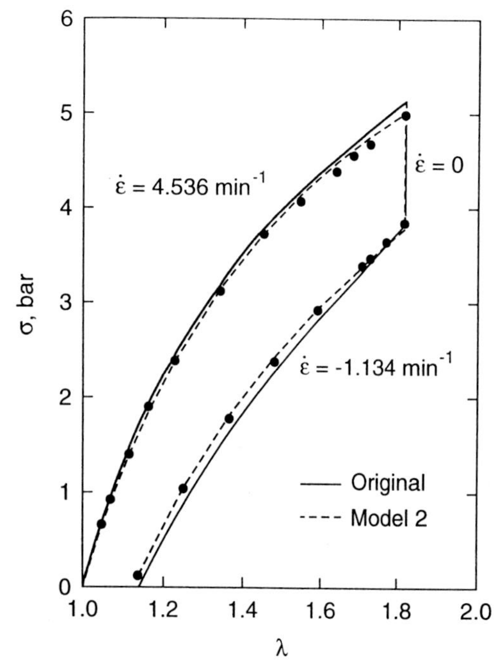

Figure 10.10 Fit found by FITSTR to the data of Bloch et al. [12] for SBR plasticized by silicone oil. The sample was stretched in an Instron tensile tester at the nominal rates shown. The stress in this figure is engineering stress (1 bar = 10^5 Pa). The curve marked Original was found by Bloch et al. Adapted from Ref. 12. Copyright © 1978 by The Society of Rheology, Inc. Reprinted by permission of the American Institute of Physics.

10.6 Program Listings

10.6.1 STRESS

Imbedded into the code are the subprograms MENU and RTF that are inherent to the execution of STRESS. Refer to Appendix A for the program dependencies.

```
      PROGRAM STRESS
C Calculates stresses arising from deformation of a fluid in various modes.
C Requires the subroutines STRMOD, RATES, and STOPP
C  and the function subprogram XL.
C Data file format:
```

```
C         1st record - title, up to 79 characters.
C         2nd record - G, B, TAU, GF, XNL
C                        G: Modulus.  For melts, about 1.0 MPa.
C                        B: Breadth parameter. Value of 0.5 gives single
C                            exponential.
C                      TAU: Relaxation time.
C                       GF: Equilibrium modulus (cross-linked).
C                      XNL: Nonlinear parameter. Value of 2.0 gives Lodge
C                            rubberlike liquid.
C         3rd record - C, ID, IJ
C                       C: 0. = simple shear; 0.5 = simple tension;
C                          1. = pure shear;   2.  = biaxial tension.
C                      ID:  1 = longitudinal stress in planar extension
C                           2 = transverse stress
C                           3 = longitudinal - transverse stress
C                      IJ:  0 = shear stress in simple shear
C                           1 = first normal stress difference
C                           2 = second normal stress difference
C                           3 = N1 - N2
C subsequent records - T, XLEN, A
C   up to 4+N, where       T: Time
C   N > 1              XLEN: Relative deformation. In simple shear, use
C                             strain.
C                          A: Path-shape parameter. Use 9999 in last record.
C                             0. = constant rate in tension; 'exponential'
C                                  shear flow;
C                             1. = constant rate in shear; constant stretch
C                                  velocity in tension.
C   (4+N+1)th record - DT, FT
C                      DT: Time added to T(I) where first calculation will
C                            be made for that block.
C                      FT: Multiplicative point spacing, e.g., 1.5.
      IMPLICIT DOUBLE PRECISION (A-H,O-Z)
      PARAMETER (NMAX=201,TINY=1.D-6)
      CHARACTER TITLE*80,ANS*1
      COMMON /BLOC1/ A(NMAX),T(NMAX),RT(NMAX),XLEN(NMAX),NP
      COMMON /BLOC2/ G,B,TAU,GF,XNL,DT,FT
      COMMON /BLOC3/ C,XN,ID,IJ
10    CALL MSHEAD
      CALL UNITS(NTER,NINT,NOUT)
      IMODE=1
      IFLAG2=0
      WRITE(NTER,*)' Run in Batch (B) or Interactive (I) mode ? (B/I)'
      READ(NTER,'(A1)')ANS
      IF(ANS.EQ.'I'.OR.ANS.EQ.'i')IMODE=2
20    READ (NINT,'(A80)',END=180)TITLE
      IF(IFLAG2.EQ.0.OR.IFLAG2.EQ.3)THEN
        READ(NINT,*)G,B,TAU,GF,XNL
        READ(NINT,*)C,XID,XIJ
        ID=INT(XID+TINY)
        IJ=INT(XIJ+TINY)
      ELSE
        READ(NINT,*)STUFF
        READ(NINT,*)STUFF
      ENDIF
      DO 30 I=1,NMAX
        READ(NINT,*)T(I),XLEN(I),A(I)
        IF(I.EQ.NMAX)CALL STOPP(NTER,NMAX-1)
        IF(A(I).GT.9000.)GOTO 40
30    CONTINUE
C Number of path points, NP.
40    NP=I
      IF(IFLAG2.EQ.0.OR.IFLAG2.EQ.3)THEN
        READ(NINT,*)DT,FT
        IF(FT.LE.1.)THEN
          WRITE(NTER,*)' WARNING! Time multiplier out of range.'
          STOP
        ENDIF
      ELSE
```

```
          READ(NINT,*)STUFF
        ENDIF
        IF(IFLAG2.EQ.0)THEN
          READ(NINT,*)XIMOD
          IMOD=INT(XIMOD+TINY)
        ELSE
          READ(NINT,*)STUFF
        ENDIF
C Calculate rates at beginning of each block.
      CALL RATES(NP,T,XLEN,A,C,RT)
      IF(IMODE.EQ.1.)WRITE(NTER,*)' Still computing!'
      WRITE(NOUT,'(///1X,71(''-''))')
      IF(IMODE.EQ.2)WRITE(NTER,'(//2X,71(''-''))')
      IBLANK=INDEX(TITLE(72:80),' ')
      IF(IBLANK.EQ.0)IBLANK=1
      WRITE(NOUT,'(A/1X,A)')TITLE(1:IBLANK+71),TITLE(IBLANK+72:80)
      IF(IMODE.EQ.2)
     & WRITE(NTER,'(1X,A/2X,A)')TITLE(1:IBLANK+71),TITLE(IBLANK+72:80)
      IF(C.LT.TINY)THEN
        IF(IJ.EQ.0)WRITE(NOUT,50)
        IF(IJ.EQ.1)WRITE(NOUT,60)
        IF(IJ.EQ.2)WRITE(NOUT,70)
        IF(IJ.EQ.3)WRITE(NOUT,80)
        IF(IMODE.EQ.2)THEN
          IF(IJ.EQ.0)WRITE(NTER,50)
          IF(IJ.EQ.1)WRITE(NTER,60)
          IF(IJ.EQ.2)WRITE(NTER,70)
          IF(IJ.EQ.3)WRITE(NTER,80)
        ENDIF
50      FORMAT(T32,'Shear Stress')
60      FORMAT(T24,'1st normal stress difference')
70      FORMAT(T24,'2nd normal stress difference')
80      FORMAT(T22,'1st - 2nd normal stress differences')
      ENDIF
      IF(ABS(C-1.D0).LT.TINY)THEN
        IF(ID.EQ.1)WRITE(NOUT, 90)
        IF(ID.EQ.2)WRITE(NOUT,100)
        IF(ID.EQ.3)WRITE(NOUT,110)
        IF(IMODE.EQ.2)THEN
          IF(ID.EQ.1)WRITE(NTER, 90)
          IF(ID.EQ.2)WRITE(NTER,100)
          IF(ID.EQ.3)WRITE(NTER,110)
        ENDIF
90      FORMAT(T27,'Longitudinal direction')
100     FORMAT(T28,'Transverse direction')
110     FORMAT(T25,'Longitudinal - Transverse')
      ENDIF
      WRITE(NOUT,'(1X,71(''-''))')
      IF(IMODE.EQ.2)WRITE(NTER,'(2X,71(''-''))')
      WRITE(NOUT,120)IMOD
      IF(IMODE.EQ.2)WRITE(NTER,120)IMOD
120   FORMAT(/28X,'Model',I3,'  results:')
      WRITE(NOUT,130)G,B,TAU,GF,XNL
      IF(IMODE.EQ.2)WRITE(NTER,130)G,B,TAU,GF,XNL
130   FORMAT(/2X,'G =',1P,E8.1,4X,'B =',0P,F5.2,4X,'TAU =',1P,E8.1,4X,
     &       'GF =',1P,E8.1,4X,'XNL =',0P,F5.2)
      WRITE(NOUT,140)
      IF(IMODE.EQ.2)WRITE(NTER,140)
140   FORMAT(//T10,'Time',T23,'Stress',T34,'Stress/Rate',T52,'Rate',
     &       T65,'Strain')
      WRITE(NOUT,'(2X,5(4X,10(''-'')))')
      IF(IMODE.EQ.2)WRITE(NTER,'(2X,5(4X,10(''-'')))')
C Get stress STRSS at times TF spaced at intervals determined by FT and DT.
      DO 170 J=1,NP-1
        IFLAG1=0
        DO 160 I=1,NMAX
          TF=T(J)+DT*FT**(I-2)
C Pick up T(J) again to check for influence of rate changes (Tanner-Simmons
C  and other strain-rate-dependent models).
```

```
          IF(I.EQ.1)TF=T(J)+TINY
          IF(I.GE.2.AND.T(J+1)-TF.LE.DT)THEN
            TF=T(J+1)
C And flag to stop calculating; this block is finished.
            IFLAG1=1
          ENDIF
          IF(I.EQ.1.AND.J.EQ.1)TF=T(1)
          CALL STRMOD(NTER,IMOD,TF,STRSS,*180)
          XLP=XL(TF,NP,T,XLEN,A,RT,C)
          IF(C.GT.TINY)XLP=LOG(XLP)
          RTFP=RTF(TF,NP,T,A,RT,C)
C Work through this again.
          IF(I.EQ.1)RTFP=RT(J)
          IF(ABS(RTFP).GT.TINY)RTFV=RTFP
C Else, keep old rate to give the viscosity decay function.
          ETA=ABS(STRSS/RTFV)
          WRITE(NOUT,150)TF,STRSS,ETA,RTFP,XLP
          IF(IMODE.EQ.2)WRITE(NTER,150)TF,STRSS,ETA,RTFP,XLP
150       FORMAT(2X,1P,5E14.4)
          IF(IFLAG1.EQ.1)GOTO 170
160     CONTINUE
170   CONTINUE
C Read in other experiments if present; if not, return here.
      GOTO 20
180   CONTINUE
C Change parameters if mode is interactive (IMODE = 2).
      IF(IMODE.EQ.2)THEN
190     WRITE(NTER,'(/2X,A)')
     &       'Change parameters (P), model (M), or quit (Q)?'
        READ(NTER,'(A1)')ANS
        IF(ANS.EQ.'Q'.OR.ANS.EQ.'q')THEN
          CALL MSTAIL(*10)
          STOP
        ELSE IF(ANS.EQ.'P'.OR.ANS.EQ.'p')THEN
          IFLAG2=1
          CALL MENU(NTER)
        ELSE IF(ANS.EQ.'M'.OR.ANS.EQ.'m')THEN
200       WRITE(NTER,*)' Enter ID for model (0 for key):'
          READ(NTER,*)IMOD
          IF(IMOD.LT.0)GOTO 200
C Fetch key if necessary (IMOD=0).
          IF(IMOD.EQ.0)CALL STRMOD(NTER,IMOD,TF,STRSS,*200)
          IFLAG2=3
        ELSE
          GOTO 190
        ENDIF
        REWIND NINT
C Return and start looping through experiments with new parameters.
        GOTO 20
      ELSE
        STOP
      ENDIF
      END
C
      SUBROUTINE MENU(NTER)
C Requests parameter values to control STRESS in interactive mode.
      IMPLICIT DOUBLE PRECISION (A-H,O-Z)
      PARAMETER (TINY=1.D-5,NRPAR=7,NIPAR=2,NPAR=NRPAR+NIPAR)
      CHARACTER C1*4(NPAR),C2*55(NPAR),CIP*4
      COMMON /BLOC2/ PAR(NRPAR)
      COMMON /BLOC3/ C,XN,IPAR(NIPAR)
      DATA (C1(I),I=1,NPAR) /'   G','   B',' TAU','  GF',' XNL','  DT',
     &                       '  FT','  IJ','  ID' /
      DATA (C2(I),I=1,NPAR)/
     &     'Modulus scale parameter                                ',
     &     'Relaxation breadth; larger => broader                  ',
     &     'Relaxation time                                        ',
     &     'Infinite-time modulus                                  ',
     &     'Nonlinear parameter; larger => stiffer                 ',
```

```
     &     'Time of first point after block starts                    ',
     &     'Time multiplier                                           ',
     &     'Picks stress: 0 = shear; 1 = N1; 2 = N2; 3 = N1-N2        ',
     &     'Planar ext.: 1=long. stress; 2=transv.; 3=long.-transv.'/
10    WRITE(NTER,20)
20    FORMAT(/T3,'ID',T7,'Name',T35,'Description',T71,'Value'
     &        /T3,'--',T7,'----',T13,55('-'),T70,9('-'))
      WRITE(NTER,30)(I,C1(I),C2(I),PAR(I),I=1,NRPAR)
30    FORMAT(T3,I2,T6,A4,T13,A55,T70,1P,G9.2)
      WRITE(NTER,40)(I,C1(I),C2(I),IPAR(I-NRPAR),I=NRPAR+1,NPAR)
40    FORMAT(T3,I2,T6,A4,T13,A55,T70,I5)
50    WRITE(NTER,'(/2X,A)')
     &            'Enter ID of parameter to be changed; 0 to return.'
      READ(NTER,'(A4)')CIP
      IP=ICHAR(CIP)-48
      IF(IP.LE.0)RETURN
      IF(IP.GT.NPAR)THEN
        WRITE(NTER,*)' >>>>>>  Parameter not applicable; reenter.'
        GOTO 50
      ENDIF
      WRITE(NTER,'(6X,A30,A4)')'Enter new parameter value for ',C1(IP)
      READ(NTER,*)TEMP
      IF((TEMP.LT.0.AND.IP.NE.5).OR.(IP.EQ.7.AND.TEMP.LE.1.))THEN
        WRITE(NTER,*)' >>>>>>  Invalid parameter value; reenter.'
      ELSE
        IF(IP.LE.NRPAR)THEN
          PAR(IP)=TEMP
        ELSE
          IPAR(IP-NRPAR)=INT(TEMP+TINY)
        ENDIF
      ENDIF
      GOTO 10
      END
C
      DOUBLE PRECISION FUNCTION RTF(U,NP,T,A,RT,C)
C Deformation rate.
      IMPLICIT DOUBLE PRECISION(A-H,O-Z)
      PARAMETER (TINY=1.D-5)
      DIMENSION T(NP),A(NP),RT(NP)
      DO 10 J=2,NP-1
        IF(U.LE.T(J))GOTO 20
10    CONTINUE
20    J=J-1
      IF(C.LT.TINY)THEN
C Simple shear:
        RTF=RT(J)*(1.D0+A(J)*ABS(RT(J))*(U-T(J)))**(1.D0/A(J)-1.D0)
      ELSE
C Extensional flows:
        RTF=RT(J)/(1.D0+A(J)*RT(J)*(U-T(J)))
      ENDIF
      RETURN
      END
```

10.6.2 RATES

```
      SUBROUTINE RATES(NP,T,XLEN,A,C,RT)
C Calculates rates at beginning of each block.
      IMPLICIT DOUBLE PRECISION(A-H,O-Z)
      DIMENSION T(NP),XLEN(NP),A(NP),RT(NP)
      PARAMETER (TINY=1.D-10)
      DO 10 J=1,NP-1
        IF(ABS(A(J)).LT.TINY)A(J)=TINY
        IF(C.LT.TINY)THEN
C Simple shear. A(J)=1 gives constant rate.
          RT(J)=((ABS(XLEN(J+1)-XLEN(J))+1.D0)**A(J)-1.D0)*
```

```
     &          SIGN(1.D0,XLEN(J+1)-XLEN(J))/(A(J)*(T(J+1)-T(J)))
        ELSE
C Extensional flow. A(J)=1 gives constant velocity; A(J)=0 constant Hencky
C  strain.
          RT(J)=((XLEN(J+1)/XLEN(J))**A(J)-1.D0)/(A(J)*(T(J+1)-T(J)))
        ENDIF
10    CONTINUE
      RETURN
      END
```

10.6.3 STRMOD

Imbedded into the code are the function subprograms STRESS*n* and CERN*n*. Specific areas have been boxed to aid custom modification.

```
      SUBROUTINE STRMOD(NTER,IMOD,TF,STRSS,*)
      IMPLICIT DOUBLE PRECISION (A-H,O-Z)
      PARAMETER (NMOD=6)
C Key for models.
      IF(IMOD.EQ.0)THEN
        WRITE(NTER,'(/8(6X,A/))')
     &  'ID                      Description                              ',
     &  '---  -----------------------------------------------------------',
     &  ' 1   Viscoelastic solid with Kohlrausch relaxation               ',
     &  ' 2   Rubberlike liquid with Kohlrausch relaxation                ',
     &  ' 3   Rubberlike liquid with Segalman relaxation                  ',
     &  ' 4   Tanner-Simmons model with Kohlrausch relaxation             ',
     &  ' 5   Papanastasiou-Scriven-Macosko with Kohlrausch relaxation',
     &  ' 6   Modified Tanner-Simmons model with Kohlrausch relaxation'
C Enter more descriptions here.
        RETURN 1
      ENDIF
C  Change value of NMOD (in PARAMETER line on incorporating new models
      IF(IMOD.LT.0.OR.IMOD.GT.NMOD)THEN
        WRITE(NTER,'(/2X,A)')'Invalid model; retry.'
        RETURN 1
      ENDIF
      IF(IMOD.EQ.1)STRSS=STRSS1(TF)
      IF(IMOD.EQ.2)STRSS=STRSS2(TF)
      IF(IMOD.EQ.3)STRSS=STRSS3(TF)
      IF(IMOD.EQ.4)STRSS=STRSS4(TF)
      IF(IMOD.EQ.5)STRSS=STRSS5(TF)
      IF(IMOD.EQ.6)STRSS=STRSS6(TF)
C Enter calls to new models here.
      RETURN
      END
C
      DOUBLE PRECISION FUNCTION STRSS1(X)
C Viscoelastic solid with Kohlrausch relaxation function.
C n strain measure (variation on S model of Chang et al. (1976)).
      IMPLICIT DOUBLE PRECISION (A-H,O-Z)
      PARAMETER (NMAX=201)
      EXTERNAL CERN1
      COMMON /BLOC1/ A(NMAX),T(NMAX),RT(NMAX),XLEN(NMAX),NP
      COMMON /BLOC2/ G,B,TAU,GF,XNL,DT,FT
      COMMON /BLOC3/ C,XN,ID,IJ
      COMMON /BLOC4/ TZ,TF,TR,XLZ,XLF,XLR
C Set up conditions for particular model.
      TF=X
      TZ=0.D0
```

```
      TR=0.D0
      XN=XNL
      XLR=XL(TR,NP,T,XLEN,A,RT,C)
      STRSS1=
C Add in integral over time TZ to TF.
     & SIMP(TZ,TF,CERN1)
      RETURN
      END
C
      DOUBLE PRECISION FUNCTION CERN1(S)
C Total integrand. Variables are carried into function using COMMON blocks.
      IMPLICIT DOUBLE PRECISION (A-H,O-Z)
      PARAMETER (NMAX=201,TINY=1.D-6)
      COMMON /BLOC1/ A(NMAX),T(NMAX),RT(NMAX),XLEN(NMAX),NP
      COMMON /BLOC2/ G,B,TAU,GF,XNL,DT,FT
      COMMON /BLOC3/ C,XN,ID,IJ
      COMMON /BLOC4/ TZ,TF,TR,XLZ,XLF,XLR
      COMMON /BLOC8/ JBLOCK
      SDN=S-TINY
      SUP=S+TINY
C Check to make sure rate calculation does not span blocks.
      IF(SDN.LT.T(JBLOCK))THEN
        SDN=S
        SUP=S+2.D0*TINY
      ENDIF
      IF(SUP.GT.T(JBLOCK+1))THEN
        SDN=S-2.D0*TINY
        SUP=S
      ENDIF
      XLDN=XL(SDN,NP,T,XLEN,A,RT,C)
      XLUP=XL(SUP,NP,T,XLEN,A,RT,C)
      CERN1=XMOD1(TF-S,G,B,TAU,GF)*
C And multiply by strain rate found with divided differences.
     &      (STRAIN(XLUP,XLR,XN,C,ID,IJ)-STRAIN(XLDN,XLR,XN,C,ID,IJ))
     &      /TINY/2.D0
      RETURN
      END
C
      DOUBLE PRECISION FUNCTION STRSS2(X)
C Rubberlike liquid with Kohlrausch relaxation function.
C n-type strain function (n = 2 is Lodge rubberlike liquid).
      IMPLICIT DOUBLE PRECISION (A-H,O-Z)
      PARAMETER (NMAX=201)
      EXTERNAL CERN2
      COMMON /BLOC1/ A(NMAX),T(NMAX),RT(NMAX),XLEN(NMAX),NP
      COMMON /BLOC2/ G,B,TAU,GF,XNL,DT,FT
      COMMON /BLOC3/ C,XN,ID,IJ
      COMMON /BLOC4/ TZ,TF,TR,XLZ,XLF,XLR
C Set up conditions for particular model.
      TF=X
      TZ=0.D0
      TR=TF
      XN=XNL
      XLZ=XL(TZ,NP,T,XLEN,A,RT,C)
      XLF=XL(TF,NP,T,XLEN,A,RT,C)
      XLR=XLF
      STRSS2=STRAIN(XLR,XLZ,XN,C,ID,IJ)*XMOD1(TF,G,B,TAU,GF)
C  Add in integral over time TZ to TF.
     &       +SIMP(TZ,TF,CERN2)
      RETURN
      END
```

```
      DOUBLE PRECISION FUNCTION CERN2(S)
C Total integrand. Variables are carried into function using COMMON blocks.
      IMPLICIT DOUBLE PRECISION (A-H,O-Z)
      PARAMETER (NMAX=201)
      COMMON /BLOC1/ A(NMAX),T(NMAX),RT(NMAX),XLEN(NMAX),NP
      COMMON /BLOC2/ G,B,TAU,GF,XNL,DT,FT
      COMMON /BLOC3/ C,XN,ID,IJ
      COMMON /BLOC4/ TZ,TF,TR,XLZ,XLF,XLR
      XLS=XL(S,NP,T,XLEN,A,RT,C)
      CERN2=XMEM1(TF-S,G,B,TAU,GF)*STRAIN(XLR,XLS,XN,C,ID,IJ)
      RETURN
      END
C
      DOUBLE PRECISION FUNCTION STRSS3(X)
C Rubberlike liquid with Segalman relaxation function.
C n-type strain function.
      IMPLICIT DOUBLE PRECISION (A-H,O-Z)
      PARAMETER (NMAX=201)
      EXTERNAL CERN3
      COMMON /BLOC1/ A(NMAX),T(NMAX),RT(NMAX),XLEN(NMAX),NP
      COMMON /BLOC2/ G,B,TAU,GF,XNL,DT,FT
      COMMON /BLOC3/ C,XN,ID,IJ
      COMMON /BLOC4/ TZ,TF,TR,XLZ,XLF,XLR
C Set up conditions for particular model.
      TF=X
      TZ=0.D0
      TR=TF
      XN=XNL
      XLZ=XL(TZ,NP,T,XLEN,A,RT,C)
      XLF=XL(TF,NP,T,XLEN,A,RT,C)
      XLR=XLF
      STRSS3=STRAIN(XLR,XLZ,XN,C,ID,IJ)*XMOD2(TF,G,B,TAU,GF)
C  Add in integral over time TZ to TF.
     &       +SIMP(TZ,TF,CERN3)
      RETURN
      END
C
      DOUBLE PRECISION FUNCTION CERN3(S)
C Total integrand. Variables are carried into function using COMMON blocks.
      IMPLICIT DOUBLE PRECISION (A-H,O-Z)
      PARAMETER (NMAX=201)
      COMMON /BLOC1/ A(NMAX),T(NMAX),RT(NMAX),XLEN(NMAX),NP
      COMMON /BLOC2/ G,B,TAU,GF,XNL,DT,FT
      COMMON /BLOC3/ C,XN,ID,IJ
      COMMON /BLOC4/ TZ,TF,TR,XLZ,XLF,XLR
      XLS=XL(S,NP,T,XLEN,A,RT,C)
      CERN3=XMEM2(TF-S,G,B,TAU,GF)*STRAIN(XLR,XLS,XN,C,ID,IJ)
      RETURN
      END
C
      DOUBLE PRECISION FUNCTION STRSS4(X)
C Viscoelastic liquid with Kohlrausch relaxation function.
C Original Tanner-Simmons model.
      IMPLICIT DOUBLE PRECISION (A-H,O-Z)
      PARAMETER (NMAX=201,TINY=1.D-9)
      EXTERNAL CERN4
      COMMON /BLOC1/ A(NMAX),T(NMAX),RT(NMAX),XLEN(NMAX),NP
      COMMON /BLOC2/ G,B,TAU,GF,XNL,DT,FT
      COMMON /BLOC3/ C,XN,ID,IJ
      COMMON /BLOC4/ TZ,TF,TR,XLZ,XLF,XLR
C Set up conditions for particular model.
```

```
      TF=X
      TZ=0.D0
      TR=TF
      XN=2.D0
C Tanner-Simmons model uses a rate-modified lower limit on integral. Use
C  the nonlinear parameter XNL to carry in information about this effect.
C  Tanner (1967) reports wide range of B; thus transform below.
      BZ=10.D0**XNL
      XLZ=XL(TZ,NP,T,XLEN,A,RT,C)
      XLF=XL(TF,NP,T,XLEN,A,RT,C)
      XLR=XLF
C Use first invariant of B strain tensor to get rupture time.
      CALL INVAR(XLZ,XLF,C,XI1,XI2)
      RATE=SQRT(XI1-3.D0)/(TF+TINY)+TINY
C If rate is very high, or BZ is low, TL will be close to TF.
C This will reduce stress.
      TL=TF-BZ/RATE
      IF(TL.LT.TZ)THEN
        STRSS4=STRAIN(XLR,XLZ,XN,C,ID,IJ)*(XMOD1(TF,G,B,TAU,GF)
     &         -XMOD1(TF-TL,G,B,TAU,GF))
C Add in integral over time TZ to TF.
     &         +SIMP(TZ,TF,CERN4)
      ELSE
        STRSS4=SIMP(TL,TF,CERN4)
      ENDIF
      RETURN
      END
C
      DOUBLE PRECISION FUNCTION CERN4(S)
C Total integrand. Variables are carried into function using COMMON blocks.
      IMPLICIT DOUBLE PRECISION (A-H,O-Z)
      PARAMETER (NMAX=201)
      COMMON /BLOC1/ A(NMAX),T(NMAX),RT(NMAX),XLEN(NMAX),NP
      COMMON /BLOC2/ G,B,TAU,GF,XNL,DT,FT
      COMMON /BLOC3/ C,XN,ID,IJ
      COMMON /BLOC4/ TZ,TF,TR,XLZ,XLF,XLR
      XLS=XL(S,NP,T,XLEN,A,RT,C)
      CERN4=XMEM1(TF-S,G,B,TAU,GF)*STRAIN(XLR,XLS,XN,C,ID,IJ)
      RETURN
      END
C
      DOUBLE PRECISION FUNCTION STRSS5(X)
C Viscoelastic liquid with Kohlrausch relaxation function.
C Variation on Papanstasiou-Scriven-Macosko (PSM) model.
      IMPLICIT DOUBLE PRECISION (A-H,O-Z)
      PARAMETER (NMAX=201,TINY=1.D-6)
      EXTERNAL CERN5
      COMMON /BLOC1/ A(NMAX),T(NMAX),RT(NMAX),XLEN(NMAX),NP
      COMMON /BLOC2/ G,B,TAU,GF,XNL,DT,FT
      COMMON /BLOC3/ C,XN,ID,IJ
      COMMON /BLOC4/ TZ,TF,TR,XLZ,XLF,XLR
C Parameter for combining first and second invariants to yield the Wagner
C  combined invariant.  This may vary with material.
      DATA BZ /.02/
C Set up conditions for particular model.
      TF=X
      TZ=0.D0
      TR=TF
      XN=2.D0
C Transformation below designed to give near-rubberlike liquid behavior
C  at XNL=2, and moderate values of ALFA for 0 < XNL < 2.
      ALFA=EXP(-XNL)/3.D0
      XLF=XL(TF,NP,T,XLEN,A,RT,C)
```

```
      XLZ=XL(TZ,NP,T,XLEN,A,RT,C)
      XLR=XLF
      CALL INVAR(XLR,XLZ,C,XI1,XI2)
      XIC=BZ*XI1+(1.D0-BZ)*XI2
      STRSS5=STRAIN(XLR,XLZ,XN,C,ID,IJ)*XMOD1(TF,G,B,TAU,GF)
C Use ALFA and XIC to give damping function.
     &        *1.D0/(1.D0+ALFA*(XIC-3.D0))
C Add in integral over time TZ to TF.
     &        +SIMP(TZ,TF,CERN5)
      RETURN
      END
C
      DOUBLE PRECISION FUNCTION CERN5(S)
C Total integrand. Variables are carried into function using COMMON blocks.
      IMPLICIT DOUBLE PRECISION (A-H,O-Z)
      PARAMETER (NMAX=201)
      COMMON /BLOC1/ A(NMAX),T(NMAX),RT(NMAX),XLEN(NMAX),NP
      COMMON /BLOC2/ G,B,TAU,GF,XNL,DT,FT
      COMMON /BLOC3/ C,XN,ID,IJ
      COMMON /BLOC4/ TZ,TF,TR,XLZ,XLF,XLR
      DATA BZ /.02D0 /
      ALFA=EXP(-XNL)/3.D0
      XLS=XL(S,NP,T,XLEN,A,RT,C)
      CALL INVAR(XLR,XLS,C,XI1,XI2)
      XIC=BZ*XI1+(1.D0-BZ)*XI2
      CERN5=XMEM1(TF-S,G,B,TAU,GF)*STRAIN(XLR,XLS,XN,C,ID,IJ)
     & *1.D0/(1.D0+ALFA*(XIC-3.D0))
      RETURN
      END
C
      DOUBLE PRECISION FUNCTION STRSS6(X)
C Viscoelastic liquid with Kohlrausch relaxation function.
C Modified Tanner-Simmons model.
      IMPLICIT DOUBLE PRECISION (A-H,O-Z)
      PARAMETER (NMAX=201,TINY=1.D-9,UNDER=158)
      EXTERNAL CERN6
      COMMON /BLOC1/ A(NMAX),T(NMAX),RT(NMAX),XLEN(NMAX),NP
      COMMON /BLOC2/ G,B,TAU,GF,XNL,DT,FT
      COMMON /BLOC3/ C,XN,ID,IJ
      COMMON /BLOC4/ TZ,TF,TR,XLZ,XLF,XLR
C Set up conditions for particular model.
      TF=X
      TZ=0.D0
      TR=TF
      XN=2.D0
C Tanner-Simmons model uses a rate-modified lower limit on integral. Use
C  the nonlinear parameter XNL to carry in information about this effect.
C  Tanner (1967) reports wide range of B; thus transform below.
C Note: -∞ < XNL < +∞, while 0 < BZ < +∞
      BZ=10.D0**XNL
C Use first invariant of B strain tensor to get rupture time.
      XLZ=XL(TZ,NP,T,XLEN,A,RT,C)
      XLF=XL(TF,NP,T,XLEN,A,RT,C)
      CALL INVAR(XLZ,XLF,C,XI1,XI2)
      RATE=SQRT(XI1-3.D0)/(TF+TINY)
C If rate is very high, TZ will be close to TF.
      IF(TF.GT.TINY.AND.BZ/TF/RATE.LT.UNDER)THEN
        TZ=TF*EXP(-BZ/TF/RATE)
      ELSE
        TZ=0.D0
      ENDIF
C Must use modified TZ here because first line of STRSS6 is uv portion of
C  integration-by-parts procedure on the integral from TZ to TF.
```

```
      XLZ=XL(TZ,NP,T,XLEN,A,RT,C)
      XLR=XLF
      STRSS6=STRAIN(XLR,XLZ,XN,C,ID,IJ)*XMOD1(TF-TZ,G,B,TAU,GF)
C Add in integral over time TZ to TF.
     &        +SIMP(TZ,TF,CERN6)
      RETURN
      END
C
      DOUBLE PRECISION FUNCTION CERN6(S)
C Total integrand. Variables are carried into function using COMMON blocks.
      IMPLICIT DOUBLE PRECISION(A-H,O-Z)
      PARAMETER (NMAX=201)
      COMMON /BLOC1/ A(NMAX),T(NMAX),RT(NMAX),XLEN(NMAX),NP
      COMMON /BLOC2/ G,B,TAU,GF,XNL,DT,FT
      COMMON /BLOC3/ C,XN,ID,IJ
      COMMON /BLOC4/ TZ,TF,TR,XLZ,XLF,XLR
      XLS=XL(S,NP,T,XLEN,A,RT,C)
      CERN6=XMEM1(TF-S,G,B,TAU,GF)*STRAIN(XLR,XLS,XN,C,ID,IJ)
      RETURN
      END
```

10.6.4 INVAR

```
      SUBROUTINE INVAR(XLR,XLU,C,XI1,XI2)
C Returns first (XI1) and second (XI2) invariants of the Cauchy relative
C  deformation tensor.
      IMPLICIT DOUBLE PRECISION (A-H,O-Z)
      PARAMETER (SMALL=1.D-3,TINY=1.D-6)
C Simple shear:
      IF(C.LT.SMALL)THEN
        XI1=3.D0+(XLR-XLU)**2
        XI2=XI1
        RETURN
      ENDIF
C Extensional flows; XLR and XLU are stretch ratios with respect to zero
C  time; their ratio is the λz in Bird et al., [4, p. 623].
      X=XLR/XLU
      X2=X**2
C Value of C controls type of deformation (analogous to b in Bird et al.).
C For simple extension, C = 0.5.
      IF(ABS(C-.5D0).LT.SMALL)THEN
        XI1=X2+2.D0/X
        XI2=1.D0/X2+2.D0*X
        RETURN
      ENDIF
C For planar extension, C = 1.
      IF(ABS(C-1.D0).LT.SMALL)THEN
        XI1=1.D0+X2+1.D0/X2
        XI2=XI1
        RETURN
      ENDIF
C For biaxial extension:
      IF(ABS(C-2.D0).LT.SMALL)THEN
        XI1=2.D0*X2+1.D0/X2**2
        XI2=2.D0/X2+X2**2
        RETURN
      ENDIF
      END
```

10.6.5 STRAIN

```
      DOUBLE PRECISION FUNCTION STRAIN(XLR,XLU,XN,C,ID,IJ)
C Finger strain (differences) for the appropriate stress (differences).
C XLR is the reference displacement; XLU is the current displacement.
      IMPLICIT DOUBLE PRECISION (A-H,O-Z)
      PARAMETER (BIG=1.D3,SMALL=1.D-3,TINY=1.D-6)
C Simple shear flows, C = 0.
      IF(C.LT.TINY)THEN
        S=XLR-XLU
C Quick return for easy cases: XN = 2 and small strain.
        IF(ABS(XN-2.).LT.TINY)THEN
          IF(IJ.EQ.0)STRAIN=S
          IF(IJ.EQ.1)STRAIN=S**2
          IF(IJ.EQ.2)STRAIN=0.D0
          IF(IJ.EQ.3)STRAIN=S**2
          RETURN
        ENDIF
        IF(ABS(S).LT.SMALL)THEN
          IF(IJ.EQ.0)STRAIN=S
          IF(IJ.EQ.1)STRAIN=S**2
          IF(IJ.EQ.2)STRAIN=(XN-2.D0)*S**2/4.D0
          IF(IJ.EQ.3)STRAIN=S**2*(2.D0+XN)/4.D0
          RETURN
        ENDIF
C Principal draw ratio, X (Note: S = X - 1/X for plane strain).
        IF(S.GT.BIG)THEN
          X=S+1.D0/S
        ELSE
          X=(S+SQRT(S**2+4.D0))/2.D0
        ENDIF
        IF(IJ.LE.1)THEN
C Shear stress and first normal stress difference; IJ=0 and IJ=1.
          IF(ABS(XN).LT.TINY)THEN
            STRAIN=4.D0*LOG(X)/(X+1.D0/X)
          ELSE
            XPN=X**XN
            STRAIN=(XPN-1.D0/XPN)/(X+1.D0/X)*2.D0/XN
          ENDIF
          IF(IJ.EQ.1)STRAIN=STRAIN*(X-1.D0/X)
C Second normal stress difference, IJ=2.
        ELSE IF(IJ.EQ.2)THEN
          X2=X**2
          IF(ABS(XN).LT.TINY)THEN
            STRAIN=2.D0*LOG(X)*(1.D0-X2)
          ELSE
            XPN=X**XN
            STRAIN=(XPN-1.D0+X2/XPN-X2)/(X2+1.D0)*2.D0/XN
          ENDIF
C Third normal stress difference, IJ=3
        ELSE IF(IJ.EQ.3)THEN
          X2=X**2
          IF(ABS(XN).LT.TINY)THEN
            STRAIN=2.D0*LOG(X)*(X2+2.*X-1.D0)/(1.D0+X)
          ELSE
            XPN=X**XN
            STRAIN=(X2-1.D0)*(XPN-1.D0/XPN)-(XPN-1.D0+X2/XPN)*2.D0/XN/
     &               (X2+1.D0)
          ENDIF
        ENDIF
      RETURN
      ENDIF
C Extensional flows; XLR and XLU are stretch ratios with respect to zero
C  time; their ratio is the $\lambda_z$ of Bird et al., [4, p. 623].
      X=XLR/XLU
C Value of C controls type of deformation (analogous to b in Bird et al.).
```

```
C ID = 2 gives transverse stress in planar extension;
C ID = 3 gives (longitudinal - transverse) stress
      IF(ABS(C-1.).LT.TINY.AND.ID.GE.2)THEN
        IF(ABS(1.D0-X).LT.TINY)THEN
C Linear limit.
          STRAIN=2.D0*(X-1.D0)
        ELSE IF(ABS(XN).LT.TINY)THEN
          STRAIN=2.D0*LOG(X)
        ELSE
          XPN=X**XN
          STRAIN=(XPN-1.D0)*2.D0/XN*XPN**(ID-3)
        ENDIF
      ELSE
C The principal stress is required.
        IF(ABS(1.D0-X).LT.SMALL)THEN
C Linear limit.
          STRAIN=2.D0*(1.D0+C)*(X-1.D0)
        ELSE IF(ABS(XN).LT.TINY)THEN
          STRAIN=2.D0*(1.D0+C)*LOG(X)
        ELSE
          STRAIN=(X**XN-X**(-C*XN))*2.D0/XN
        ENDIF
      ENDIF
      RETURN
      END
```

10.6.6 XL

```
      DOUBLE PRECISION FUNCTION XL(U,NP,T,XLEN,A,RT,C)
C Stretch ratio in extension or strain in shear at time U.
      IMPLICIT DOUBLE PRECISION (A-H,O-Z)
      PARAMETER (SMALL=1.D-4,TINY=1.D-6)
      DIMENSION T(NP),XLEN(NP),A(NP),RT(NP)
C Find correct block; 1 ≤ J ≤ NP-1 no matter what.
      DO 10 J=2,NP-1
        IF(T(J).GE.U)GOTO 20
10    CONTINUE
20    J=J-1
      IF(C.LT.TINY)THEN
C Shear:
        X=ABS(RT(J))*(U-T(J))
        XSIGN=DSIGN(1.D0,RT(J))
        IF(ABS(A(J)-1.D0).LT.TINY.OR.ABS(X).LT.SMALL)THEN
          XL=XLEN(J)+XSIGN*X
        ELSE IF(ABS(A(J)).LT.TINY)THEN
          XL=XLEN(J)+XSIGN*(EXP(X)-1.D0)
        ELSE
          XL=XLEN(J)+XSIGN*((1.D0+A(J)*X)**(1.D0/A(J))-1.D0)
        ENDIF
      ELSE
C Extension:
        X=RT(J)*(U-T(J))
        IF(ABS(X).LE.SMALL.OR.ABS(A(J)-1.D0).LT.TINY)THEN
          XL=XLEN(J)*(1.D0+X)
        ELSE
          IF(ABS(A(J)).LT.SMALL)THEN
            XL=XLEN(J)*EXP(X)
          ELSE
            XL=XLEN(J)*(1.D0+A(J)*X)**(1.D0/A(J))
          ENDIF
        ENDIF
      ENDIF
      RETURN
      END
```

10.6.7 XMEM*n*

```
      DOUBLE PRECISION FUNCTION XMEM1(T,G,B,TAU,GF)
C Memory function, Kohlrausch relaxation.
      IMPLICIT DOUBLE PRECISION (A-H,O-Z)
      PARAMETER (SMALL=1.D-10,UNDER=158)
      X=T/TAU
      B2=.5D0/B
      XB=X**B2
C Avoid singularity.
      IF(X.LT.SMALL)X=SMALL
C Avoid underflow arithmetic.
      IF(XB.GT.UNDER)XB=UNDER
      XMEM1=(G-GF)*B2/TAU*XB/X*EXP(-XB)
      RETURN
      END
C
      DOUBLE PRECISION FUNCTION XMEM2(T,G,B,TAU,GF)
C Memory function.
      IMPLICIT DOUBLE PRECISION(A-H,O-Z)
      PARAMETER (SMALL=1.D-10)
      X=T/TAU
      A=1.D0-.5D0/B
C Avoid singularity.
      IF(X.LE.SMALL)X=SMALL
      XMEM2=(XMOD2(T,G,B,TAU,GF)-GF)*(1.D0+A/X)/TAU
      RETURN
      END
```

10.6.8 XMOD*n*

```
      DOUBLE PRECISION FUNCTION XMOD1(T,G,B,TAU,GF)
C Relaxation function, Kohlrausch type. B=0.5 gives single exponential.
      IMPLICIT DOUBLE PRECISION (A-H,O-Z)
      PARAMETER (UNDER=158)
      X=T/TAU
      B2=.5D0/B
      XB=X**B2
C Avoid underflow arithmetic.
      IF(XB.LT.UNDER)THEN
        XMOD1=(G-GF)*EXP(-XB)+GF
      ELSE
        XMOD1=GF
      ENDIF
      RETURN
      END
C
      DOUBLE PRECISION FUNCTION XMOD2(T,G,B,TAU,GF)
C Segalman relaxation function (Bird et al., [4, p. 285]).
C Requires the routine GAMMLN from Numerical Recipes (1986, 1992).
      IMPLICIT DOUBLE PRECISION (A-H,O-Z)
      PARAMETER (SMALL=1.D-10,TINY=1.D-15)
      SAVE GAM,BSAVE,A
      DATA BSAVE /0.D0/
      X=T/TAU
C Save coefficients for next call to save time.
      IF(B.NE.BSAVE)THEN
        BSAVE=B
C Transform 0.5 < B < large to 0 < A < 1. B=0.5, A=0 is single exponential.
        A=1.D0-.5D0/B
        PIA=3.1415926536D0*A
        IF(PIA.LT.SMALL)THEN
          GAM=1.D0
        ELSE
```

```
          GAM=EXP(GAMMLN(1.D0+A))*SIN(PIA)/PIA
        ENDIF
      ENDIF
C Avoid singularity at X = 0.
      IF(X.LE.0.D0)X=SMALL
      IF(A.LT.TINY)A=TINY
      XMOD2=(G-GF)*GAM*X**(-A)*EXP(-X)+GF
      RETURN
      END
```

10.6.9 SIMP

```
      DOUBLE PRECISION FUNCTION SIMP(X1,X2,CX)
C Controls integration. For the Romberg method, it is best to avoid
C  discontinuities in the derivative of the function to be integrated.
C Requires the subroutine QROMB2.
      IMPLICIT DOUBLE PRECISION (A-H,O-Z)
      PARAMETER (EPSS=1.D-6,SMALL=1.D-1,TINY=1.D-9,NMAX=201)
      COMMON /BLOC1/ A(NMAX),T(NMAX),RT(NMAX),XLEN(NMAX),NP
      COMMON /BLOC8/ J
      DIMENSION SEST(NMAX),EPS(NMAX)
      EXTERNAL CX
      IF(X1.EQ.X2)THEN
        SIMP=0.D0
        RETURN
      ENDIF
C Find beginning and ending blocks.
      JS=1
      DO 10 J=1,NP
        IF(T(J).LE.X1)JS=J
        IF(T(J).GE.X2)GOTO 20
10    CONTINUE
20    JF=J-1
C Integrate over each block and sum.
      SIMP=0.D0
      JEND=1
      IF(X1.LT.T(1))THEN
        J=1
        JEND=2
        CALL QROMB2(CX,X1,T(1),SI,EPSS)
        SIMP=SI
      ENDIF
C Estimate value of integral and relative contribution of each interval.
      IF(JF-JS.GE.1)THEN
        SSEST=ABS(SIMP)
        DO 30 J=JS,JF
          SEST(J)=ABS(CX((MAX(T(J),X1)+MIN(T(J+1),X2))/2.)
     &              *(MAX(T(J),X1)-MIN(T(J+1),X2)))+TINY
          SSEST=SSEST+SEST(J)
30      CONTINUE
        SMEAN=SSEST/REAL(JF-JS+JEND)
        DO 40 J=JS,JF
          EPS(J)=EPSS*SMEAN/SEST(J)
          IF(EPS(J).GT.SMALL)EPS(J)=SMALL
          IF(EPS(J).LT.EPSS)EPS(J)=EPSS
40      CONTINUE
      ELSE
        EPS(JS)=EPSS
      ENDIF
      DO 50 J=JS,JF
        CALL QROMB2(CX,MAX(T(J),X1),MIN(T(J+1),X2),SI,EPS(J))
        SIMP=SIMP+SI
50    CONTINUE
      RETURN
      END
```

10.6.10 FITSTR

Imbedded into the code are the subprograms CCEROR and FUNC inherent to the execution of FITSTR. Refer to Appendix A for the program dependencies.

```
      PROGRAM FITSTR
C Fits time-strain-stress data with a nonlinear constitutive model.
C Six models listed in Table 10.1 are programmed in STRMOD (Sect. 10.6.3).
C Requires the subroutines STRMOD, RATES, and STOPP and the subroutine
C  POWELL from Numerical Recipes, by Press et al., Cambridge University
C  Press, Cambridge, 1986.
C Data file format:
C         1st record - Title, up to 79 characters.
C         2nd record - C, ID, IJ
C                        C: 0.   = simple shear
C                           0.5 = simple tension
C                           1.   = pure shear
C                           2.   = biaxial tension.
C                       ID: 1    = longitudinal stress in planar extension
C                           2    = transverse stress
C                           3    = longitudinal - transverse stress
C                       IJ: 0    = shear stress in simple shear
C                           1    = first normal stress difference
C                           2    = second normal stress difference
C                           3    = N1 - N2
C subsequent records - T, XLEN, STRSS
C                           T    = time
C                        XLEN    = deformation
C                       STRSS    = stress
      IMPLICIT DOUBLE PRECISION (A-H,O-Z)
      PARAMETER (NACC=11,NDIM=5,NLARGE=500,NMAX=201,NMOD=6,TINY=1.D-6)
      CHARACTER ANS*1,TITLE*80,PROP(NDIM)*3,PROPI(NDIM)*3
      DIMENSION X(NDIM),XI(NDIM,NDIM)
      DIMENSION BLO(NDIM),BHI(NDIM),Q(NDIM),PLOI(NDIM),PHII(NDIM)
      EXTERNAL FUNC
      COMMON /BLOC1/ A(NMAX),T(NMAX),RT(NMAX),XLEN(NMAX),NP
      COMMON /BLOC2/ Q
      COMMON /BLOC3/ C,XN,ID,IJ
      COMMON /BLOC5/ STRSS(NMAX),STRCAL(NMAX),RES(NMAX)
      COMMON /BLOC6/ IDROP(NDIM),IMOD,IWRITE,ND
      COMMON /BLOC7/ PI(NDIM,NLARGE),PHI(NDIM),PLO(NDIM),SSI(NLARGE),
     &               NCALL1,NCALL2
C Initial direction vector for POWELL.
      DATA XI /1.,0.,0.,0.,0.,0.,1.,0.,0.,0.,0.,0.,1.,0.,0.,0.,0.,0.,1.,
     &         0.,0.,0.,0.,0.,1./
      DATA (PROPI(I),I=1,NDIM) /'G  ','B  ','TAU','GF ','XNL'/
      IDROP(4)=1
      Q(4)=0.D0
      PLOI(4)=0.D0
10    CALL MSHEAD
      CALL UNITS(NTER,NINT,NOUT)
      READ(NINT,'(A80)')TITLE
C Read in characteristics of deformation.
      READ(NINT,*)C,XID,XIJ
      ID=INT(XID+TINY)
      IJ=INT(XIJ+TINY)
C Read in time, deformation, stress records.
      DO 20 I=1,NMAX
        READ(NINT,*,END=30)T(I),XLEN(I),STRSS(I)
        IF(I.EQ.NMAX)CALL STOPP(NTER,NMAX-1)
20    CONTINUE
30    NP=I-1
C Compute path shapes at each point.
```

```
      DO 40 I=1,NP-2
        IF(C.LT.TINY)THEN
          Y1=(XLEN(I+1)-XLEN(I))/(T(I+1)-T(I))
          Y2=(XLEN(I+2)-XLEN(I))/(T(I+2)-T(I))
        ELSE
          Y1=(LOG(XLEN(I+1)/XLEN(I)))/(T(I+1)-T(I))
          Y2=(LOG(XLEN(I+2)/XLEN(I)))/(T(I+2)-T(I))
        ENDIF
        SL=(Y2-Y1)/(T(I+2)-T(I+1))
        DEN=(Y1-SL*(T(I+1)-T(I)))**2
        IF(ABS(DEN).GT.TINY)THEN
          IF(C.LT.TINY)THEN
            A(I)=1.D0-2.D0*SL/DEN
          ELSE
            A(I)=-2.D0*SL/DEN
          ENDIF
        ELSE
          A(I)=1.D0
        ENDIF
        IF(ABS(A(I)).LT.TINY)A(I)=0.D0
        IF(ABS(A(I)-1.D0).LT.TINY)A(I)=1.D0
40    CONTINUE
      A(NP-1)=A(NP-2)
      A(NP)=A(NP-2)
      DO 50 I=1,NP,MAX(1,NP/6)
        WRITE(NTER,'(6X,I3,12X,F10.2)')I,A(I)
50    CONTINUE
      WRITE(NTER,60)
60    FORMAT(/2X,'Above are approximate path shapes. Computational',
     &        1X,'speed can be improved'/2X,'by setting to 0.0 or 1.0;',
     &        2X,'if path shapes are close to one of these it'/
     &        2X,'will probably be beneficial to set them all to this',
     &        1X,'value. To do this,'/
     &        2X,'enter 0.0 or 1.0; otherwise a large number.')
      READ(NTER,*)AA
      IF(AA.LE.1.D0)THEN
        DO 70 I=1,NP
          A(I)=AA
70      CONTINUE
      ENDIF
C Compute deformation rates at beginning of each block.
      CALL RATES(NP,T,XLEN,A,C,RT)
      WRITE(NTER,'(/A80/)')TITLE
      WRITE(NOUT,'(/A80/)')TITLE
      WRITE(NTER,*)'  Number of points =',NP
80    WRITE(NTER,*)'  Enter ID for model (0 for key):'
      READ(NTER,*)IMOD
      IF(IMOD.LT.0.OR.IMOD.GT.NMOD)GOTO 80
C Fetch key if necessary (IMOD=0).
      IF(IMOD.EQ.0)CALL STRMOD(NTER,IMOD,TF,STR,*80)
90    DO 120 I=1,NDIM
C Exclude GF specifically.
        IF(I.EQ.4)GOTO 120
        IDROP(I)=0
100     WRITE(NTER,110)PROPI(I)
110     FORMAT(6X,'Enter low and high limits for ',A3)
        READ(NTER,*)PLOI(I),PHII(I)
        IF(ABS(PHII(I)-PLOI(I)).LT.TINY)IDROP(I)=1
        IF(PHII(I).LT.PLOI(I).OR.(I.NE.5.AND.PLOI(I).LT.TINY))THEN
          WRITE(NTER,*)' >>>>> Check entries!'
          GOTO 100
        ENDIF
120   CONTINUE
      J=0
      DO 130 I=1,NDIM
        IF(IDROP(I).EQ.1)THEN
          Q(I)=PLOI(I)
        ELSE
```

```
          J=J+1
          PROP(J)=PROPI(I)
          IF(I.EQ.1.OR.I.EQ.3.OR.I.EQ.4)THEN
            PLO(J)=LOG10(PLOI(I))
            PHI(J)=LOG10(PHII(I))
          ELSE
            PLO(J)=PLOI(I)
            PHI(J)=PHII(I)
          ENDIF
        ENDIF
130   CONTINUE
      ND=J
      DO 140 J=1,ND
        X(J)=0.D0
140   CONTINUE
      IWRITE=1
      WRITE(NTER,'(/2X,A)')'Computing center point:'
      DUMMY=FUNC(X)
      WRITE(NTER,'(/2X,A)')'Guess new starting values? (Y/N)'
      READ(NTER,'(A1)')ANS
      IF(ANS.EQ.'Y'.OR.ANS.EQ.'y')GOTO 90
C Set up for tracking of parameter and function values.
      NCALL1=0
      NCALL2=0
C Call Powell's method minimization.
      DO 200 K=1,NACC,2
        FTOL=10.D0**(-K)
        IWRITE=0
        WRITE(NTER,*)' Display parameters at each function call? (Y/N)'
        READ(NTER,'(A1)')ANS
        IF(ANS.EQ.'Y'.OR.ANS.EQ.'y')IWRITE=1
        WRITE(NTER,*)' Computing!'
        CALL POWELL(X,XI,ND,NDIM,FTOL,ITER,FRET)
        DF=FLOAT(NP-ND)
C SEE will be relative standard error of estimate, based on the residual
C divided by the stress at that point.
        SEE=SQRT(FRET/DF)
        WRITE(NTER,150)FRET,SEE
150     FORMAT(/12X,'SS = ',1P,E10.3,4X,'SEE =',E10.3//)
C Set starting values for confidence limits.
        DO 160 J=1,ND
          BLO(J)=PHI(J)
          BHI(J)=PLO(J)
160     CONTINUE
        CALL CCEROR(ND,DF,BLO,BHI,ICOUNT)
        DO 170 J=1,ND
          PP=PLO(J)+(PHI(J)-PLO(J))*EXP(X(J))/(1.D0+EXP(X(J)))
          IF(PROP(J).EQ.'G   '.OR.PROP(J).EQ.'TAU')THEN
            WRITE(NTER,230)PROP(J),10.D0**PP,10.D0**BLO(J),10.D0**BHI(J)
          ELSE
            WRITE(NTER,240)PROP(J),PP,BLO(J),BHI(J)
          ENDIF
170     CONTINUE
        WRITE(NTER,260)
        DO 180 I=2,NP,MAX(1,NP/6)
          RES(I)=(STRSS(I)-STRCAL(I))/STRCAL(I)
          WRITE(NTER,270)T(I),XLEN(I),STRSS(I),STRCAL(I),RES(I)
180     CONTINUE
        WRITE(NTER,190)NLARGE*NCALL2+NCALL1,ICOUNT
190     FORMAT(/2X,'Number of function calls =',I6
     &         /9X,'Calls in 95% C.I. =',I6)
        IF(K.LT.NACC) THEN
          WRITE(NTER,'(/2X,A)')'Go for higher accuracy? (Y/N)'
          READ(NTER,'(A1)')ANS
          IF(ANS.EQ.'N'.OR.ANS.EQ.'n')GOTO 210
        ENDIF
200   CONTINUE
210   WRITE(NOUT,220)IMOD
220   FORMAT(/26X,'Model',I3,' results:'/)
```

```
      DO 250 J=1,ND
        PP=PLO(J)+(PHI(J)-PLO(J))*EXP(X(J))/(1.D0+EXP(X(J)))
        IF(PROP(J).EQ.'G  '.OR.PROP(J).EQ.'TAU')THEN
          WRITE(NOUT,230)PROP(J),10.D0**PP,10.D0**BLO(J),10.D0**BHI(J)
230       FORMAT(11X,A3,' =',1P,E10.3,';   C.I. =',E10.3,'  to',E10.3)
        ELSE
          WRITE(NOUT,240)PROP(J),PP,BLO(J),BHI(J)
240       FORMAT(11X,A3,' =',F10.3,';   C.I. =',F10.3,'  to',F10.3)
        ENDIF
250   CONTINUE
      WRITE(NOUT,260)
260   FORMAT(//T6,'Time',T20,'Strain',T34,'Stress',T46,'Calc. Stress',
     &       T63,'Residual'/
     &       T4,8('-'),T18,9('-'),T32,10('-'),T46,12('-'),T62,10('-'))
      RES(1)=0.D0
      WRITE(NOUT,270)T(1),XLEN(1),STRSS(1),STRCAL(1),RES(1)
      DO 280 I=2,NP
        RES(I)=(STRSS(I)-STRCAL(I))/STRCAL(I)
        WRITE(NOUT,270)T(I),XLEN(I),STRSS(I),STRCAL(I),RES(I)
270     FORMAT(F11.2,F15.3,1P,E15.3,E15.3,E15.3)
280   CONTINUE
      WRITE(NTER,'(/3(2X,A/))')'Results written to file.',
     &     'Try another model (M) or starting point (S)?',
     &     '(Enter Q to Quit)'
      READ(NTER,'(A1)')ANS
      IF(ANS.EQ.'S'.OR.ANS.EQ.'s')GOTO 90
      IF(ANS.EQ.'M'.OR.ANS.EQ.'m')GOTO 80
      CALL MSTAIL(*10)
      STOP
      END
C
      SUBROUTINE CCEROR(ND,DF,BLO,BHI,ICOUNT)
C Returns estimate of the 95% confidence limits on the parameters.
      IMPLICIT DOUBLE PRECISION (A-H,O-Z)
      PARAMETER (NDIM=5,NLARGE=500)
      DIMENSION BLO(NDIM),BHI(NDIM),CCHI(6)
      COMMON /BLOC7/ PI(NDIM,NLARGE),PHI(NDIM),PLO(NDIM),SSI(NLARGE),
     &              NCALL1,NCALL2
C Values for Δχ for degrees of freedom = 1 to 6, 95% confidence level.
      DATA (CCHI(I),I=1,6)/4.00, 6.17, 8.02, 9.07, 11.3, 12.8/
C Find lowest value of SS to use as estimate of intrinsic error.
      SSBEST=SSI(NCALL1)
      IBEST=NCALL1
      IF(NCALL2.EQ.0)THEN
        NC=NCALL1
      ELSE
        NC=NLARGE
      ENDIF
      DO 10 I=1,NC
        IF(SSI(I).LT.SSBEST)THEN
          SSBEST=SSI(I)
          IBEST=I
        ENDIF
10    CONTINUE
C Find SS values within critical χ (CCHI) range.
      ICOUNT=0
      DO 30 I=1,NC
        IF(SSI(I)/SSBEST.LT.1.D0+CCHI(ND)/DF)THEN
          ICOUNT=ICOUNT+1
C Find if parameter values are lower or higher than those found before.
          DO 20 J=1,ND
            IF(PI(J,I).GT.BHI(J))BHI(J)=PI(J,I)
            IF(PI(J,I).LT.BLO(J))BLO(J)=PI(J,I)
20        CONTINUE
        ENDIF
30    CONTINUE
      RETURN
      END
```

```
      DOUBLE PRECISION FUNCTION FUNC(X)
C Supplies the function value to be minimized to the the Numerical Recipes
C  function F1DIM as required by POWELL (Press et al., Cambridge University
C  Press, Cambridge, 1986, 1992).
C Requires the subroutines STRMOD and UNITS.
      IMPLICIT DOUBLE PRECISION (A-H,O-Z)
      PARAMETER (NDIM=5,NLARGE=500,NMAX=201)
      DIMENSION X(NDIM),P(NDIM),Q(NDIM)
      CHARACTER*1 RESS(NMAX)
      COMMON /BLOC1/ A(NMAX),T(NMAX),RT(NMAX),XLEN(NMAX),NP
      COMMON /BLOC2/ Q
      COMMON /BLOC3/ C,XN,ID,IJ
      COMMON /BLOC5/ STRSS(NMAX),STRCAL(NMAX),RES(NMAX)
      COMMON /BLOC6/ IDROP(NDIM),IMOD,IWRITE,ND
      COMMON /BLOC7/ PI(NDIM,NLARGE),PHI(NDIM),PLO(NDIM),SSI(NLARGE),
     &               NCALL1,NCALL2
      CALL UNITS(NTER,NINT,NOUT)
      NCALL1=NCALL1+1
C Transform within allowed range.
      DO 10 J=1,ND
         XEX=EXP(X(J))
         P(J)=PLO(J)+(PHI(J)-PLO(J))*XEX/(1.D0+XEX)
10    CONTINUE
C Assign to correct material properties.
      J=0
      DO 20 I=1,NDIM
C Leave material properties untouched if they have been dropped.
        IF(IDROP(I).EQ.1)GOTO 20
        J=J+1
        Q(I)=P(J)
        IF(I.EQ.1.OR.I.EQ.3.OR.I.EQ.4)Q(I)=10.D0**Q(I)
20    CONTINUE
      SS=0.D0
      DO 30 I=2,NP
        RESS(I)='+'
        CALL STRMOD(NTER,IMOD,T(I),STRCAL(I),*70)
        IF(STRCAL(I).GT.STRSS(I))RESS(I)='-'
        SS=SS+((STRCAL(I)-STRSS(I))/STRSS(I))**2
30    CONTINUE
      FUNC=SS
      IF(IWRITE.EQ.1)THEN
        WRITE(NTER,40)Q(1),Q(2),Q(3),Q(5),SS
40      FORMAT(2X,'G,B,TAU,XNL =',1P,E12.4,0P,F10.5,1P,E12.4,0P,F10.5,
     &            '; SS =',1P,E12.4)
        WRITE(NTER,50)(RESS(I),I=2,NP)
50      FORMAT(8X,32(' ',A1)/)
      ENDIF
      IF(NCALL1.GT.NLARGE)THEN
C In this case write new values at top of array over the old ones.
        NCALL1=1
        NCALL2=NCALL2+1
      ENDIF
      SSI(NCALL1)=SS
      DO 60 J=1,ND
        PI(J,NCALL1)=P(J)
60    CONTINUE
      RETURN
70    WRITE(NTER,*)' Model selected is unavailable.'
      STOP
      END
```

10.7 Sample Listings of INPUT/OUTPUT Files

10.7.1 Data File for STRESS

Shown here is a data file, in the format required for running STRESS, to calculate the classical stress-growth function for a Lodge rubberlike liquid stretched at 1 s^{-1}. The precision of the deformations and the point spacing are necessary for accurate results in exponential flow.

```
Simple extension, constant-rate startup experiment.
   1.0  0.5  1.  0.  2.          G, B, TAU, GF, XNL
   0.5  1    1                   C, ID, IJ
   0.0  1.   0.                  TIME, DEFORMATION, PATH SHAPE
  10.   22026.46579  9999.       TIME, DEFORMATION, PATH SHAPE
   0.01 3.16227766               FIRST POINT; POINT SPACING MULTIPLIER
   2                             MODEL NUMBER
```

10.7.2 Data File for STRESS Using Batch Mode

This sample data file has seven calculations stacked to run in STRESS in batch mode. The results are shown in Table 10.5.

```
Simple shear, constant-rate, shear stress, high strain.
   1. .5  1.  0.  2.             G, B, TAU, GF, XNL
   0.  1  0                      C, ID, IJ
   0.  0. 1.                     TIME, DEFORMATION, PATH SHAPE
1000.   100.  9999.              TIME, DEFORMATION, PATH SHAPE
1000.  2.                        FIRST POINT; POINT SPACING MULTIPLIER
  2                              MODEL NUMBER
Simple shear, constant-rate, shear stress, high strain.
   1. .5  1.  0.  2.
   0.  1  0
   0.  0. 1.
1000.  1000.  9999.
1000.  2.
  2
Simple shear, constant-rate, shear stress, high strain.
   1. .5  1.  0.  2.
   0.  1  0
   0.  0. 1.
1000.  10000. 9999.
1000.  2.
  2
Simple shear, constant-rate, shear stress, high strain.
   1. .5  1.  0.  2.
   0.  1  0
   0.  0. 1.
1000.  100000. 9999.
1000.  2.
  2
Simple shear, constant-rate, shear stress, high strain.
   1. .5  1.  0.  2.
   0.  1  0
   0.  0. 1.
1000. 1000000. 9999.
1000.  2.
  2
```

10.7.3 Data File for **FITSTR**

Shown here is a data file in the format required for running FITSTR.

```
Simple shear, constant-rate startup, shear stress.
   0.   0    0                                         C, ID, IJ
   0.   0.   0.
 .1000E+00  .1000E+00  .9516E-01
 .2000E+00  .2000E+00  .1813E+00
 .4000E+00  .4000E+00  .3297E+00
 .8000E+00  .8000E+00  .5507E+00
 .1600E+01  .1600E+01  .7981E+00
 .3200E+01  .3200E+01  .9592E+00
 .6400E+01  .6400E+01  .9983E+00
 .1000E+02  .1000E+02  .1000E+01
```

10.7.4 Terminal Dialog with **FITSTR**

Reproduced here is the terminal dialog for running FITSTR with the data file FITSTR.DAT listed in Section 10.7.3. The string of + signs refer to the signs of the residuals at the starting point and indicate that the starting parameter values lead to stresses that are all lower than the data.

```
>FITSTR
 Enter name of data file (N, if unavailable).
FITSTR.DAT
 Results to be written to NEW file: FITSTR.RES
 OK (Y/N)?
Y
       1                  1.00
       2                  1.00
       3                  1.00
       4                  1.00
       5                  1.00
       6                  1.00
       7                  1.00
       8                  1.00
       9                  1.00

 Above are approximate path shapes. Computational speed can be improved
 by setting to 0.0 or 1.0;  if path shapes are close to one of these it
 will probably be beneficial to set them all to this value. To do this,
 enter 0.0 or 1.0; otherwise, a large number.
1.0

 Simple shear, constant-rate startup, shear stress.

  Number of points =              9
  Enter ID for model (0 for key):
2
     Enter low and high limits for G
0.1  2.0
     Enter low and high limits for B
0.3  1.0
     Enter low and high limits for TAU
0.3  1.5
     Enter low and high limits for XNL
1.0  3.0

 Computing center point:
```

```
 G,B,TAU,XNL =  4.4721E-01  .65000  6.7082E-01  2.00000;  SS =  3.2196E+00
        + + + + + + + +

 Guess new starting values? (Y/N)
N
 Display parameters at each function call? (Y/N)
N
 Computing!

          SS = 1.928E-02       SEE = 6.210E-02

          G   = 1.199E+00;   C.I. = 1.179E+00  to 1.230E+00
          B   =      .720;   C.I. =      .650  to      .777
          TAU = 6.801E-01;   C.I. = 6.631E-01  to 7.174E-01
          XNL =     1.994;   C.I. =     1.959  to     2.000

     Time          Strain          Stress        Calc. Stress      Residual
   --------      ---------       ----------      ------------     ----------
       .10          .100         9.516E-02        1.028E-01       -7.411E-02
       .20          .200         1.813E-01        1.874E-01       -3.266E-02
       .40          .400         3.297E-01        3.241E-01        1.728E-02
       .80          .800         5.507E-01        5.177E-01        6.373E-02
      1.60         1.600         7.981E-01        7.401E-01        7.837E-02
      3.20         3.200         9.592E-01        9.235E-01        3.866E-02
      6.40         6.400         9.983E-01        1.014E+00       -1.523E-02
     10.00        10.000         1.000E+00        1.031E+00       -3.002E-02

 Number of function calls =     91
        Calls in 95% C.I. =     56

 Go for higher accuracy? (Y/N)
N
```

10.7.5 Terminal Display for **FITSTR** Data in Section 10.7.3

The results using a more favorable starting point than that shown in Section 10.7.4. `SS` is the sum of squares of the relative residuals, `SEE` is the standard error of estimate of the relative stress, and `C.I.` is the 95% confidence interval.

```
         SS =  2.227E-08       SEE = 6.673E-05

         G   =  1.000E+00;   C.I. = 9.999E-01  to 1.000E+00
         B   =       .500;   C.I. =      .500  to      .501
         TAU =  1.001E+00;   C.I. = 1.001E+00  to 1.002E+00
         XNL =      1.996;   C.I. =     1.996  to     1.996

     Time          Strain          Stress        Calc. Stress      Residual
   --------      ---------       ----------      ------------     ----------
       .10          .100         9.516E-02        9.517E-02       -7.062E-05
       .20          .200         1.813E-01        1.813E-01        1.124E-04
       .40          .400         3.297E-01        3.297E-01       -2.189E-05
       .80          .800         5.507E-01        5.507E-01       -3.122E-05
      1.60         1.600         7.981E-01        7.981E-01        2.927E-06
      3.20         3.200         9.592E-01        9.592E-01        2.441E-05
      6.40         6.400         9.983E-01        9.983E-01       -4.296E-05
     10.00        10.000         1.000E+00        1.000E+00        2.736E-05

 Number of function calls =    404
        Calls in 95% C.I. =     54
```

10.7.6 Output File from **FITSTR** Data with 1% Random Noise

Reproduced here is an output file from FITSTR with the results for a problem similar to the one in Section 10.7.3, but with 1% random noise added. The noise shows up in larger residuals and poorer fit.

```
Simple shear, constant-rate startup, shear stress, 1% noise.

                         Model  2  results:

          G   = 8.979E-01;   C.I. = 8.483E-01  to 9.821E-01
          B   =      .419;   C.I. =      .361  to      .535
          TAU = 1.175E+00;   C.I. = 9.865E-01  to 1.272E+00
          XNL =     2.001;   C.I. =     2.001  to     2.001

     Time         Strain        Stress      Calc. Stress     Residual
   --------     ---------    -----------   ------------    ----------
        .00          .000    0.000E+00       0.000E+00      0.000E+00
        .10          .100    8.378E-02       8.765E-02     -4.418E-02
        .20          .200    1.798E-01       1.700E-01      5.745E-02
        .40          .400    3.208E-01       3.176E-01      1.002E-02
        .80          .800    5.435E-01       5.477E-01     -7.680E-03
       1.60         1.600    8.054E-01       8.114E-01     -7.414E-03
       3.20         3.200    9.605E-01       9.685E-01     -8.252E-03
       6.40         6.400    9.965E-01       9.941E-01      2.436E-03
      10.00        10.000    1.003E+00       9.944E-01      8.626E-03
```

10.7.7 Output File from **FITSTR** Using Published Data

Results for data on the stretching of SBR gum in an Instron machine (see Figure 4 of Reference 10). An indication of a good fit is the six changes in sign of the residuals.

```
SBR extension, Chang, Bloch and Tschoegl (J. Polym. Sci., PPE, 1977)

                         Model  2  results:

          G   = 4.733E+05;   C.I. = 4.732E+05  to 4.762E+05
          B   =     2.492;   C.I. =     2.492  to     2.500
          TAU = 3.168E+00;   C.I. = 3.118E+00  to 3.169E+00
          XNL =     1.357;   C.I. =     1.333  to     1.357

     Time         Strain        Stress      Calc. Stress     Residual
   --------     ---------    -----------   ------------    ----------
        .00         1.000    0.000E+00       0.000E+00      0.000E+00
      90.53         1.066    2.043E+04       1.877E+04      8.842E-02
     259.00         1.190    3.633E+04       3.644E+04     -3.038E-03
     419.30         1.307    4.560E+04       4.713E+04     -3.247E-02
     558.80         1.409    5.288E+04       5.424E+04     -2.501E-02
     783.10         1.573    6.103E+04       6.296E+04     -3.070E-02
    1025.00         1.751    7.032E+04       7.018E+04      1.955E-03
    1239.00         1.907    7.480E+04       7.495E+04     -2.001E-03
    1378.00         2.009    7.798E+04       7.772E+04      3.366E-03
    1506.00         2.103    8.082E+04       7.998E+04      1.051E-02
    1748.00         2.280    8.380E+04       8.357E+04      2.740E-03
    1989.00         2.457    8.620E+04       8.668E+04     -5.547E-03
    2226.00         2.630    8.834E+04       8.915E+04     -9.129E-03
    2475.00         2.812    9.150E+04       9.146E+04      4.822E-04
    2619.00         2.918    9.324E+04       9.273E+04      5.467E-03
    2753.00         3.016    9.517E+04       9.374E+04      1.528E-02
```

10.7.8 Data File for **FITSTR** Using Published Data

Data for the stretching and relaxation of an SBR gum plasticized with silicone oil (see Figure 8 of Reference 11). From about 11 to 611 s (first column), the specimen is held fixed at a stretch ratio of about 1.8 (second column), and then the stretch ratio is decreased.

```
SBR + silicone oil, Fig 8, Bloch, Chang and Tschoegl (J. Rheol., 1978)
0.5  0   0
0.          1.0         0.
 .6253E+00  .1047E+01  .6880E+05
 .9070E+00  .1069E+01  .9869E+05
 .1511E+01  .1114E+01  .1558E+06
 .2152E+01  .1163E+01  .2209E+06
 .3011E+01  .1228E+01  .2927E+06
 .4520E+01  .1342E+01  .4186E+06
 .5978E+01  .1452E+01  .5404E+06
 .7228E+01  .1546E+01  .6300E+06
 .8430E+01  .1637E+01  .7186E+06
 .9011E+01  .1681E+01  .7662E+06
 .9595E+01  .1725E+01  .8077E+06
 .1081E+02  .1817E+01  .9067E+06
 .6108E+03  .1817E+01  .6989E+06
 .6133E+03  .1769E+01  .6447E+06
 .6156E+03  .1726E+01  .5988E+06
 .6167E+03  .1705E+01  .5773E+06
 .6228E+03  .1591E+01  .4653E+06
 .6287E+03  .1479E+01  .3518E+06
 .6347E+03  .1365E+01  .2423E+06
 .6408E+03  .1250E+01  .1300E+06
 .6468E+03  .1137E+01  .1351E+05
```

References

1. Shaw, M. T.; Miller, J. C.: "Simulation of Extrusion Coating by Tape Drawing," *Polym. Eng. Sci.* **19**, 1098-1103 (1979).
2. Seth, B. R.: "Generalized Strain and Transition Concepts for Elastic–Plastic Deformation–Creep and Relaxation," *Proceedings of the XI International Congress of Applied Mechanics.* (H. Görtler, ed.), Springer-Verlag, New York, 1969, pp. 383-389.
3. Kohlrausch, R.: "Theorie des Elestrischen Ruckstandes in der Leidener Flasche," *Pogg. Ann. Physik.* **IV-91**, 56-82, 179-214 (1854).
4. Bird, R. B.; Armstrong, R. C.; Hassager, O.: *Dynamics of Polymeric Liquids.* Wiley-Interscience, New York, 1987, Vol. 1.
5. Larson, R. G.: *Constitutive Equations for Polymer Melts and Solutions.* Butterworths, Boston, 1988.
6. Dealy, J. M.; Wissbrun, K. F.: *Melt Rheology and Plastics Processing.* Van Nostrand Reinhold, New York, 1990.

7. Williams, G.; Watts, D. C.: "Non-Symmetrical Dielectric Relaxation Behavior Arising from a Simple Empirical Decay Function," *Trans. Faraday Soc.* **66**, 80-85 (1971).
8. Knoff, W. F.; Hopkins, I. L.; Tobolsky, A. V.: "Studies on the Stress Relaxation of Polystyrenes in the Rubbery Flow Region. II," *Macromol.* **4**, 750-754 (1971).
9. Blatz, P. J.; Sharda, S. C.; Tschoegl, N. W.: "Strain Energy Function for Rubberlike Materials Based on a Generalized Measure of Strain," *Trans. Soc. Rheol.* **18**, 145-161 (1974).
10. Chang, W. V.; Bloch, R.; Tschoegl, N. W.: "On the Theory of the Viscoelastic Behavior of Soft Polymers in Moderately Large Deformations," *Rheol. Acta* **15**, 367-378 (1976).
11. Chang, W. V.; Bloch, R.; Tschoegl, N. W.: "Study of the Viscoelastic Behavior of Uncrosslinked (Gum) Rubbers in Moderately Large Deformations," *J. Polym. Sci., Polym. Phys. Ed.* **15**, 923-944 (1977).
12. Bloch, R.; Chang, W. V.; Tschoegl, N. W.: "The Behavior of Rubberlike Materials in Moderately Large Deformations," *J. Rheol.* **22**, 1-32 (1978).
13. Shaw, M. T.: "Application of the Seth Generalized Strain Function to Polyethylene Melt Rheology," *Rheol. Acta* **20**, 231-239 (1981).
14. Tanner, R. I.; Simmons, J. M.: "Combined Simple and Sinusoidal Shearing in Elastic Liquids," *Chem. Eng. Sci.* **22**, 1803-1815 (1967).
15. Papanastasiou, A. C.; Scriven, L. E.; Macosko, C. W.: "An Integral Constitutive Equation for Mixed Flows: Viscoelastic Characterization," *J. Rheol.* **27**, 387-410 (1983).
16. Wagner, M. H.: "Zur Netzwerktheorie von Polymer-Schmelzen," *Rheol. Acta* **18**, 33-50 (1979).
17. Press, W. H.; Flannery, B. P.; Teukolsky, S. A.; Vetterling, W. T.: *Numerical Recipes in* FORTRAN: *The Art of Scientific Computing.* Cambridge University Press, Cambridge, 1986.
18. Shaw, M. T.; Lin, F. B.: "Material Functions in Extension Using Arbitrary Deformation Programs," *Current Topics in Polymer Science Volume II: Rheology and Polymer Processing/Multiphase Systems.* R. M. Ottenbrite, L. A. Utracki, S. Inoue, eds.), Hanser, New York, 1987, pp. 137-148.

Appendix 10.1

Path Definition[a] (XL.FOR)

	Shear	Extension
General	$\gamma = \gamma_i + \left[(1 + a_i\lvert\dot{\gamma}_i\rvert t)^{1/a_i} - 1\right] \mathrm{SGN}(\dot{\gamma}_i)$ $\lvert\dot{\gamma}_i\rvert t << 1, \quad \gamma = \gamma_i + \lvert\dot{\gamma}_i\rvert t\, \mathrm{SGN}(\dot{\gamma}_i)$	$\lambda = \lambda_i (1 + a_i\dot{\varepsilon}_i t)^{1/a_i}$ $\dot{\varepsilon}_i t << 1,$ $\lambda = \lambda_i(1 + \dot{\varepsilon}_i t)$
$a_i = 1$	$\gamma = \gamma_i + \lvert\dot{\gamma}_i\rvert t\, \mathrm{SGN}(\dot{\gamma}_i)$	$\lambda = \lambda_i (1 + \dot{\varepsilon}_i t)$
$a_i = 0$	$\gamma = \gamma_i + (e^{\lvert\dot{\gamma}_i\rvert t} - 1)\, \mathrm{SGN}(\dot{\gamma}_i)$	$\lambda = \lambda_i\, e^{\dot{\varepsilon}_i t}$
$a_i = -1$	$\gamma = \gamma_i + \left[\dfrac{1}{1 - \lvert\dot{\gamma}_i\rvert t} - 1\right] \mathrm{SGN}(\dot{\gamma}_i)$	$\lambda = \lambda_i \dfrac{1}{1 - \dot{\varepsilon}_i t}$

[a] Here, t refers to time in excess of t_i, and is always positive.

Path Definition, Initial Rates (RATES.FOR)

	Shear	Extension
General	$\dot{\gamma}_i = \dfrac{\left[(\lvert\gamma_{i+1} - \gamma_i\rvert + 1)^{a_i} - 1\right] \mathrm{SGN}(\gamma_{i+1} - \gamma_i)}{a_i(t_{i+1} - t_i)}$	$\dot{\varepsilon}_i = \dfrac{(\lambda_{i+1}/\lambda_i)^a - 1}{a_i\,(t_{i+1} - t_i)}$
$a_i = 1$	$\dot{\gamma}_i = \dfrac{\lvert\gamma_{i+1} - \gamma_i\rvert\, \mathrm{SGN}(\gamma_{i+1} - \gamma_i)}{t_{i+1} - t_i}$	$\dot{\varepsilon}_i = \dfrac{\lambda_{i+1}/\lambda_i - 1}{t_{i+1} - t_i}$
$a_i = 0$	$\dot{\gamma}_i = \dfrac{\ln(\lvert\gamma_{i+1} - \gamma_i\rvert + 1)\mathrm{SGN}(\gamma_{i+1} - \gamma_i)}{t_{i+1} - t_i}$	$\dot{\varepsilon}_i = \dfrac{\ln(\lambda_{i+1}/\lambda_i)}{t_{i+1} - t_i}$
$a_i = -1$	$\dot{\gamma}_i = \dfrac{1 - 1/(\lvert\gamma_{i+1} - \gamma_i\rvert + 1)}{t_{i+1} - t_i}\, \mathrm{SGN}(\gamma_{i+1} - \gamma_i)$	$\dot{\varepsilon}_i = \dfrac{1 - \lambda_i/\lambda_{i+1}}{t_{i+1} - t_i}$

Appendix A

Program Dependencies

The programs listed in this volume conform to the ANSI standard X3.9-1978 and should be portable to any compiler conforming to this standard. All the programs have been tested using Microsoft® FORTRAN (Version 5.1) in a PC-DOS environment and VS FORTRAN (Release 2 Version 5) in the Conversational Monitor System environment (CMS 5, VM/HPO Release 5 Level 50). The program dependencies, if any, for the routines alphabetically listed in the first column are shown here. The 16 main programs described in this volume are set in boldface type. In the second column, the bracketed characters denote where the source code for the routine in the first column may be found. A number represents the corresponding chapter, the letter `B` is for Appendix B, `NR` is for *Numerical Recipes* (© 1986, 1992), and `IM` is for the IMSL libraries (Versions 1.0 - 2.0). The program dependencies start from the third column onward in alphabetical order. A routine inherent to the execution of a program is placed within parentheses as an indication that the code is imbedded within the code of the program listed in the first column.

Eight subprograms in the first column call a module that varies depending on the application. For each subprogram, we have given the modules distinct names but all starting with the same letter. To illustrate, `GAUSSN` is used by seven main programs but calls a different module for each one. Under `GAUSSN`, we refer to these modules collectively as `F*`. When the main program `CAPSEQ` calls `GAUSSN`, the relevant `F*` module, listed under `CAPSEQ` and using its full name, is `FXCR`.

We introduce here four service routines, found in Appendix B, that were not explained in the text. The first three routines service *all* the main programs in the volume and, for brevity, are not listed as dependents. The routines `MSHEAD` and `MSTAIL` are used to set up program execution and define the input/output (I/O) files for programs compiled using Microsoft® FORTRAN. For testing the programs on an IBM ES/9000 mainframe with VS FORTRAN, we deactivated the program statements calling `MSHEAD` and `MSTAIL`. The third routine, `UNITS`, parameterizes three external unit devices for I/O operations. Thus, there is the option of defining the unit specifiers for the terminal (keyboard and screen), input (data), and output (results) files, as may be required when porting programs from one computer operating system to another. When a subprogram is dependent on `UNITS`, it is specified in the list. The last routine, `STOPP`, is listed here as a dependent because it services all but the Chapter 7 programs. `STOPP` issues a warning that the maximum number of allowable records for the data file is being exceeded and terminates program execution.

AKIMA	[B]						
ALINR	[B]						
AMAX	[B]						
AMAXI	[B]						
AMIN	[B]						
AMINI	[B]						
ASINH	[9]						
BRENT	[NR]	F1DIM					
CALTEN	[1]						
CAPCON	[8]	STOPP	UNITS				
CAPSEQ	[8]	ALINR	CAPCON	EQSOLN	(FSV)	FXCR	GAUSSN
		SLOPE	SORT2				
CAPSIM	[8]	ALINR	CAPCON	DBCONF	EQSOLN	ERROR	(FCAP)
		(FVPC)	FXCR	GAUSSN	SLOPE	SORT2	SPECC
CCEROR	[10]						
CERN1	[10]	STRAIN	XL	XMOD1			
CERN2	[10]	STRAIN	XL	XMEM1			
CERN3	[10]	STRAIN	XL	XMEM2			
CERN4	[10]	STRAIN	XL	XMEM1			
CERN5	[10]	INVAR	STRAIN	XL	XMEM1		
CERN5A	[7]	XMEM1					
CERN6	[10]	STRAIN	SL	XMEM1			
CERN6A	[7]	XMEM1					
CHATH3	[3]	ALINR	AMIN	CTIN	CTOUT	GAUSSN	(FCT3)
		POLYN	SLOPE	SPECC			
CHATH4	[3]	ALINR	AMAX	AMIN	CTIN	CTOUT	(FCT4)
		GAUSSN	POLYN	SLOPE	SPECC		
CONVRT	[5]	AKIMA	STOPP				
CONVSI	[1]						
CSTEDY	[7]						
CTIN	[3]	STOPP					
CTOUT	[3]						
DBCONF	[IM]	F*					
DCURVE	[B]						
DFDHES	[IM]	F*					
DRNNOA	[IM]						
DUMINF	[IM]	FWLF					
DUVMIF	[IM]	FAT1					
EQSOLN	[B]	F*					
ERROR	[B]	DFDHES	F*	MATINV			
EVEN	[4]	AKIMA	AMAX	AMIN	(LNTER)		
FASR	[8]						
FAT	[5]	POLYN	STAT				
FAT1	[5]	FAT					
FCAP	[8]	EQSOLN	(FASR)	(STR)			
FCT3	[3]						
FCT4	[3]						
FITSTR	[10]	(CCEROR)	(FUNC)	POWELL	RATES	STOPP	STRMOD

```
FMR      [ 2]
FOPEN    [ 6]
FORCE    [ 6]
FSLOPE   [ 7]
FSS      [ 1]
FSTR     [ 2]  EQSOLN   (FMR)
FSV      [ 8]
FUNC     [10]  STRMOD   UNITS
FVPC     [ 8]
FWLF     [ 5]
FXBU     [ 9]
FXCA     [ 9]
FXCR     [ 9]
FXEY     [ 9]  (ASINH)
FXGR     [ 9]
FXSU     [ 9]  (ASINH)
F1DIM    [NR]  FUNC
GAMMLN   [NR]
GAUSSI   [ 7]  MTS      PSM
GAUSSN   [ B]  F*       MATINV   MATMUL
INVAR    [10]
LINMIN   [NR]  BRENT    F1DIM    MNBRAK
INTER    [ 4]
LUBKSB   [NR]
LUDCMP   [NR]
MATINV   [ B]  LUBKSB   LUDCMP
MATMUL   [ B]
MEDFIL   [ 1]  (XMED)
MENU     [10]
MNBRAK   [NR]  F1DIM
MPROVE   [NR]  LUBKSB
MSHEAD   [ B]  UNITS
MSTAIL   [ B]  UNITS
MTS      [ 7]  (CERN6A) QROMB2   XMOD1
MWD      [ 4]  ALINR    AMAXI    AMINI    DCURVE   (EVEN)   POLYN
               SLOPE    SORT2    STOPP    (VALUE)
NOISE    [ 5]  DRNNOA
POLINT   [NR]
POLYN    [ B]  LUBKSB   LUDCMP   MPROVE
POWELL   [NR]  FUNC     LINMIN
PSM      [ 7]  (CERN5A) QROMB2   XMOD1
QROMB2   [ B]  C*       TRAPZD   POLINT
RATES    [10]
RESFIL   [ 6]
RHOR     [ 5]
RTF      [10]
SHIFT    [ 5]  AKIMA    ALINR    DUVMIF   ERROR    (FAT)    (FAT1)
               (SUMS)
```

SHIFTT	[5]	DUMINF	ERROR	(FWLF)	NOISE	(RHOR)	(SHIFT)
		STAT	STOPP				
SIMP	[10]	C*	QROMB2				
SLOPE	[B]	POLYN					
SLOPEV	[6]						
SMOOTH	[1]	(MEDFIL)					
SORT	[1]						
SORT2	[B]						
SPECC	[B]						
SQUEEZ	[6]	ALINR	DCURVE	(FOPEN)	(FORCE)	(RESFIL)	(SLOPEV)
		STOPP					
START	[1]	MEDFIL					
STAT	[B]						
STEADY	[7]	(CSTEDY)	QROMB2	(VIS)			
STOPP	[B]						
STR	[8]						
STRAIN	[10]						
STRECH	[2]	ALINR	DBCONF	EQSOLN	ERROR	(FSTR)	SORT2
		STOPP					
STRESS	[10]	(MENU)	RATES	(RTF)	STOPP	STRMOD	XL
STRMOD	[10]	(STRSS1)	(STRSS2)	(STRSS3)	(STRSS4)	(STRSS5)	(STRSS6)
STRSS1	[10]	(CERN1)	SIMP	XL			
STRSS2	[10]	(CERN2)	SIMP	STRAIN	XL	XMOD1	
STRSS3	[10]	(CERN3)	SIMP	STRAIN	XL	XMOD2	
STRSS4	[10]	(CERN4)	INVAR	SIMP	STRAIN	XL	XMOD1
STRSS5	[10]	(CERN5)	INVAR	SIMP	STRAIN	XL	XMOD1
STRSS6	[10]	(CERN6)	INVAR	SIMP	STRAIN	XL	XMOD1
SUMS	[5]						
TENSIL	[1]	AMAXI	(CALTEN)	(CONVSI)	(FSS)	GAUSSN	POLYN
		(SMOOTH)	SPECC	(START)	STAT	STOPP	
TRANST	[7]	(GAUSSI)	(MTS)	(PSM)			
TRAPZD	[NR]	C*					
TWIST	[7]	ALINR	AMAX	AMIN	GAUSSN	(FSLOPE)	FXBU
		FXEY	SORT2				
UNITS	[B]						
VALUE	[4]						
VIS	[7]	ASINH					
VISFIT	[9]	(FXBU)	(FXCA)	(FXCR)	(FXEY)	(FXGR)	(FXSU)
		GAUSSN	SORT2	SPECC	STOPP		
XL	[10]						
XMED	[1]	(SORT)					
XMEM1	[10]						
XMEM2	[10]						
XMOD1	[10]						
XMOD2	[10]	GAMMLN					

Appendix B

Subprogram Listings

Twenty-two subprograms are called to several other routines in the text and are listed in alphabetical order. Refer to Appendix A for program dependencies or to comment lines immediately following the program statement.

B.1 AKIMA

```
      SUBROUTINE AKIMA(NP,X,Y,XI,YI)
C Interpolation based on Akima algorithm from Ruckdeshel, "Basic Scientific
C  Subroutines," Vol.II, p 309. Search method based on bisection.  Modified
C  to allow mild extrapolation. XI is abscissa for interpolation (input);
C  YI is the interpolated value (output) from NP pairs of X,Y data.
      DOUBLE PRECISION S(5),X(NP),Y(NP),Z(2),TINY,XI,YI
      PARAMETER (TINY=1.D-20)
C If only two points, use linear interpolation.
      IF(NP.EQ.2)THEN
        YI=(XI-X(1))*(Y(2)-Y(1))/(X(2)-X(1))+Y(1)
        RETURN
      ENDIF
C Prepare for extrapolation.
      IF(XI.LT.X(1))THEN
        II=1
        GOTO 30
      END IF
      IF(XI.GE.X(NP))THEN
        II=NP-1
        GOTO 30
      END IF
C Find the two data points which bracket XI.
      I1=1
      I2=NP-1
      NIT=IFIX(SQRT(FLOAT(NP)))+1
      DO 10 I=1,NIT
        IMID=(I1+I2)/2
        IF(XI.GE.X(IMID))THEN
          I1=IMID
          IF((XI-X(I1))/(X(I1+1)-X(I1)).LT.TINY)THEN
            YI=(XI-X(I1))*(Y(I1+1)-Y(I1))/(X(I1+1)-X(I1))+Y(I1)
            RETURN
          ENDIF
          IF(XI.LT.X(I1+1))GOTO 20
        ELSE
          I2=IMID
        END IF
10    CONTINUE
20    II=I1
C Set up five slopes.
30    DO 40 I=MAX(II-2,1),MIN(II+2,NP-1)
        S(3+I-II)=(Y(I+1)-Y(I))/(X(I+1)-X(I))
40    CONTINUE
      DO 50 I=NP,II+2
        J=3+I-II
        S(J)=2.D0*S(J-1)-S(J-2)
50    CONTINUE
```

```
      DO 60 I=0,II-2,-1
        J=3+I-II
        S(J)=2.D0*S(J+1)-S(J+2)
60    CONTINUE
C Calculate weighted slopes.
      DO 70 I=1,2
        A=ABS(S(I+3)-S(I+2))
        B=ABS(S(I+1)-S(I))
        IF(A+B.EQ.0.D0)THEN
          Z(I)=(S(I+2)+S(I+1))/2.D0
        ELSE
          Z(I)=(A*S(I+1)+B*S(I+2))/(A+B)
        ENDIF
70    CONTINUE
      A=XI-X(II)
      B=X(II+1)-X(II)
      YI=Y(II)+Z(1)*A+(3.D0*S(3)-2.D0*Z(1)-Z(2))*A*A/B
     & +(Z(1)+Z(2)-2.D0*S(3))*A*A*A/B/B
      RETURN
      END
```

B.2 ALINR

```
      SUBROUTINE ALINR(NP,X,Y,A)
C Performs linear regression analysis for set of NP pairs of X,Y data.
      IMPLICIT DOUBLE PRECISION (A-H,O-Z)
      DIMENSION X(NP),Y(NP),A(11)
      SX=0.D0
      SY=0.D0
      SXX=0.D0
      SYY=0.D0
      SXY=0.D0
      DO 10 I=1,NP
        SX=SX+X(I)
        SY=SY+Y(I)
        SXX=SXX+X(I)*X(I)
        SYY=SYY+Y(I)*Y(I)
        SXY=SXY+X(I)*Y(I)
10    CONTINUE
      XM=FLOAT(NP)*SXX-SX*SX
      XN=FLOAT(NP)*SYY-SY*SY
      XP=NP*SXY-SX*SY
C Compute the parameter estimates: A(1) is the estimated intercept.
      A(1)=(XM*SY-XP*SX)/XM/FLOAT(NP)
C A(2) is the estimated slope.
      A(2)=XP/XM
C Compute the means of the X, A(3), and Y, A(4), data.
      A(3)=SX/FLOAT(NP)
      A(4)=SY/FLOAT(NP)
C Compute the variances of the X, A(5), and Y, A(6) data.
      A(5)=XM/FLOAT(NP)/FLOAT(NP-1)
      A(6)=XN/FLOAT(NP)/FLOAT(NP-1)
C Compute the sum of squares for error, A(7).
      A(7)=0.D0
      DO 20 I=1,NP
        RES=Y(I)-A(1)-A(2)*X(I)
        A(7)=A(7)+RES*RES
20    CONTINUE
C Compute the standard error of estimate, A(8).
C Written to handle NP=2 in the case of sparse capillary flow data.
      IF(NP.EQ.2)THEN
        A(8)=0.D0
      ELSE
        A(8)=SQRT(A(7)/FLOAT(NP-2))
      ENDIF
```

```
C Compute standard error for parameter A(1)
      A(9)=A(8)*SQRT(SXX/XM)
C  and A(2).
      A(10)=A(8)/SQRT(XM/FLOAT(NP))
C Compute the sample correlation coefficient, r.
      A(11)=SQRT(XP*XP/XM/XN)
      RETURN
      END
```

B.3 AMAX

```
      DOUBLE PRECISION FUNCTION AMAX(NP,X)
C Returns maximum value from the array X(NP).
      IMPLICIT DOUBLE PRECISION (A-H,O-Z)
      DIMENSION X(NP)
      AMAX=X(1)
      DO 10 I=1,NP
        IF(X(I).GT.AMAX)AMAX=X(I)
10    CONTINUE
      RETURN
      END
```

B.4 AMAXI

```
      SUBROUTINE AMAXI(NP,X,XMAX,IXMAX)
C Returns maximum value XMAX and index IXMAX from the array X(NP).
      IMPLICIT DOUBLE PRECISION (A-H,O-Z)
      DIMENSION X(NP)
      XMAX=X(1)
      IXMAX=1
      DO 10 I=1,NP
        IF(X(I).GT.XMAX)THEN
          XMAX=X(I)
          IXMAX=I
        ENDIF
10    CONTINUE
      RETURN
      END
```

B.5 AMIN

```
      DOUBLE PRECISION FUNCTION AMIN(NP,X)
C Returns minimum value from the array X(NP).
      IMPLICIT DOUBLE PRECISION (A-H,O-Z)
      DIMENSION X(NP)
      AMIN=X(1)
      DO 10 I=1,NP
        IF(X(I).LT.AMIN)AMIN=X(I)
10    CONTINUE
      RETURN
      END
```

B.6 AMINI

```
      SUBROUTINE AMINI(NP,X,XMIN,IXMIN)
C Returns minimum value XMIN and index IXMIN from the array X(NP).
      IMPLICIT DOUBLE PRECISION (A-H,O-Z)
      DIMENSION X(NP)
      XMIN=X(1)
      IXMIN=1
```

```
      DO 10 I=1,NP
        IF(X(I).LT.XMIN)THEN
          XMIN=X(I)
          IXMIN=I
        ENDIF
10    CONTINUE
      RETURN
      END
```

B.7 DCURVE

```
      SUBROUTINE DCURVE(NP,Y,W,EMP)
C Performs three-point smoothing on equally spaced points.  Minimizes the
C  second derivative defined by three points by moving the center point
C  closer to the arithmetic average depending on input parameter EMP
C  while not affecting longer-range curvature.
      IMPLICIT DOUBLE PRECISION (A-H,O-Z)
      PARAMETER (NPASS=5)
      DIMENSION Y(NP),W(NP)
      A=(1.D0-EMP)/(1.D-7+EMP)
      DO 10 I=1,NP
        W(I)=Y(I)
10    CONTINUE
      DO 30 J=1,NPASS
        WS1=W(1)
        WS2=WS1
        W(1)=(2.D0*W(2)-W(3)+A*Y(1))/(1.D0+A)
        DO 20 I=2,NP-1
          WS1=WS2
          WS2=W(I)
          W(I)=(2.D0*WS1+2.D0*W(I+1)+A*Y(I))/(4.D0+A)
20      CONTINUE
        W(NP)=(2.D0*WS2-WS1+A*Y(NP))/(1.D0+A)
30    CONTINUE
      RETURN
      END
```

B.8 EQSOLN

```
      SUBROUTINE EQSOLN (F,N,A,Y,XI,ER,X)
C Finds X given Y = F(N,X,A) using Newton's method. Evaluates the
C  derivative, DER. Input are the set of parameters, A(N); initial
C  estimate of solution X, XI; and relative error for solution, ER.
      IMPLICIT DOUBLE PRECISION(A-H,O-Z)
      PARAMETER (NITER=50,TINY=1.D-8)
      DIMENSION A(N)
      X=XI
      DO 20 I=1,NITER
        XT=X
        FX=F(N,A,X)
        DX=X*TINY
10      CONTINUE
        DER=(F(N,A,X+DX)-FX)/DX
        IF(DER.EQ.0.)THEN
          DX=1.D1*DX
          GOTO 10
        ENDIF
        X=X+(Y-FX)/DER
        IF(ABS((X-XT)/XT).LT.ER.AND.ABS((FX-Y)/DER/X).LT.ER)RETURN
20    CONTINUE
      ER=1.
      RETURN
      END
```

B.9 ERROR

```
      SUBROUTINE ERROR(FCN,N,B,BSCALE,FX,H,HI,COV)
C Returns N diagonal elements of the covariance matrix COV estimated from
C  the curvature matrix HI.  Uses the IMSL optimization service routine
C  DFDHES to approximate the Hessian matrix H using forward differences at
C  the function value FX.
C Requires the routine MATINV and the IMSL subroutine DFDHES.
      IMPLICIT DOUBLE PRECISION (A-H,O-Z)
      PARAMETER (NPD=10,EPSFCN=0.2221D-15)
      DIMENSION INDX(NPD)
      DIMENSION B(N),BSCALE(N),COV(N),H(N,N),HI(N,N)
      EXTERNAL FCN
C Set up call to IMSL DFDHES.
      LDH=N
      CALL DFDHES(FCN,N,B,BSCALE,FX,EPSFCN,H,LDH)
C Construct the upper triangular part of the Hessian matrix.
      DO 10 I=1,N
      DO 10 J=1,N
        H(I,J)=H(J,I)
10    CONTINUE
C Calculate the curvature matrix from the Hessian matrix.
      DO 20 I=1,N
      DO 20 J=1,N
        H(I,J)=H(I,J)/2.D0
20    CONTINUE
C Find the inverse of the curvture matrix.
      CALL MATINV(N,N,INDX,H,HI)
      DO 30 I=1,N
        COV(I)=HI(I,I)
30    CONTINUE
      RETURN
      END
```

B.10 GAUSSN

```
      SUBROUTINE GAUSSN(FX,CC,N,NP,X,Y,B,A,AT,R,ERR,SEE)
C Fits a quasilinear regression model by the Gauss-Newton iterative method.
C  User-supplied routine FX(N,X,B,F,ERR) returns function value F and
C  partial first derivatives ERR of model with respect to each parameter
C  B(I=1,N) evaluated at X.  CC is input as the convergence criterion.
C  Returns standard error for each parameter as ERR and the standard error
C  of estimate for the model as SEE.
C Requires the routines MATINV and MATMUL.
      IMPLICIT DOUBLE PRECISION (A-H,O-Z)
      PARAMETER (NPD=10)
      DIMENSION X(NP),Y(NP),B(N),ERR(N)
      DIMENSION A(NP,NPD),AT(NPD,NP),ATA(NPD,NPD),ATAI(NPD,NPD)
      DIMENSION ATR(NPD,1),DB(NPD,1),INDX(NPD),R(NP,1)
C Zero the arrays.
      DO 30 I=1,NPD
        ATR(I,1)=0.D0
        DO 10 J=1,NP
           A(J,I)=0.D0
          AT(I,J)=0.D0
10      CONTINUE
        DO 20 K=1,NPD
           ATA(I,K)=0.D0
          ATAI(I,K)=0.D0
20      CONTINUE
30    CONTINUE
C  Set the number of iterations, NIT.
      NIT=2*N**2+2
```

```
      DO 90 K=1,NIT
        DO 40 I=1,NP
          CALL FX(N,X(I),B,F,ERR)
          R(I,1)=Y(I)-F
        DO 40 J=1,N
          A(I,J)=ERR(J)
40      CONTINUE
        DO 50 I=1,N
        DO 50 J=1,NP
          AT(I,J)=A(J,I)
50      CONTINUE
C MATMUL will multiply matrix AT by matrix A to return ATA.
        CALL MATMUL(NPD,NPD,NP,AT,A,ATA)
C MATINV will invert matrix ATA to return ATAI.
        CALL MATINV(N,NPD,INDX,ATA,ATAI)
C MATMUL will multiply matrix AT by matrix R to return ATR.
        CALL MATMUL(1,NPD,NP,AT,R,ATR)
C MATMUL will multiply matrix ATAI by matrix ATR to return DB.
        CALL MATMUL(1,NPD,NPD,ATAI,ATR,DB)
C Determine the parameters B(I) and check for convergence: IFLAG=1.
        IFLAG=1
        DO 60 I=1,N
          B(I)=B(I)+DB(I,1)
          IF(ABS(DB(I,1)/B(I)).GT.CC)IFLAG=2
60      CONTINUE
        IF(IFLAG.EQ.1)THEN
C Determine respective errors ERR(I) in the parameters B(I).
          DO 70 I=1,N
            ERR(I)=SQRT(ATAI(I,I))
70        CONTINUE
C Determine the standard error of estimate, SEE.
          SUM=0.D0
          DO 80 I=1,NP
            SUM=SUM+R(I,1)*R(I,1)
80        CONTINUE
          SEE=SQRT(SUM/FLOAT(NP-N))
          RETURN
        ENDIF
90    CONTINUE
      PAUSE ' The maximum number of iterations was exceeded.'
      STOP
      END
```

B.11 MATINV

```
      SUBROUTINE MATINV(N,NPD,INDX,A,B)
C Inverts matrix A of logical size N, maximum physical array size NPD=10,
C  to return B.  Requires the routines LUDCMP and LUBKSB from Numerical
C  Recipes by Press et al., Cambridge University Press (1986, 1992).
      IMPLICIT DOUBLE PRECISION (A-H,O-Z)
      DIMENSION A(NPD,NPD),B(NPD,NPD),INDX(NPD)
C Set up identity matrix.
      DO 20 I=1,N
        DO 10 J=1,N
          B(I,J)=0.D0
10      CONTINUE
        B(I,I)=1.D0
20    CONTINUE
C Decompose the matrix.
      CALL LUDCMP(A,N,NPD,INDX,D)
C Find inverse by columns.
      DO 30 J=1,N
        CALL LUBKSB(A,N,NPD,INDX,B(1,J))
30    CONTINUE
      RETURN
      END
```

B.12 MATMUL

```
      SUBROUTINE MATMUL(L,M,N,A,B,C)
C Multiplies matrix A(M,N) by matrix B(N,L) to return C(M,L).
      IMPLICIT DOUBLE PRECISION (A-H,O-Z)
      DIMENSION A(M,N),B(N,L),C(M,L)
      DO 30 K=1,L
        DO 20 I=1,M
          SUM=0.D0
          DO 10 J=1,N
C Compute the dot product for element C(I,K).
            SUM=SUM+A(I,J)*B(J,K)
10        CONTINUE
          C(I,K)=SUM
20      CONTINUE
30    CONTINUE
      RETURN
      END
```

B.13 MSHEAD

```
      SUBROUTINE MSHEAD
C Handles data file opening for programs compiled using Microsoft FORTRAN.
C Requires the subroutine UNITS to specify Input/Output units.
      LOGICAL FEXIST,REXIST,FOPEN
      PARAMETER (NMAX=10000)
      CHARACTER ANS*1,CROW*80,CSTAT*3,FDAT*33,FRES*33
      CALL UNITS(NTER,NINT,NOUT)
10    WRITE(NTER,'(2X,A)')
     &     'Enter name of data file (N, if not applicable).'
      READ(NTER,'(A33)')FDAT
      IF((FDAT(1:1).EQ.'N'.OR.FDAT(1:1).EQ.'n').AND.(FDAT(2:2).EQ.' '))
     & GOTO 100
      INQUIRE(FILE=FDAT,EXIST=FEXIST)
      IF(FEXIST)THEN
        OPEN(NINT,FILE=FDAT,STATUS='OLD')
      ELSE
        WRITE(NTER,*)' Cannot find the file ',FDAT
20      WRITE(NTER,*)' Retry (R), Keyboard (K) entry or Quit (Q)?'
        READ(*,'(A1)')ANS
        IF(ANS.EQ.'Q'.OR.ANS.EQ.'q')STOP
        IF(ANS.EQ.'R'.OR.ANS.EQ.'r')GOTO 10
        IF(ANS.EQ.'K'.OR.ANS.EQ.'k')THEN
          OPEN(NINT,FILE=FDAT,STATUS='NEW')
          DO 50 I=1,NMAX
            WRITE(NTER,30)I
30          FORMAT(6X,'Enter data for record:',I5,'; 9999 to quit.')
            READ(NTER,'(A80)')CROW
            IF(CROW(2:4).EQ.'999')THEN
              WRITE(NTER,40)I-1,FDAT
40            FORMAT(6X,I5,' records added to file ',A33)
              REWIND NINT
              GOTO 60
            ELSE
              WRITE(NINT,'(1X,A79)')CROW
            ENDIF
50        CONTINUE
        ELSE
          GOTO 20
        ENDIF
      ENDIF
60    II=INDEX(FDAT,'.')
      IF(FDAT(II+1:II+3).EQ.'RES'.OR.FDAT(II+1:II+3).EQ.'res')GOTO 100
      IF((II).EQ.0)II=INDEX(FDAT,' ')
```

```
      FRES=FDAT(1:II-1)//'.RES'
70    INQUIRE(FILE=FRES,EXIST=REXIST,OPENED=FOPEN)
      IF(FOPEN)THEN
        WRITE(NTER,*)' The file ',FRES(1:INDEX(FRES,' ')),'  is already
     &attached; retry.'
      ELSE
        IF(REXIST)THEN
          CSTAT='OLD'
        ELSE
          CSTAT='NEW'
        ENDIF
        WRITE(NTER,*)' Results to be written to ',CSTAT,' file: ',FRES
        WRITE(NTER,*)' OK? (Y/N)'
        READ(NTER,'(A1)')ANS
        IF(ANS.EQ.'Y'.OR.ANS.EQ.'y')THEN
          OPEN(NOUT,FILE=FRES,STATUS=CSTAT)
          IF(CSTAT.EQ.'OLD')THEN
            DO 80 I=1,NMAX
              READ(NOUT,'(A80)',END=90)CROW
80          CONTINUE
90          BACKSPACE NOUT
          ENDIF
          RETURN
        ENDIF
      ENDIF
100   WRITE(NTER,*)' Enter name for results file.'
      READ(NTER,'(A33)')FRES
      GOTO 70
      END
```

B.14 MSTAIL

```
      SUBROUTINE MSTAIL(*)
C Concludes the execution of a program compiled using Microsoft FORTRAN.
C Requires the subroutine UNITS to specify Input/Output units.
      CHARACTER ANS*1,FNAME*33,FRES*33
      CALL UNITS(NTER,NINT,NOUT)
      INQUIRE (UNIT=NINT,NAME=FNAME)
      INQUIRE (UNIT=NOUT,NAME=FRES)
      CLOSE(NINT)
      CLOSE(NOUT)
      WRITE(NTER,10)FRES
10    FORMAT(/2X,'The results are in ',A33
     &       /2X,'Run another set of data? (Y/N)')
      READ(NTER,'(A1)')ANS
      IF(ANS.EQ.'Y'.OR.ANS.EQ.'y')THEN
        RETURN 1
      ELSE
        RETURN
      ENDIF
      RETURN
      END
```

B.15 POLYN

```
      SUBROUTINE POLYN(N,NP,XP,YP,B,ERR,RES,SEE)
C Uses LU decomposition to fit a (N-1)th-order polynomial to NP pairs of
C  XP, YP data.  Returns N coefficients, corresponding standard errors
C  and residuals in vectors B, ERR and RES, respectively; and the
C  standard error of estimate SEE.
C Requires the LU matrix decomposition routines from Numerical Recipes
C  by Press et al., Cambridge University Press (1986, 1992), Chapter 2:
C  LUDCMP, LUBKSB and MPROVE.
```

```
      IMPLICIT DOUBLE PRECISION (A-H,O-Z)
      PARAMETER (NPD=10)
      INTEGER INDX(NPD)
      DIMENSION A(NPD,NPD),AO(NPD,NPD),AI(NPD),BO(NPD),B(N),ERR(N)
      DIMENSION RES(NP),XP(NP),YP(NP),DA(2*NPD),DB(NPD)
C Zero the arrays.
      DO 10 I=1,N
        DB(I)=0.D0
10    CONTINUE
      DO 20 I=1,2*N-1
        DA(I)=0.D0
20    CONTINUE
      DO 50 I=1,NP
        P=XP(I)
C Determine column vector.
        DB(1)=DB(1)+YP(I)
        DO 30 J=2,N
          DB(J)=DB(J)+YP(I)*P**(J-1)
30      CONTINUE
C Determine sum of powers XP(I).
        DA(1)=DA(1)+1.D0
        DO 40 J=2,2*N-1
          DA(J)=DA(J)+P**(J-1)
40      CONTINUE
50    CONTINUE
      DO 70 I=1,N
        B(I)=DB(I)
C Preserve original vector B(I) as BO(I,J) to be input in MPROVE.
        BO(I)=DB(I)
        DO 60 J=1,N
          A(I,J)=DA(I+J-1)
C Preserve original matrix A(I,J) as AO(I,J) to be input in MPROVE.
          AO(I,J)=DA(I+J-1)
60      CONTINUE
70    CONTINUE
C Solve the linear set of equations by LU decomposition.
      CALL LUDCMP(A,N,NPD,INDX,D)
      CALL LUBKSB(A,N,NPD,INDX,B)
C Improve solution B by iteration method (just once).
      CALL MPROVE(AO,A,N,NPD,INDX,BO,B)
C Determine residuals RES and standard error of estimate SEE.
      SUM=0.D0
      DO 90 I=1,NP
        YC=B(1)
        DO 80 J=1,N-1
          YC=YC+B(J+1)*XP(I)**J
80      CONTINUE
        RES(I)=YP(I)-YC
        SUM=SUM+RES(I)*RES(I)
90    CONTINUE
      SEE=SQRT(SUM/FLOAT(NP-N))
C Extract covariance matrix AI; calculate parameter errors ERR.
      DO 110 I=1,N
        DO 100 J=1,N
          AI(J)=0.D0
100     CONTINUE
        AI(I)=1.D0
        CALL LUBKSB(A,N,NPD,INDX,AI)
        ERR(I)=SEE*SQRT(AI(I))
110   CONTINUE
      RETURN
      END
```

B.16 QROMB2

QROMB2 is a slightly modified version of the routine QROMB available from *Numerical Recipes: The Art of Scientific Computing* and published by Cambridge University Press. Reproduced by permission. Copyright © 1986, 1992 Numerical Recipes Software.

```
      SUBROUTINE QROMB2(FUNC,A,B,SS,EPS)
C Integrates a function FUNC from A to B using Romberg's method.
C  Slightly modified from the version found in Numerical Recipes: The Art
C  of Scientific Computing. Reproduced by permission. Copyright (C) 1986,
C  1992 Numerical Recipes Software.
C Requires the routines TRAPZD and POLINT from Numerical Recipes.
C Note that the allowed error (EPS) is an argument; not a parameter.
      IMPLICIT DOUBLE PRECISION (A-H,O-Z)
      PARAMETER (JMAX=30,JMAXP=JMAX+1,K=5,KM=K-1,ZERO=0.D0)
      DIMENSION S(JMAXP),H(JMAXP)
      EXTERNAL FUNC
      IF(A.EQ.B)THEN
        SS=0.D0
        RETURN
      ENDIF
      H(1)=1.D0
      DO 10 J=1,JMAX
        CALL TRAPZD(FUNC,A,B,S(J),J)
        IF(J.GE.2.AND.ABS(S(J)).EQ.ZERO)THEN
          SS=S(J)
C Dangerous to return on absolute value, but OK for the integrals in
C  this application; means strain is zero or memory very low.
          RETURN
        ENDIF
        IF(J.GE.K)THEN
          L=J-KM
          CALL POLINT(H(L),S(L),K,ZERO,SS,DSS)
          IF(ABS(DSS).LE.EPS*ABS(SS))RETURN
        ENDIF
        S(J+1)=S(J)
        H(J+1)=0.25D0*H(J)
10    CONTINUE
      PAUSE ' Too many steps in QROMB2.'
      RETURN
      END
```

B.17 SLOPE

```
      SUBROUTINE SLOPE(NP,X,Y,NST,N,S,RES)
C Returns slope S for NST non-evenly spaced points in a moving strip using
C  a (N-1)th order polynomial fit for NP pairs of X,Y data.  Maximum order
C  for polynomial function is 4.
C Requires the subroutines POLYN and UNITS.
      IMPLICIT DOUBLE PRECISION(A-H,O-Z)
      PARAMETER (NPD=5)
      DIMENSION A(NPD),AERR(NPD),RES(NP),S(NP),X(NP),Y(NP)
      CALL UNITS(NTER,NINT,NOUT)
      IF(N.GT.NPD)THEN
        WRITE(NTER,*)' Order too high; maximum is fourth order.'
        RETURN
      ENDIF
      IF(NST.GT.NP)THEN
        WRITE(NTER,*)' Strip has more points than allowed.'
        RETURN
      ENDIF
```

```
      IF(NST.EQ.2)THEN
C Just do finite differences.
        DO 10 I=1,NP-1
          S(I)=(Y(I+1)-Y(I))/(X(I+1)-X(I))
10      CONTINUE
        S(NP)=S(NP-1)
        RETURN
      ENDIF
      IF(N.GE.NST)THEN
        WRITE(NTER,*)
     &       ' Polynomial order must be LT number of points in strip)'
        RETURN
      ENDIF
      NS2=NST/2
C Do case when strip is as large as the total number of points;
C  just one polynomial.
      IF(NP-NS2.LT.NS2+1)THEN
        CALL POLYN(N,NP,X,Y,A,AERR,RES,SEE)
        DO 30 I=1,NP
          S(I)=A(2)
          DO 20 J=3,N
            S(I)=S(I)+(J-1)*A(J)*X(I)**(J-2)
20        CONTINUE
30      CONTINUE
        RETURN
      ENDIF
      DO 90 I=NS2+1,NP-NS2
        CALL POLYN(N,NST,X(I-NS2),Y(I-NS2),A,AERR,RES,SEE)
        S(I)=A(2)
        DO 40 J=3,N
          S(I)=S(I)+(J-1)*A(J)*X(I)**(J-2)
40      CONTINUE
        IF(I.EQ.NS2+1)THEN
          DO 60 K=1,NS2
            S(K)=A(2)
            DO 50 J=3,N
              S(K)=S(K)+(J-1)*A(J)*X(K)**(J-2)
50          CONTINUE
60        CONTINUE
        ENDIF
        IF(I.EQ.NP-NS2)THEN
          DO 80 K=NP-NS2+1,NP
            S(K)=A(2)
            DO 70 J=3,N
              S(K)=S(K)+(J-1)*A(J)*X(K)**(J-2)
70          CONTINUE
80        CONTINUE
          RETURN
        ENDIF
90    CONTINUE
      RETURN
      END
```

B.18 SORT2

```
      SUBROUTINE SORT2(NP,X,Y,U,V)
C Returns sorted array X and corresponding rearrangement of array Y in
C  arrays U and V via Box-Lacey Combsort routine: BYTE Magazine, (McGraw-
C  Hill, April 1991, p. 315) plus a revision for IGAP calculation by
C  Hills, BYTE, (August 1991, p. 18).
      IMPLICIT DOUBLE PRECISION (A-H,O-Z)
      DIMENSION X(NP),Y(NP),U(NP),V(NP)
      IGAP=NP
```

```
      DO 10 I=1,NP
        U(I)=X(I)
        V(I)=Y(I)
10    CONTINUE
      DO 30 K=1,NP
C Hills revision, integer math.
        IGAP=MAX(IGAP*8/11,1)
        ISWIT=0
        DO 20 I=1,NP-IGAP
          J=I+IGAP
          IF(U(I).GT.U(J))THEN
            HOLDU=U(I)
            HOLDV=V(I)
            U(I)=U(J)
            V(I)=V(J)
            U(J)=HOLDU
            V(J)=HOLDV
            ISWIT=ISWIT+1
           ENDIF
20      CONTINUE
        IF(ISWIT.EQ.0.AND.IGAP.EQ.1)RETURN
30    CONTINUE
      RETURN
      END
```

B.19 SPECC

```
      SUBROUTINE SPECC(NTER,CC)
C Provides option of specifying convergence criterion.
C Input is NTER - the unit specified for the terminal device.
      CHARACTER ANS*1
      DOUBLE PRECISION CC
      CC=1.D-4
      WRITE(NTER,*)
     &     ' Use the default convergence criterion, 1.E-4? (Y/N)'
      READ(NTER,'(A1)')ANS
      IF(ANS.EQ.'N'.OR.ANS.EQ.'n') THEN
        WRITE(NTER,*)' Specify convergence criterion.'
        READ(NTER,*)CC
      ENDIF
      RETURN
      END
```

B.20 STAT

```
      SUBROUTINE STAT(NP,Y,AVE,SDY)
C Returns the average (AVE) and standard deviation (SDY) of vector Y(NP).
      IMPLICIT DOUBLE PRECISION (A-H,O-Z)
      DIMENSION Y(NP)
      AVE=0.D0
      DO 10 I=1,NP
        AVE=AVE+Y(I)
10    CONTINUE
      AVE=AVE/FLOAT(NP)
      SS=0.D0
      DO 20 I=1,NP
        SS=SS+(Y(I)-AVE)**2
20    CONTINUE
      SDY=SQRT(SS/FLOAT(NP-1))
      RETURN
      END
```

B.21 STOPP

```
      SUBROUTINE STOPP(NTER,NMAX)
C Stops program execution when allotted number of records for the data
C  file NMAX is exceeded.
C Input is NTER - the unit specified for the terminal device.
      WRITE(NTER,10)NMAX
10    FORMAT(1X,'The maximum allowable number of data points (',I4,
     &          ') was exceeded.')
      STOP
      END
```

B.22 UNITS

```
      SUBROUTINE UNITS(NTER,NINT,NOUT)
C Specifies the unit to which a file is connected.
      INTEGER NTER,NINT,NOUT
C NTER is the unit specified for the keyboard and screen terminals.
      NTER=0
C NINT is the unit specified for the input file.
      NINT=1
C NOUT is the unit specified for the output file.
      NOUT=2
      RETURN
      END
```

Author Index

Subject Index

This diskette contains the software discussed in the book *Computer Programs for Rheologists* by Glenn V. Gordon and Montgomery T. Shaw, Carl Hanser Verlag, 1994.

The purchase of this book is a license to install the accompanying software on a single, DOS-based computer. The three executable files in the diskette (EXEC.EXE, SOURCE.EXE, and DATA.EXE) contain the files from the book in a compressed format and must be copied onto a hard disk of a DOS-based computer and run before these files can be used. When expanded, the files will require approximately 1.6 megabytes of hard-disk storage space. To effect the expansion, simply enter the appropriate name (i.e., EXEC, SOURCE, or DATA) at the DOS prompt (do not include the extension EXE). Running EXEC will produce 16 executable files, running SOURCE will produce 51 source (FORTRAN) files, and running DATA will produce 13 sample data files. Of these, only the files from EXEC are needed to run your own data.

The 16 executable programs available here were compiled with Microsoft FORTRAN (Version 5.1) in a DOS environment with the floating-point option which uses a math coprocessor if one is installed in your computer. Imbedded by permission in the executable codes are routines from the Numerical Recipes Software (Copyright © 1986, 1992) and the IMSL Libraries (Copyright © 1992 Visual Numerics, Inc.).

The 51 source codes here do not include the published routines from the book *Numerical Recipes in* FORTRAN: *The Art of Scientific Computing,* by Press et al., Cambridge University Press (Copyright © 1986, 1992), or the routines from IMSL Libraries (Copyright © 1992 Visual Numerics, Inc.). The codes conform to the American National Standards Institute (ANSI) standard X3.9-1978 and are compatible with IBM VS FORTRAN.

To serve as examples, 13 data files are provided. The file MWD.DAT is adapted by permission from a publication by W. Pfandl and F. R. Schwarzl: "Relation Between Rheological Properties and the Molecular Weight Distribution of a Technical Polystyrene," *Colloid Polym. Sci.,* **263**, 328-334 (1985). Copyright © 1985 by Springer-Verlag. The file SHIFTT.DAT is adapted in cooperation with John Wiley & Sons, Inc. from a publication by J.J. Aklonis and W.J. MacKnight: *Introduction to Polymer Viscoelasticity.* Wiley-Interscience, New York, 1983. Copyright © 1983 by John Wiley & Sons, Inc.

The diskette is warranted to the original purchaser to be free from defects in materials and workmanship under normal use for 90 days from the date of purchase. Hanser's entire liability for a defect occurring in the original diskette shall be a replacement of the diskette returned to Hanser with a copy of the proof of purchase. A replacement will be made for a nominal fee if the diskette is damaged after the warranty period expires.

While most of the programs have been tested over the years with a variety of data, the authors occasionally encounter sets that will not execute properly. The user can expect similar occurrences. While the authors will be pleased to make arrangements to